# Insect Conservation and Australia's Inland Waters

Tim R. New

# Insect Conservation and Australia's Inland Waters

Springer

Tim R. New
Department of Ecology
Environment and Evolution
La Trobe University
Bundoora, VIC, Australia

ISBN 978-3-030-57007-1 ISBN 978-3-030-57008-8 (eBook)
https://doi.org/10.1007/978-3-030-57008-8



This Springer imprint is published by the registered company Springer Nature Switzerland AG
The registered company address is: Gewerbestrasse 11, 6330 Cham, Switzerland

# Preface

Freshwater is a critical resource in Australia. The country's reputation as 'a land of droughts and flooding rains' encapsulates the uncertainty of water supply and the unpredictable outcomes for agricultural production and human welfare that so largely depend on more reliable rainfall and water availability, and that can be affected dramatically by either excess or lack. When I first drafted this preface in early autumn 2019, people in large areas of Queensland were attempting to recover from devastating floods in which vast numbers of cattle and sheep perished, and numerous livelihoods rendered uncertain, large amounts of water were moving inland toward the normally dry Lake Eyre in the centre of the continent, stretches of the Darling River in the southeast were dry, and Melbourne's normally adequate water reserves were down to about half their possible capacity. Six months later, the ramifications of 'water shortage' remained at the forefront of political and economic concern (and increasingly linked to concerns over climate change), with no prospect of resolution in sight and continuing calls for ever-increasing government recompense to primary producers and affected rural communities, who faced a continuing major crisis as lack of rain persisted and prolonged a drought that had already continued for around 8 years in some places. By the start of autumn 2020, the bushfires that devastated much of eastern and southern Australia over the summer months had given way to localised torrential rainfall and flooding, perhaps exacerbating soil and debris run-off into many waterways. Such vagaries in supply illustrate the extreme varieties of impact on inland waters and their enveloping environments, and with which people and aquatic biota alike must contend.

More typically, large arid and semi-arid 'desert areas' of Australia contrast with the northern monsoon tropics with more predictable seasonal deluges, and also with the uncertainties of rain over most of the more densely settled and cropping areas of the island continent. Water supply is both a critical ecological need for native biota, human settlement, crop and pasture industries, and a powerful political tool affecting the management of Australia's economy. The latter is reflected in the allocations of water from major catchment systems, notably the Murray-Darling Basin of the south and central east, in which energetic political and economic debate continues over the reality that upstream withdrawal of water for irrigation affects supply for people and environmental needs nearer the river mouth—in this case, the rivers of this complex system flow through several States with separate jurisdictions essentially competing for (and dependent on) water from the same system, and attempts continue to formulate a harmonious integrated plan to satisfy the varying needs across these different political entities and priority interests. Outcomes of a recent government allocation of Au$ 13 billion towards a unifying management scheme for the Murray-Darling Basin have met (at July 2019) with considerable criticisms for inadequacy, fostering private interests rather than a common good, and poor planning and monitoring. The history of the Murray-Darling Basin Plan, described as 'tortured' by Boon (2020), displays many shortcomings despite abundant evidence of the urgent need for such an integrated scheme and long-term advocacy from aquatic ecologists, as displayed in references in this book. But, as Boon (2020, p. 33) put it 'aquatic ecologists are often the couriers of those sad and inconvenient and, to politicians especially, mostly unwanted messages'.

In short, the unpredictable uncertainties voiced in Dorothea Mackellar's (1908) poem, cited above, affect the livelihoods of many people, the nation's agricultural and pastoral economy with significant international trade impacts, patterns of human settlement, and the ecology of the entire country. A large number of native aquatic biota are unwitting but key components of the inland water systems that participate in this scenario and are affected by the varying impacts of changes.

This book is the last of a trilogy in which I try to summarise the major needs and prospects for conservation of Australia's remarkable insect fauna in a rapidly changing world, drawing on the mass of information available from studies elsewhere to augment perspective from the generally more incomplete picture available from the less-documented Australian taxa and ecosystems. The two earlier volumes focused on two major terrestrial biomes—forests (New 2018) and grasslands (New 2019). Concerns for insects associated with—and in many cases restricted to—Australia's freshwaters or, more embracingly, 'inland waters' are no less diverse and complex. A high proportion of species are endemic. Many are also very localised in incidence and are deemed both rare and vulnerable. Many of the host aquatic ecosystems have suffered immense changes from human activities over the past century or so, and concerns for their native denizens parallel those for terrestrial taxa.

Freshwater ecosystems have been considered widely to be among the most globally threatened environments, with impacts on both the biota that depend on them, and on their central contributions to ecosystem goods and services on which much of terrestrial life also depends. Concerns over the health of inland water environments and the susceptibility of their biota to pollution, direct loss of habitats, and a variety of associated threats have led to considerable study to characterise the affected fauna, clarify features of the environments on which they depend, elucidate the nature and causes of changes, and clarify the needs for practical protection, conservation and restoration. The wide variety of these environments encompasses permanent, seasonal and more intermittent or temporary water bodies, natural and anthropogenic constructions, both lotic (running water) and lentic (still water) systems, and sizes from tiny pools (such as treeholes, phytotelmata) to large lakes and rivers. Recent studies of aquifers and other subterranean waters have revealed unsuspected high levels of resident biodiversity. 'Freshwater biology' has for long been a well-established discipline in ecology and human welfare, and this book focuses on a small, but important, suite of themes vital for the future of Australia's notable endemic aquatic invertebrate biota.

Insects are among the most abundant, diverse, ecologically important and recognisable components of that biota. They are also vulnerable. Australia's native aquatic insects span groups that are reasonably well known (notably Odonata, dragonflies and damselflies, for which most species are named) to others that are far less tractable to name or to enable studies of individual species or assemblages. The background to the major taxa, including recognition and diagnosis, ecology, importance to humanity, and needs for conservation are summarised in later chapters, but the focused information in major texts such as 'The Insects of Australia' (Naumann 1991), and guides such as those produced through the Murray-Darling Freshwater Research Centre augments this considerably. Wider synopses of aquatic invertebrates (such as for the Malaysian region: Yule and Yong 2004), and peer-reviewed papers in a host of specialist journals—many referred to in the references for later chapters here—continually add to the formidable literature on ecology and conservation of insects of inland waters. The diverse taxonomy and ecology of aquatic insects is the theme of several important texts, such as those by Williams and Feltmate (1992) and Lancaster and Downes (2013), but neither of these emphasises insect conservation. The second edition of the major text on Australian freshwater ecology (Boulton et al. 2014) is an indispensable overview and source of information on these ecosystems. The last paragraph of the introduction to that volume commences 'Our aquatic ecosystems are precious', and the many reasons to endorse that comment include the remarkable suites of unusual, endemic and in many instances threatened insects that they harbour and which are a key feature of Australia's insect biodiversity. I have here drawn from the voluminous reviewed publications, and also from the 'grey literature' on Australian aquatic ecology, encountered up to late 2019 to demonstrate the variety of themes and concerns relevant to insect conservation and aquatic ecosystem management. My focus here is to provide an overview of the ecology and conservation of inland water insects and their relationships with host ecosystems, that can be used by non-entomologist stewards seeking to conserve and wisely manage Australia's inland aquatic ecosystems and assure the survival of the numerous little-known species that depend on those ecosystems, and are vulnerable to the widespread anthropogenic changes inflicted. Most entomologists (myself included) are not limnologists—and most limnologists are not entomologists; further, many conservation managers are neither. My hope here is to provide sufficient information for each of these cohorts to participate effectively in enhancing both policy and management for conservation of insects and other invertebrates in Australia's inland water environments.

The first chapters (Chaps. 1–3) introduce the scope of need for the variety of Australia's inland water environments and their main characteristics, and how their invertebrate faunas may be appraised (Chap. 4). Changes and threats to those environments and biota (Chaps. 5, 6) set a scene for appraising the diversity and ecology of Australia's aquatic insects and their conservation needs (Chaps. 7, 8) with some notable species treated further in Chap. 9. The final two chapters expand on the management and key aspects of conservation of inland water insects. Because each chapter is required to be read independently, and so independently referenced, some overlaps are inevitable as key themes emerge in different contexts: page cross-references throughout the text facilitate broader use.

Melbourne, Australia Tim R. New

## References


Boon PI (2020) The environmental history of Australian rivers: a neglected field of opportunity. Mar Freshw Res 71: 1–45

Boulton AJ, Brock MA, Robson BJ, Ryder DS, Chambers JM, Davis JA (2014) Australian freshwater ecology. Processes and management, 2nd edn. Wiley-Blackwell, Oxford

Lancaster J, Downes B (2013) Aquatic entomology. Oxford University Press, Oxford

Mackellar D (1908) Poem: 'Core of my heart', initially published in London Spectator; then published (1911) as 'My Country' in 'The closed door and other verses', Melbourne

Naumann ID (chief ed) (1991) The insects of Australia (2nd edn). Melbourne University Press, Carlton

New TR (2018) Forests and insect conservation in Australia. Springer, Cham

New TR (2019) Insect conservation and Australia's grasslands. Springer, Cham

Williams DD, Feltmate BW (1992) Aquatic insects. CAB International, Wallingford

Yule CM, Yong HS (eds) (2004) Freshwater invertebrates of the Malaysian region. Academy of Sciences Malaysia, Kuala Lumpur

# Acknowledgements

I am very grateful to the following publishers, organisations and individuals for granting permission to use materials to which they hold the copyright. Every effort has been made to obtain such permissions, and the publishers would appreciate notice of any inadvertent omissions or corrections that should be included in any future imprints or editions of this book. Most figures have been redrawn to ensure standardised lettering and some have been modified, as indicated in individual legends. Thanks are extended to CAB International, Wallingford; CSIRO Publishing, Melbourne; Dr. Tarmo Raadik (Department of Environment, Land, Water and Planning, Victoria); Melbourne Water; Museum Victoria; Centre for Environmental Studies, University of Tasmania; Great Lakes Entomologist; South Australian Department of Health; European Pond Conservation Network; Taylor & Francis Ltd; John Wiley and Sons; Elsevier; Oxford University Press, Oxford.

It is difficult to overstate my indebtedness to the Springer team in Dordrecht, with whom it has been a privilege to work over many years and whose professionalism is a model for emulation. Zuzana Bernhart (as commissioning editor) gratifyingly accepted my suggestion for this book, and the continuing encouragement and enthusiastic support from her colleague Mariska van der Stigchel are again appreciated greatly. More recently, Ineke Ravesloot's considerable help in completing formalities for the book's completion greatly eased those final stages. My very grateful thanks for this support. Later production stages were overseen by Mr. Rajan Muthu (Project Coordinator), and I am especially grateful to my Project Manager, Ms. Ritu Chandwani, for her careful attention to checking and processing the manuscript.

# Contents

# Chapter 1
# Introduction: Aquatic Insects in Australia's Environments

## 1.1 Introduction: The Background to Concern

The Australian environment is complex, reflecting the size of the country, the span from tropical to cool temperate climates, and from mesic to arid and semi-arid regions—so that water resources across the country range from reasonably assured to highly uncertain and unpredictable. That variety is reflected in the range of lotic (running water: streams and rivers), lentic (still water: lakes and ponds) and subterranean aquatic ecosystems, each with characteristic biota that may be highly localised, and scarce. Continuing investigations demonstrate the unique nature of these biota and environments but, as elsewhere, conservation of invertebrates is commonly secondary to concerns for more charismatic taxa, notably vertebrates—with realisation that numerous taxa are disappearing or becoming more vulnerable as anthropogenic changes proliferate. The few semiaquatic mammals, in particular the Platypus (*Ornithorhynchus anatinus*), have far higher public profile and sympathy and one of the two semiaquatic rodents, the False water rat ('Yirkoo', *Xeromys myoides*) is also of significant conservation concern. Both are significant ambassadors (or 'flagships') for conservation of aquatic environments, with their wellbeing helping to sustain environments needed by many other taxa. Endemic freshwater fish also garner conservation concerns and attention. However, as for those vertebrates, conservation attention to aquatic invertebrates has flowed largely from concerns over individual species, with a number of insects and others signaled in various ways as 'threatened', as their habitats have changed. Development of that interest since the mid twentieth century has led to considerable interest in conserving Australia's aquatic fauna (Michaelis 1986) and demonstrated the severe problems in doing so in an essentially rather poorly documented fauna. Thus, Michaelis commented on the paucity of taxonomic, distributional and ecological information as a severe obstacle to preparing meaningful lists of threatened insects and other invertebrates, so that progress must be pursued with, often, limited and sometimes highly subjective information and with considerable caution needed in defining conservation status of the numerous poorly known species. That restriction persists. The wider perspective of


T. R. New, *Insect Conservation and Australia's Inland Waters*,
https://doi.org/10.1007/978-3-030-57008-8_1

a 'habitat approach' is perhaps even more complex but acceptance that (1) habitat changes are the most universal and pervasive causes of conservation need, and (2) many insects have very restricted resource requirements and depend on geographically and physically restricted habitats, implies a considerable variety of patterns of diversity and species' distributions across the range of environments available. That variety is introduced below.

Even for the best-known fauna of aquatic insects, that of Britain, Foster (1991) was able to comment that 'the main threat to conservation of aquatic insects is ignorance'. He recognised four categories of ignorance: (1) lack of enlightenment on the part of the land owner or developer; (2) lack of enlightenment on the part of the conservation manager; (3) low public awareness; and (4) ignorance of entomologists. All transfer easily to Australia, with the last a considerable barrier to species-level studies and assessments, and a driver toward according greater priority to conservation of sites and 'habitats' signalled as in some way important for particular taxa or ecological features. One important impediment is simply inability to predict the impacts and consequences of many environmental changes.

Australia is not only the driest of all inhabited continents, but much of the country also has the most unpredictable rainfall that, almost inevitably, leads to substantial fluctuations in water supply and demand, and the quality—even, existence—of many waterbodies. Much of the continent is regarded as 'arid' or 'semi-arid' (Fig. 1.1). Humphreys (2006) noted that two-thirds of mainland Australia is within the 'arid zone' where rainfall is indeed infrequent and unpredictable. Central Australia, even with average rainfall of only <300mm/year $^{-1}$ (Brim-Box et al. 2014), supports a range of freshwater habitats that collectively 'provide refuges for relict, endemic and widely dispersed species' and the greatest values accorded to long-lasting or permanent waterbodies. Changes in invertebrate faunas have mostly been documented from rather recent comparative surveys, whilst 'pressures' on waterbodies continue to proliferate. Aquatic invertebrates surveyed on three occasions (1986, 1994, 2008) from 16 sites in central Australia, by the time of the most recent survey mostly in national parks and representing two major drainage divisions, comprised a total of 118 taxa. These were dominated by insects: Coleoptera (35 taxa), and Odonata (19) were followed by smaller numbers of Hemiptera (10), Ephemeroptera (3), Trichoptera (6) and Diptera (10), and representation also of Lepidoptera (two aquatic larvae) and Neuroptera (1) (Brim-Box et al. 2014). Many species found in the 2008 survey were also represented earlier and are likely to have been continually present. In addition to reserved status of the sites in national parks, most were remote and difficult to access—features possibly also contributing to the persistence of the fauna, and low levels of externally-imposed disturbances.

Globally, fresh waters occupy less than 1% of the world's land area—but have been estimated to contain about 10% of all species, including about a third of all vertebrates (Strayer and Dudgeon 2010). However, one conclusion reached by Dudgeon et al. (2006) is that freshwater systems are experiencing declines in biodiversity that are far greater than in most terrestrial ecosystems; indeed, they noted that they may be the world's most endangered ecosystems, in part reflecting the increased demands by human society and burgeoning populations since the early

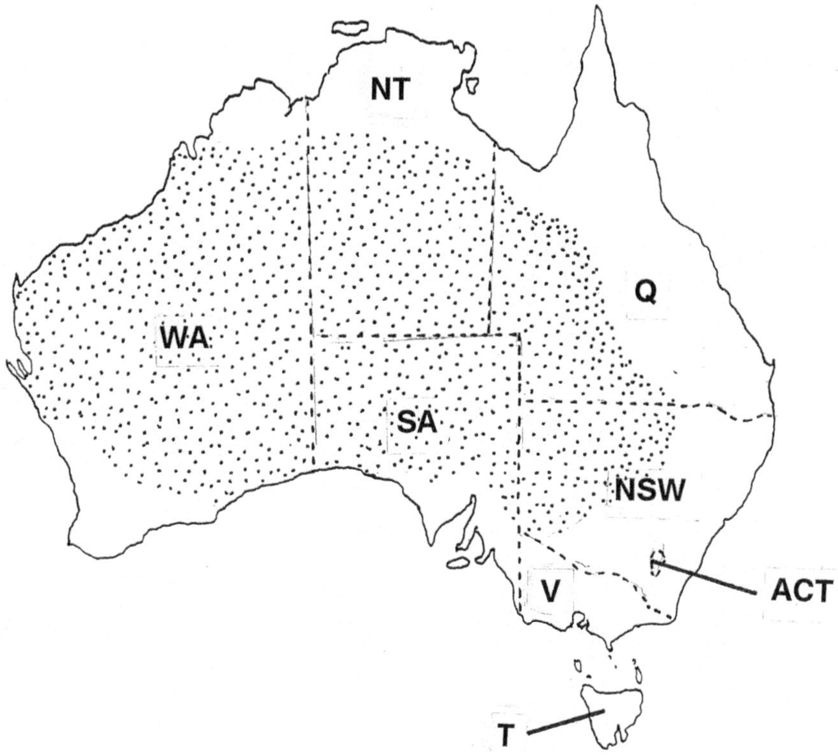


**Fig. 1.1** The political divisions of Australia, and the extent of the arid/semi-arid zone (dotted) (WA, Western Australia; SA, South Australia; NT, Northern Territory; Q, Queensland; NSW, New South Wales; V, Victoria: ACT, Australian Capital Territory; T, Tasmania)

twentieth century (Strayer and Dudgeon 2010). Dudgeon et al.'s severe prognosis for global freshwater ecosystem losses has worsened. A vertebrate-based assessment, the 'Living Planet Index' (WWF 2016), showed that freshwater species are indeed declining more steeply than either marine or terrestrial taxa (Reid et al. (2019); there is no reason to suppose that invertebrates are any more secure. Freshwater habitats are proclaimed widely as the world's most threatened environments (Dijkstra et al. 2014) and, together with agricultural conversions, are 'the most extensively and rapidly altered ecosystems on the planet' (Carpenter et al. 2011). And, as Darwall et al. (2018) commented, the practices of water use, viewing water primarily as a resource for humanity, have led to 'catastrophic declines in freshwater species' and global loss continues largely unabated.

Thus, there are strong suggestions that aquatic insect biodiversity is declining to an equivalent—or perhaps even greater—extent than the better-appraised terrestrial fauna. For example, stoneflies (Plecoptera) in the Czech Republic have declined substantially since surveys over 1955–1960. Bojkova et al. (2014) found that three-quarters of the 77 species then recorded had declined in frequency of occurrence across 175 streams by 2006–2011, about half of them by >30%. Ten species found earlier were not rediscovered in the later samples. Lowland or submontane streams were most affected, and in those systems losses included pollution-sensitive species and habitat specialists. In some places those losses were masked by colonisations of common generalist species.

The above inferences are supported widely elsewhere. As well as freshwater environments themselves, many of their inhabitants are vulnerable to changes. Although dealing with non-insect groups in temperate region freshwater bodies, the projected extinction rates for these in North America summed to a mean estimate that extinction rates for freshwater fauna were about five times those for terrestrial fauna (Ricciardi and Rasmussen 1999). Other workers (such as Richter et al. 1997) have also remarked on the proportionately greater threats to aquatic fauna over terrestrial species and, again, implied the great variety of impacts from this trend. They wrote, soberingly: 'A quiet crisis is taking place beneath the surface of the world's rivers and lakes'. With evidence for invertebrate extinctions largely arising from documentation of crustaceans and unionid molluscs, there is no compelling reason to exclude insects from undergoing parallel losses and contributing significantly to the large proportion of freshwater invertebrates that are in some way becoming more vulnerable to extinctions. Evidence of declines and losses comes largely from studies in temperate regions, especially in so-called 'first world countries' in which documented anthropogenic changes are high. Although claimed commonly that freshwater biodiversity is higher in the tropics than in temperate regions, this inference (other than for fish) suffers from relative lack of studies in many tropical areas (Pearson and Boyero 2009), so that contrasting opinions have been published over whether such latitudinal gradients occur—and, if so, how they are governed and differ across taxa. Thus, it seemed that no latitudinal gradients in diversity occurred for Trichoptera, whilst distinct gradients for Ephemeroptera and Plecoptera were associated with greater diversity at higher latitudes. Conversely, significant gradients for Odonata were toward higher diversity in the tropics. The long terrestrial phase of dragonflies perhaps affords greater opportunity for habitat partitioning in the tropics, resulting in considerable specialisation and speciation there (Pearson and Boyero 2009), capitalising also on the wide range of aquatic habitats available.

As noted above, the fate of freshwater invertebrates has commonly been overshadowed by conservation attention to vertebrates, as a symptom of the more general bias in animal conservation toward more popular taxa that garner greater public and political support (Darwall et al. 2011)—but it is sobering that even basic documentation of many aquatic vertebrate groups is still far from complete in many parts of the world. Dudgeon et al. (2006) noted, for example, that numerous fish and amphibian species had been described over the previous decade, and that estimates of fish species richness in some major systems, such as the Mekong discharge of south-east Asia, have been wildly below the reality now emerging from more recent surveys. Major radiations of fish (and of many groups of invertebrates) remain to be described and evaluated effectively. But, simply reflecting the greater information available, declines of aquatic vertebrates appear to be substantial, but incomplete knowledge of all aquatic taxa suggests that rates of species loss may be higher than any current

**Table 1.1** The five challenges facing freshwater invertebrate conservation, and that may mould the best approaches to pursue this (Strayer 2006)

| |
|---|
| 1. Many species of freshwater invertebrates around the world may already be extinct or endangered |
| 2. Human pressures on freshwater resources are intense and will increase in coming decades and put more species at risk |
| 3. Scientific knowledge of freshwater invertebrates, although still substantial for many groups, is far less than for vertebrates that have largely stimulated development of modern conservation biology |
| 4. Because freshwater ecosystems are downhill from and embedded in their watersheds, conservation must usually address entire watersheds rather than the small local sites from which endangered species are known |
| 5. Society spends rather modest amounts of money on freshwater invertebrate conservation |

estimates imply. The lack of taxonomic information is especially evident for many invertebrate groups, particularly as more remote and lesser-explored habitats are investigated. The subterranean aquifers of Australia, for example, have yielded high numbers of small, locally endemic waterbeetles as a major suite of radiations in these isolated habitats (p. 194), which have considerably increased the richness of Dytiscidae known from the country. Taxonomic and ecological knowledge of the different aquatic insect groups is very uneven, and hampers assessments and comparisons of diversity and species richness. Thus, Balian et al. (2008) contrasted the largely reliable and relatively complete documentation of Odonata (p. 178) with the much poorer and more uneven information available on some Diptera families, whose diversity can at present be inferred only with considerable caution.

However, wherever waterbodies and invertebrates occur and whatever specific threats apply, 'Freshwater invertebrate conservation faces huge challenges' (Strayer 2006), a statement that remains entirely valid. Strayer identified a series of such challenges (Table 1.1), collectively posing the even greater unifying challenge of what the best approaches to conserving freshwater invertebrate diversity may be. Fundamentally, rapid declines in global freshwater biodiversity continue, and can be reduced only by well-considered and effectively promoted actions (Turak and Linke 2011), within the certainty that pressures on freshwater systems and needs for water will continue and intensify, logistic resources for their alleviation will be outpaced by need, and further novel threats will arise. Turak and Linke noted that, in principle, the broad actions needed at any particular location can often be defined, but moving to implement these may become very complex and even impracticable. Thus, actions projected or needed include restoring or maintaining (1) migration or dispersal routes, promoting connectivity; (2) refuges; (3) natural flow regimes; (4) good water quality; (5) food supplies and sources; (6) habitat structure and integrity; (7) preventing invasions, and suppressing or eradicating alien species; (8) preventing over-exploitation; (9) reducing pollution; (10) reducing nutrient enrichment; and (11) reducing sedimentation. The relative needs for these will clearly differ across sites, so that local priorities differ markedly, and also reflect (1) cost and feasibility under local conditions, and (2) potential benefits of the local actions for wider-scale freshwater biodiversity. The broader perspective is that, whatever levels of threat to biodiversity are evident, the most threatened individual species of insects and their relatives, even when designated as 'critically endangered' gain very low priority in relation to sustained economic and social needs for water. Planning for their specific conservation needs must usually take place within that wider scenario.

Three characteristics of freshwater habitats impinge directly on invertebrate conservation (Strayer 2006). These are (1) freshwater habitats are scarce compared with other major ecosystems, and are markedly less in extent than either terrestrial or marine biomes; (2) all freshwater habitats are in some sense 'islands' in a sea of dry land and salt water, and each is more-or-less isolated from other such habitats: in many cases their inhabitants have only very limited opportunity to disperse to other 'islands', and in each such island particular features (such as riffles, pools, substrates, vegetation and many others) are further restricted and isolated; and (3) freshwater habitats are 'downhill', so that many of their characteristics are moulded by the drainage basins and the human activities there—such as land clearing, settlement, agriculture and other industry—which in many cases lead to degradation of the waterbodies and their enveloping environments. Thus, much of the world's agriculture was developed on lowland plains, and most major urban settlements founded on lakes or rivers. Disturbances anywhere in the catchments may pose threats to freshwater biota, and few such bodies remain free of human interventions of some form in interacting strongly with land use. Essentially, this means that quality of numerous freshwater habitats depends directly on management of the landscapes that envelop and affect them.

The high level of habitat isolation ensures that many invertebrate inhabitants have small distribution ranges—many are deemed narrow range endemics—and are susceptible to local disturbances and changes. Groups with limited dispersal powers can show high levels of endemism and local diversification—so that taxa that cannot withstand desiccation and have no aerial or resting stages, such as many non-insect invertebrates, can exhibit higher endemism than winged insects. There is some tendency for 'hotspots' of freshwater insect diversity to also be hotspots of local endemism, so that assemblages of geographically restricted taxa are not uncommon. Within Australia, many insects characterise such assemblages and many species are both highly restricted in distribution and ecologically specialised. Local vulnerability is also common.

The peculiarities of Australian inland water ecosystems and biota give them globally unique significance but can be difficult to appraise. Much of the general understanding of limnology has stemmed from studies in the northern hemisphere: Williams (1988) noted that the historical perspective on limnology was founded in Europe and north eastern North America, where 'the founding fathers of limnology resided'. Not surprisingly, those faunas are by far the most thoroughly documented and understood. Application of some of those ideas to Australia then needed to be re-considered. Beyond those northern regions, Williams considered that limnology then lacked 'a cohesive intellectual framework upon which local studies may be based in a scientifically satisfactory manner'. However, the principles of the more restricted ambit of aquatic insect ecology may be rather more universal and studies from outside Australia furnish numerous examples and contexts spanning threats and dynamics that are common to many freshwater ecosystems. Conservation approaches and needs have common relevance but can differ in the extent to which they can be undertaken based on logistics and soundness of the background information available. Individual contexts define the measures needed, but the fundamental biology of many Australian aquatic insects differs rather little from that of their relatives elsewhere.

## References

Balian EV, Segers H, Leveque C, Martens K (2008) The freshwater animal diversity assessment: an overview of the results. Hydrobiologia 595:627–637

Bojkova J, Radkova V, Soldamn T, Zahradkova S (2014) Trends in species diversity of lotic stoneflies (Plecoptera) in the Czech Republic over five decades. Insect Conserv Divers 7:252–262

Brim-Box J, Davis J, Strehlow K, McBurnie G, Duguid A (and four other authors) (2014) Persistence of central Australian aquatic invertebrate communities. Mar Freshw Res 65: 562–572

Carpenter SR, Stanley EH, Vander Zanden MJ (2011) State of the world's freshwater ecosystems: physical, chemical and biological changes. Annu Rev Environ Resour 36:75–99

Darwall WRT, Holland RA, Smith KG, Allen D, Brooks EGE (and 14 other authors) (2011) Implications of bias in conservation research and investment for freshwater species. Conserv Lett 4: 474–482

Darwall W, Bremerich V, De Wever A, Dell AI, Freyhof J (and 18 other authors) (2018) The Alliance for Freshwater Life: a global call to unite efforts for freshwater biodiversity science and conservation. Aquat Conserv Mar Freshw Ecosyst 28:1015–1022

Dijkstra K-DB, Monaghan MT, Pauls SU (2014) Freshwater biodiversity and aquatic insect diversification. Annu Rev Entomol 59:143–163

Dudgeon D, Arthington AH, Gessner MO, Kawabara Z-I, Knowler DJ, Leveque C (and five other authors) (2006) Freshwater biodiversity: importance, threats, status and conservation challenges. Biol Revs 81: 163–182

Foster GN (1991) Conserving insects of aquatic and wetland habitats, with special reference to beetles. In: Collins NM, Thomas JA (eds) The conservation of insects and their habitats. Academic Press, London, pp 237–262

Humphreys WF (2006) Aquifers: the ultimate groundwater-dependent ecosystems. Aust J Bot 54:115–132

Michaelis FB (1986) Conservation of Australian aquatic fauna. In: De Dekker P, Williams WD (eds) Limnology in Australia. CSIRO, Melbourne/W Junk, Dordrecht, pp 599–613

Pearson RG, Boyero L (2009) Gradients in regional diversity of freshwater taxa. J N Am Benthol Soc 28:504–514

Reid AJ, Carlson AK, Creed IF, Eliason EJ, Gell PA (and 11 other authors) (2019) Emerging threats and persistent conservation challenges for freshwater biodiversity. Biol Revs 94: 849–873

Ricciardi A, Rasmussen JB (1999) Extinction rates of North American freshwater fauna. Conserv Biol 13:1220–1222

Richter BD, Braun DP, Mendelson MA, Master LL (1997) Threats to imperiled freshwater fauna. Conserv Biol 11:1081–1093

Strayer DL (2006) Challenges for freshwater invertebrate conservation. J N Am Benthol Soc 25:271–287

Strayer DL, Dudgeon D (2010) Freshwater biodiversity conservation: recent progress and future challenges. J N Am Benthol Soc 29:344–358
Turak E, Linke S (2011) Freshwater conservation planning: an introduction. Freshw Biol 56:1–5
Williams WD (1988) Limnological imbalances: an antipodean viewpoint. Freshw Biol 20:407–420
WWF (2016) Living Planet Report 2016. Risk and resilience in a new era, WWF International, Gland, Switzerland

# Chapter 2
# Major Habitats

## 2.1 Introduction

The two major categories of freshwater habitats are both highly diverse but, simplistically, may be thought of as 'still water' (ponds, lakes) and 'running water' (streams, rivers). The main features of these environments differ, and many workers have drawn attention to the differing characteristics and complements of the biota that typify them, and human impacts on the conditions they provide, together with the wider ecological ramifications of those impacts. Those consequences are reflected in many of the contexts and examples discussed in later chapters, and this short introduction to the major features of these systems outlines some of the major considerations that determine and drive their suitability for aquatic insects. The greater variety of inland water bodies are largely subdivisions of these two major categories and are outlined in the next chapter.

## 2.2 Ponds and Lakes

'Ponds' have been defined as small still water bodies between 1 $m^2$ and 2 ha in area, which normally hold water for at least four months of the year (Britain: Pond Action, see Biggs et al. 1998). Definitions vary somewhat, and the subtle (and sometimes subjective) gradations between ponds and lakes provide abundant opportunities for semantic interferences! Lakes are larger but depth is, perhaps, a fundamental parameter for differentiation—with ponds shallow enough to allow rooted vegetation to occur on the substrate, whilst lakes are deeper and do not enable rooted vegetation to occur throughout. However, the sentiment expressed by Bayley and Williams (1973), namely '… we suggest a pond is a body of standing water so termed by a good limnologist!', allows for much variation to be considered. Precise definition is arbitrary (Wissinger et al. 2016). In alpine and subalpine systems, for example, standing water bodies range from large deep lakes to small shallow ponds, with many structural


T. R. New, *Insect Conservation and Australia's Inland Waters*,
https://doi.org/10.1007/978-3-030-57008-8_2

and biological features shared through much of the gradient between these. A further, occasionally confusing, distinction sometimes drawn is between ponds (permanent) and 'pools' (temporary), with many such transitions used inconsistently. Essentially, standing water bodies are both numerous and diverse. Cereghino et al. (2014) quoted estimates of a global 277 400,000 ponds less than a hectare in area, plus 24 120,000 water bodies of 1–10 ha, collectively comprising >90% of the global 304 million standing water bodies estimated by Downing et al. (2006). On a more local scale, Biggs et al. (2005) noted 'approximately 400,000' ponds in the United Kingdom.

Ponds may vastly outnumber lakes, and can also show greater ecological amplitude than lakes and rivers. The sheer number of ponds in any given area is assuredly welcome, but also renders assessments of conservation values more complex, because individual ponds can differ considerably in their character, and sampling to represent regional conservation significance becomes laborious if it is sufficient to adequately reflect that variety on any regional scale. In the past, many evaluations of ponds for conservation values based on their invertebrate complements have involved only a single 'snapshot' visit (Jeffries 2005), so not revealing aspects of temporal/seasonal/phenological changes, and likely also to obscure reliable spatial relationships between ponds, even when they are inspected at the same season: both temporal and spatial variations across ponds influence interpretation of species distributions and richness. Interactions of invertebrates between pond sites, reflecting dispersal (p. 231), can occur over substantial scales. So-called 'spatial autocorrelation', whereby sites closer together have more similar species composition than found in sites further apart, was endorsed by Briers and Biggs (2005). Their surveys of invertebrates of ponds in Oxfordshire, UK, were over an area of 60 × 60 km, so encompassing distances greater than the likely normal natural dispersal ranges of most species present. Large-scale positive autocorrelation supports the principle that pond networks should be adopted over larger scales than immediately local, and incorporate wider, regional scale conservation targets. Thus, Briers and Biggs (2003) recommended selecting a suite of macroinvertebrate indicator taxa and using their species richness as a tested surrogate for all other taxa. For Oxfordshire ponds they enumerated species at the family level and noted that Coleoptera and Odonata had earlier been used as criteria for selecting Sites of Special Scientific Interest. Data from 130 ponds—about 5% of ponds in Oxfordshire—represented 256 macroinvertebrate species, and the pool of families for potential indicators was selected as those with at least four species represented. Correlation coefficients were calculated across taxa, leading to a considerable range of values (Fig. 2.1). From this summary, Coenagrionidae and Limnephilidae both had highly significant cross-taxon correlations, suggesting that variation in their species richness is strongly representative of variations in all other taxa. The proportion of the 256 species in the wider pool that would be represented in selecting sites on information from these two families, confirmed that both are potentially useful indicators of richness of all other taxa present. Nevertheless, Briers and Biggs councelled that other aspects may also be relevant—low richness ponds may support rare species absent from most others, for example, and a range of different metrics and criteria may be needed to hone selection of the 'most valuable' sites from the array available. Values of ponds in understanding aquatic

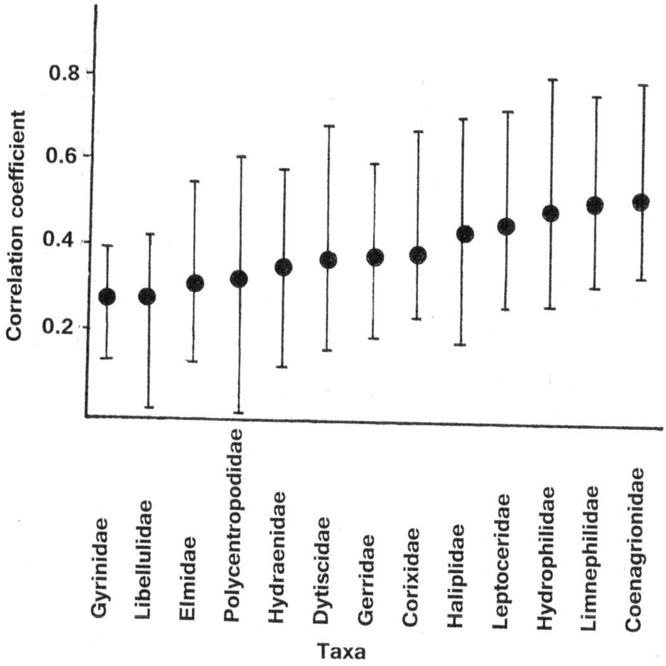


**Fig. 2.1** The mean and range of cross-taxon correlations for potential indicator taxa, arranged in sequence of increasing mean cross-taxon correlations. (from Briers and Biggs 2003, non-insect groups omitted; Pearson correlation coefficients of original omitted)

biodiversity (Cereghino et al. 2008) make them very relevant in conserving that diversity, not least as model systems for investigating a wide range of associations with (and influences of) ecological gradients of many kinds—such as hydroperiod, surface area, salinity, vegetation and inter-pond connectivity. As Cereghino et al. wrote 'if these associations are causal, it is clear that the conservation of such environmental gradients at the landscape scale is essential'.

For the Tatra Mountains of central Europe, Hamerlik et al. (2014) used a threshold area of 1 ha to effectively distinguish small shallow ponds from larger deeper lakes. They showed that ponds had (1) lower local diversity; (2) higher among-site diversity; and (3) similar regional diversity to lakes. That small ponds can support rare invertebrate species within considerable species richness renders them important considerations in sustaining local biota. In that comparison, involving benthic invertebrates from 25 ponds and 34 lakes, only 23 of 69 taxa occurred in both categories. Twenty one taxa were found only in ponds, and 25 only in lakes. Chironomidae were by far the richest group (23 taxa in ponds and 20 in lakes), followed by Trichoptera (nine species in each). The ponds and lakes were regarded here as qualitatively different systems, with differences in local diversity reflecting habitat heterogeneity and overall environmental 'harshness'. In general, however, the ponds had lower local diversity and higher species turnover than lakes in the same area.

Many ponds are natural, but numerous others have been created for human needs—in Australia, farm dams for water storage and watering livestock, ponds for aquaculture, and ornamental ponds in parks and gardens are all familiar features, with a wide variety of amenity and landscape features encompassing more traditional and more contemporary roles. However, many ponds are also lost with declines of some traditional uses, by direct destruction by urban or other developments (p. 126), or by neglect allowing increased sedimentation, succession and reductions in water depth and area. Many constructed ponds are stocked with fish, some of them alien species (p. 111), for ornamental or recreational angling purposes. They, together with the vegetation present, can strongly influence the invertebrate communities present. Throughout the world, farm ponds (whether natural or constructed) are important habitats for insects. For example, they support >30% of the Irish water beetle species and especially in intensively farmed or cleared areas are among the most important aquatic habitats (Gioria et al. 2010). Their complement of vegetation is a key influence, and the richest water beetle assemblages were in permanent ponds with swamp zones dominated by Bull rushes (*Typha latifolia*) or Bur weed (*Sparganium erectum*). In contrast, temporary ponds did not support any unique species, and Gioria et al. found that plant species composition in farmland ponds was related to that of water beetles—so that wetland plants might be a useful surrogate group for identifying ponds with high conservation value for the beetles, so implying that preliminary investigation of the vegetation might be useful in appraising wider conservation value.

Artificial ponds can contribute to conservation through increasing the area of occupancy for species and so reduce risks of their extinction. The roles of artificial ponds as water storages for irrigation can essentially provide additional relatively permanent habitat that could otherwise be regionally scarce—as in the Cape Floristic Region of South Africa (Apinda Legnouo et al. 2014). Surveys of aquatic Hemiptera and Coleoptera in 18 artificial ponds in that region, both orders with high local endemism, endorsed those values. Bugs were the more abundant group (with 14,953 of the total 17,814 individuals), but this survey showed that the assemblages of both orders were as rich as those reported elsewhere. Not all the ponds were permanent, and the scale of 'permanency' used is likely to be paralleled in Australia (Table 2.1) and may be used as a measure of habitat condition and availability. Nevertheless, the artificial ponds were collectively important conservation areas, and greater consideration for their conservation, increasing regional diversity and security for bugs, beetles and probably other taxa at present undocumented there for South Africa includes attention to

**Table 2.1** The four categories of stream 'permanency' used to compare invertebrate assemblages of artificial ponds in South Africa (Apinda Legnouo et al. 2014)

| |
|---|
| 1. Temporary pool |
| 2. Pool becomes stagnant in summer |
| 3. Water level drops by more than 20 cm |
| 4. Water level fluctuates by no more than 20 cm |

macrophytes and controls of local agricultural runoff. Some merit management as 'temporary pools', fostering some species not usually found in permanent ponds.

Although temporary ponds, forming in depressions that are periodically (often, seasonally) filled, support fewer macroinvertebrate species than found in many permanent ponds (Britain: Collinson et al. 1995), they can still have considerable conservation value by supporting assemblages containing rare and threatened species. A number of possible reasons for this were noted—for examples, the shallowness of many temporary ponds allows them to warm up quickly, enhancing growth in some species; lack of large fish predators may benefit open water species; and the ponds can be very stable and fill with sediment only very slowly. Collinson et al. contrasted these with permanent ponds in which organic breakdown may be relatively slow—but, conversely to such putative benefits, many temporary ponds are indeed small and susceptible to human activities. In Oxfordshire, two insects were indicators of temporary ponds—the caddisfly *Limnephilus auricula* and a lesser waterboatman *Callicorixa praeusta*. The latter frequents open water, and might benefit from absence of predatory fish. Conversely, again, Odonata and Ephemeroptera were largely confined to permanent ponds. Four of the five highest Species Rarity Score occurrences (p. 276) were from temporary or semi-permanent ponds. Conservation lessons from this survey, likely to have far wider relevance, were (1) that careful survey of temporary ponds should be undertaken before any are sacrificed or management work undertaken, especially for long-established ponds or where management involves measures such as deepening or removal of vegetation; and (2) protection of the water regime and quality, such as by use of semi-natural buffer zones and preventing increased drainage or water extraction may prevent undue changes.

Ponds are usually thought of as discrete isolated entities. Many are, but many others, including those designed for practical uses such as water retention (p. 277) have some form of inflow and/or outflow to facilitate use of the water for industrial or agricultural purposes, so have features of both lentic and lotic environments. Wood and Barker (2000) showed some of the resulting differences in invertebrate communities from surveys of mill ponds in Britain. Surveys of old industrial woolen mill ponds, many of them stocked with fish for local recreational angling, confirmed that unmanaged ponds had higher macroinvertebrate diversity (Wood et al. 2001), whilst managed ponds were dominated by burrowing Chironomidae and Oligochaeta. However, neglected ('unmanaged') ponds were at greater risk of loss through drainage and urban development—a fate of many earlier ponds in the region. Such ponds may appear unimportant to many observers, but they can support high invertebrate diversity and their conservation—reflecting the high quality resources they furnish—is considered by many people to be important.

A pond can have considerable conservation values even if it is small, so that preservation of existing ponds and construction of additional ones to enhance their connectivity and abundance over landscapes is an active conservation focus. Within this, even very small ponds merit attention. Indeed, some insects (such as the beetle *Agabus bipustulatus* in Europe) were associated with smaller ponds in a comparative survey across different pond sizes (Oertli et al. 2002). Using measures of species richness and conservation value, pond size was important for adult Odonata, but not for most beetles, gastropods, amphibians and flora. Seven of the 17 species of Odonata were associated with the largest ponds, and large ponds can support species not found in smaller ponds; some species, however, occurred across different sizes. Conservation value was calculated as the sum for all species in a pond scored according to their degree of rarity, an approach possible only for such a well-documented fauna. Thus, for the relatively well-known aquatic invertebrates of Britain, each species can be given a numerical conservation score depending on any formal recognition of conservation significance and frequency of occurrence in samples. These categories (Table 2.2) range from '10' (for formally recognised endangered species) to '1' (for very common species found in a high proportion of samples). Accurate identification is critical and, even for Britain, some groups remain intractable and cannot be included reliably in comparative indices. A scheme for British ponds reflects similar logic and approach but has slightly different categorisations (Table 2.3) (Biggs et al. 2005), drawing on a Species Rarity Index that incorporates assessment of rarity at a site (alpha diversity, the richness at each site) and regional level (gamma diversity, the total species richness of the study sites) (Williams et al. 2004).

Studies in Europe imply that ponds contain more species, including more rare species, than found in lakes, rivers or streams (Williams et al. 2004, for southern England), but the unique features of ponds have only recently been distinguished from the presumption that they are simply 'small lakes' (Cereghino et al. 2012). Reasons for the richness of ponds continue to be debated, but Biggs et al. (2005) suggested that their catchments reflect very local environmental variations, so that ponds are collectively more varied chemically and physically than larger water bodies with larger catchments that might 'average out' the various conditions encountered. Such small catchments, however, can pose conservation dilemmas—they are

**Table 2.2** Scheme for assessing Conservation Score (CS) for freshwater invertebrates in Britain (Chadd and Extence 2004)

| CS | Definition |
|---|---|
| 10 | Red Data Book 1 (Endangered) |
| 9 | Red Data Book 2 (Vulnerable) |
| 8 | Red Data Book 3 (Rare) |
| 7 | Notable (but not Red Data Book status) |
| 6 | Regionally notable |
| 5 | Local |
| 4 | Occasional (species not in categories 10–5, which occur in up to 10% of all samples from similar habitats) |
| 3 | Frequent (species not in categories 10–5, which occur in >10–25% of all samples from similar habitats) |
| 2 | Common (species not in categories 10–5, which occur in >25–50% of all samples from similar habitats) |
| 1 | Very common (species not in categories 10–5, which occur in >50–100% of all samples from similar habitats) |

**Table 2.3** The provisional categories of macroinvertebrate assemblages used to assess conservation value of ponds in Britain (after Biggs et al. 2005) (SRI, Species Rarity Index, see text)

| Assemblage conservation value | Qualifying characteristics |
|---|---|
| Low | Few invertebrate species (0–10) and no local species (SRI = 1.00) |
| Moderate | Below average number of invertebrate species (11–30), or SRI of 1.01–1.19 |
| High | Above average number of invertebrate species (31–50), or SRI of 1.20–1.49. No Nationally Scarce or Red Data Book species |
| Very high | Supports one or more Nationally Scarce or Red Data Book species, or SRI of 1.50 or more, or an exceptionally rich invertebrate assemblage (50 or more species) |

easily degraded, with changes not 'diluted' in their impacts and, conversely, may be completely protected from land-based disturbances to which larger water bodies are almost inevitably exposed.

The importance of conservation of 'pondscapes', as major contributors to regional aquatic biodiversity, is recognised increasingly as encompassing the collective variety supported by individual ponds in an area. Hill et al. (2018) attributed the significance of ponds to three main factors, as (1) the small catchments of individual ponds leading to local variations and collective habitat heterogeneity; (2) the variety of artificial ponds (such as farm dams, p. 270) increasing the overall habitat area; and (3) the provision of refuge habitats for aquatic communities in progressively altered landscapes. They commented that current conservation policy globally is failing to conserve much aquatic biodiversity in pondscapes, and that these scenarios should be incorporated more effectively into conservation policy at all levels.

## 2.3 Streams and Rivers

Running waters provide milieux through which transport by water flow is usual, so that conditions, disturbances and changes to any upstream area can potentially affect all downstream environments. Simplistically, each stream or river is a continuum of connected linear systems along the entire length from source to sea, with gradients in both physical and biological features reflecting changed environments along that length. The continuing and varied drives to harness and sustain Australia's rivers focus largely on human needs, superimposing numerous anthropogenic changes on these ecosystems.

Since the 'River Continuum Concept' (RCC) was advanced to help understand the natural changes along a river (Vannote et al. 1980), an enormous variety of studies have assessed its ramifications for interpreting the distribution and ecological patterns of lotic biodiversity (Stendera et al. 2012; Tornwall et al. 2015). Benthic macroinvertebrates are amongst the most frequently and intensively studied biota in those endeavours. The RCC emphasised that changes along a longitudinal river gradient can be both predictable and quantifiable and, in Tornwall et al.'s words, 'has inspired generations of studies in lotic ecosystems'. It has stimulated much further understanding—with varying levels of support or dissent—by providing a hypothesis on ecosystem changes along a river continuum from headwaters to mouth, and aiding definition of the conditions against which human impacts can be appraised. Later evaluations (such as the Riverine Ecosystem Synthesis of Thorp et al. 2006) recognise the limitations of a continuum perspective and attempt to predict how patterns of species distributions and ecological processes can be expected to vary over a range of spatial and temporal scales in relation to physical differences throughout river/stream networks. The varied patterns of rivers—the arrangements of channels in the landscape—are determined largely by structure and slope (Twidale 2004).

The RCC was developed specifically in reference to natural stream ecosystems and was considered to accommodate many anthropogenic disturbances that essentially change either nutrient loads (Vannote et al. noted nutrient enrichment, organic pollution, changes in riparian vegetation) or transport (such as high sediment load or impoundment). They postulated that such changes could be considered 're-set mechanisms' that cause the continuum response to be shifted either seaward or toward the headwaters. The RCC thereby incorporated a range of biological strategies along a river and related to the gradient of physical conditions. Those physical variables form a continuous dynamic set of limiting conditions, and the assemblages of species present respond in their occurrence and composition to the gradient stages they encounter.

The postulated correspondences can also include functional relationships. For example, in Vannote et al.'s (1980) statement topics such as energy input, organic matter storage, transport and use by macroinvertebrate functional feeding groups (p. 166) may be regulated largely by physical processes. They appraised a river as a gradient from heterotrophic headwater environments to a more seasonal autotrophic regime in the middle reaches and downstream return to more heterotrophic processes.

A rather broader perspective of lotic ecology recognised that many rivers are not simply single channels to which an idealised RCC can apply but, rather, have multiple channels, floodplains and/or groundwater aquifers, so that heterogeneity is increased considerably. Ward and Tockner (2001) recognised a range of major environments in river corridors that contribute biodiversity to lotic ecosystems (Fig. 2.2), with complex interplays between the three major systems. In essence, the longitudinal gradient picture on which the RCC was founded extends to the more complex environmental gradients along that length through lateral inputs and vertical influences from subsurface waters. Rivers expand or contract in any dimension as response to flow regimes.

Continuously flowing rivers thereby present series of gradients along their length, with many of their features influenced from upstream inputs and any point sources of disturbance perhaps affecting substantial downstream reaches, in some cases with gradual 'dilution' of impacts with distance downstream. Invertebrate communities are thus not as clearly isolated as those in more discrete ponds, and changes may

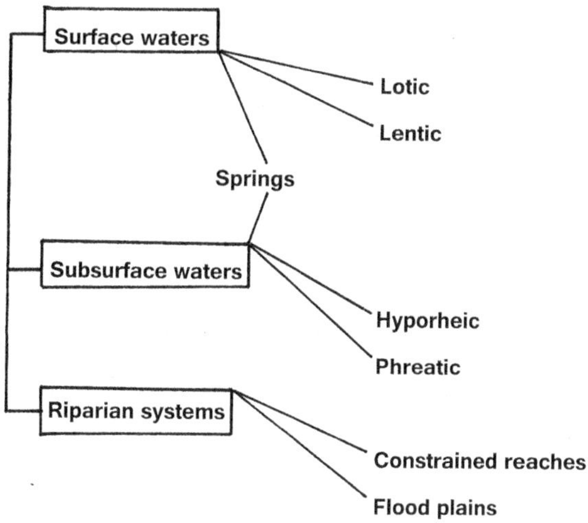


**Fig. 2.2** The major environments within river corridors that contribute biodiversity to lotic ecosystems (Ward and Tockner 2001)

occur from along-stream dispersal (such as drift or flight). The latter (p. 132) is clearly a primary mode of replenishment in still water bodies but is important also in rivers and streams—in which longitudinal movements by flying insects can avoid them traversing inhospitable terrestrial terrains as they seek breeding sites.

## References

Apinda Legnouo EA, Samways MJ, Simaika JP (2014) Value of artificial ponds for aquatic beetle and bug conservation in the Cape Floristic region biodiversity hotspot. Aquat Conserv Mar Freshw Ecosyst 24:522–535

Bayley IAE, Williams WD (1973) Inland waters and their ecology. Longmans, Sydney

Biggs J, Fox G, Nicolet P, Walker D, Whitfield M, Williams P (1998) A guide to the methods of the National Pond Survey. Pond Action, Oxford

Biggs J, Williams P, Whitfield M, Nicolet P, Weatherby A (2005) 15 years of pond assessment in Britain: results and lessons learned from the work of Pond Conservation. Aquat Conserv Mar Freshw Ecosyst 15:693–714

Briers RA, Biggs J (2003) Indicator taxa for the conservation of pond invertebrate diversity. Aquat Conserv Mar Freshw Ecosyst 13:323–330

Briers RA, Biggs J (2005) Spatial patterns in pond invertebrate communities: separating environmental and distance effects. Aquat Conserv Mar Freshw Ecosyst 15:549–557

Cereghino R, Biggs J, Oertli B, Declerck S (2008) The ecology of European ponds: defining the characteristics of a neglected freshwater habitat. Hydrobiologia 597:1–6

Cereghino R, Oertli B, Bazzabti M, Coccia C, Compin A, (and six other authors), (2012) Biological traits of European pond macroinvertebrates. Hydrobiologia 689:51–61

Cereghino R, Boix D, Cauchie H-M, Martens K, Oertli B (2014) The ecological role of ponds in a changing world. Hydrobiologia 723:1–6

Chadd R, Extence C (2004) The conservation of freshwater macroinvertebrate populations; a community-based classification scheme. Aquat Conserv Mar Freshw Ecosyst 14:597–624

Collinson NH, Biggs J, Corfield A, Hodson MJ, Walker D, Whitfield M, Williams PJ (1995) Temporary and permanent ponds: an assessment of the effects of drying out on the conservation value of aquatic macroinvertebrate communities. Biol Conserv 74:125–133

Downing JA, Prairie YT, Cole JJ, Duarte CM, Tranvik LJ, (and six other authors), (2006) The global abundance and size distribution of lakes, ponds and impoundments. Limnol Oceanogr 51:2388–2397

Gioria M, Schaffers A, Bacaro G, Feehan J (2010) The conservation value of farmland ponds: predicting water beetle assemblages using vascular plants as a surrogate group. Biol Conserv 143:1125–1133

Hamerlik L, Svitok M, Novikmec M, Ocadlik M, Bitusik P (2014) Local, among-site and regional diversity patterns of benthic macroinvertebrates in high altitude waterbodies: do ponds differ from lakes? Hydrobiologia 723:41–52

Hill MJ, Hassall C, Oertli B, Fahrig L Robson BJ (and six other authors) (2018) New policy directions for global pond conservation. Conserv Lett 11:e12447. https://doi.org/10.1111/conl.12447

Jeffries M (2005) Small ponds and big landscapes: the challenge of invertebrate spatial and temporal dynamics for European pond conservation. Aquat Conserv Mar Freshw Ecosyst 15:541–547

Oertli B, Joye DA, Castella E, Juge R, Cambin D, Lachavanne J-B (2002) Does size matter? The relationship between pond area and biodiversity. Biol Conserv 104:59–70

Stendera S, Adrian R, Bonada N, Canedo-Arguelles M, Hugueny B, (and three other authors), (2012) Drivers and stressors of freshwater biodiversity patterns across different ecosystems and scales; a review. Hydrobiologia 696:1–28

Thorp JH, Thoms MC, Delong MC (2006) The riverine ecosystem hypothesis: biocomplexity in river networks across space and time. River Res Appl 22:123–147

Tornwall B, Sokol E, Skelton J, Brown BL (2015) Trends in stream biodiversity research since the River Continuum Concept. Diversity 7:16–35

Twidale CR (2004) River patterns and their meaning. Earth Sci Revs 67:159–218

Vannote RL, Minshall GW, Cummins KW, Sedell JR, Cushing CE (1980) The River continuum concept. Can J Fish Aquat Sci 37:130–137

Ward JV, Tockner K (2001) Biodiversity: towards a unifying theme for river ecology. Freshw Biol 46:807–819

Williams P, Whitfield M, Biggs J, Bray S, Fox G, Nicolet P, Sear D (2004) Comparative biodiversity of rivers, streams, ditches and ponds in an agricultural landscape in southern England. Biol Conserv 115:329–341

Wissinger SA, Oertli B, Rosset V (2016) Invertebrate communities of alpine ponds. In: Batzer DP, Boix D (eds) Invertebrates in freshwater wetlands: an international perspective on their ecology. Springer, Switzerland, pp 55–103

Wood PJ, Barker S (2000) Old industrial mill ponds: a neglected ecological resource. Appl Geogr 20:65–81

Wood PJ, Greenwood MT, Barker SA, Gunn J (2001) The effects of amenity management for angling on the conservation value of aquatic invertebrate communities in old industrial ponds. Biol Conserv 102:17–29

# Chapter 3
# Australian Inland Waters

## 3.1 Introduction

The great variety of aquatic ecosystems of Australia includes some of especial global significance, none more so than the unique range of waterbody types within the vast arid biome of the continent's interior (Table 3.1), but with the continent's major drainage basins spanning both endorheic and exorheic features. 'Inland waters' are generally delimited as the 'surface and subsurface water resources not associated with the seas', thus excluding estuaries and coastal lagoons. Most are freshwater but some are markedly saline, and their variety also spans permanent and temporary waterbodies (Fairweather and Napier 1998). The 12 major drainage basins in Australia (Fig. 3.1, note that some authors recognised 13 basins, separating the South-East Coastal into two state-based units, as indicated) collectively include 456 river basins, amongst which most management and conservation interest has focused on the most densely populated regions of the South-East Coast Drainage Division and the Murray-Darling Basin Drainage Division. The 12 drainage basins include ten that are exorheic (draining to the sea), one is endorheic (draining to the inland) and one is arheic (with no well-defined drainage). Of the first category, the enormous Murray-Darling Basin complex in the southeast has received the greatest political and ecological attention as the water source for a large proportion of the country's people and economic activity. It covers about one seventh of Australia's land area. This and the South-East Coastal Basin supply the areas in which most of Australia's people live and so provide water for much of the populace. Their biotas are the best-documented of all the drainage areas. Collectively, the exorheic systems include more than 98% of water drainage from the continent, although comprising under half the land area (Williams 1983). The North-East Coast Basin has the greatest annual discharge of the twelve.

Reflecting the generally low relief of Australia, alpine and other upland waters are restricted to the south east, and support numerous endemic biota, many of them known from only small areas and of particular conservation concern. Isolation is also a feature of some inland waters. Some depend wholly on ground water, and



T. R. New, *Insect Conservation and Australia's Inland Waters*,
https://doi.org/10.1007/978-3-030-57008-8_3

**Table 3.1** Australia's drainage divisions, with approximate area of each (from Boulton et al. 2014) (see Fig. 3.1, for a somewhat different terminology)

| Drainage division | Area (1000 $km^2$) |
|---|---|
| Pilbara Gascoyne | 478 |
| North-Western Plateau | 716 |
| Tanami-Timor Sea Coast | 1162 |
| Carpentaria Coast | 647 |
| North-East Coast | 451 |
| Lake Eyre Basin | 1200 |
| Murray-Darling Basin | 1061 |
| South-East Coast (New South Wales) | 129 |
| South-East Coast (Victoria) | 135 |
| Tasmania | 68 |
| South Australian Gulf | 118 |
| South-Western Plateau | 1093 |
| South-West Coast | 326 |

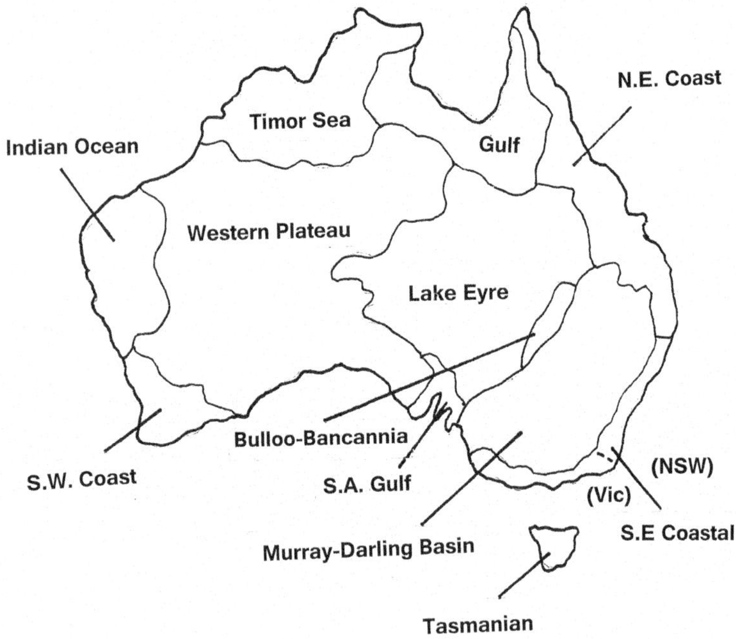


**Fig. 3.1** The major drainage basins of Australia

some are connected with riverine networks (Davis et al. 2016). Others, such as rockholes, are supplied mainly or solely by surface water, and infrequent rainfall allows development of temporary river networks through which isolation may be reduced over short periods. Many such connections are highly transient, and unpredictable in occurrence. Hydrologically isolated communities often support more distinctive invertebrate faunas than do closely grouped riverine communities, and Davis et al. suspected that distributions of some families indicated diversification within the sites, with ancient dispersal processes and very limited recent dispersal. Distributions of some low-mobility taxa (they noted psephenid beetles, p. 193) in inland Australia may be related to ancient past aridification. Amongst the most isolated inland waters, rockholes in the Great Victoria Desert were dominated by taxa that have desiccation-resistant eggs that may remain dormant for many years (some crustaceans), a feature shared by some weakly dispersing insect groups (notably some Diptera: Chironomidae, Culicidae).

Despite the extreme aridity of much of central Australia, a considerable variety of waterbodies occur—Davis (1997) commented that 'the only waterbodies not represented are permanently flowing streams and rivers'. In one complex site, in the West MacDonnell Ranges, Northern Territory (noted as 'perhaps the most isolated freshwater habitats in Australia'), the 75 macroinvertebrate taxa recorded were dominated by insects (58 species identified), with Odonata and Coleoptera the richest orders—each with 16 species. Many of the dragonflies were widespread, even nomadic, species but a few were more localised and Davis noted that *Hemicordulia flava* (the Desert emerald, Corduliidae) appeared to be endemic to the Central Ranges region. Plecoptera were absent and several species of Coleoptera and Hemiptera were noted to occur only in waterbodies from which fish were absent. Increasing tourism and expanding pastoral activity were potential threats to these systems, with risk that human uses and needs for water could compete with natural ecosystems for the very limited water supply. Davis recommended that overpumping of ground water should not be allowed.

The following examples demonstrate the great variety of inland water ecosystems present as habitats for insects in Australia.

## 3.2 Waterfalls

The ability of some insects to resist water flow is most fully demonstrated by the assemblages that occur on the vertical rock faces of waterfalls, where species of Plecoptera, Coleoptera and Trichoptera can be well represented (Rackemann et al. 2013). Comparison of the 131 insect taxa found on the rock face itself, directly below the waterfall and in a nearby riffle of 12 waterfalls in western Victoria in autumn and winter samples showed riffles to support higher richness—but also that some taxa occurred only on the vertical face, where highest occurrences were in the most highly vegetated areas, with up to about 30% moss cover. As a likely refuge for such specialised fauna on the rock faces, conservation of moss cover may be an important conservation focus in such extreme habitats. Rackemann et al. pondered whether waterfalls may provide a refuge for some rheophilic species over periods of low water flow and noted also the range of specialised morphological and behavioural traits that enabled them to live in these torrential environments. Some such taxa are rare. The flightless stonefly *Dinotoperla walkeri*, for example, has been recorded only from two waterfalls and one adjacent riffle area in Victoria (Dean and St Clair 2006), and the more recent surveys there also yielded a likely new species of the related genus *Trinotoperla*.

## 3.3 Subterranean Aquifers

The complexities of Australia's vast and still far under-explored subterranean waters are reflected in the then 55 families of fauna associated with them, of which 44 families are crustaceans, and only one represents insects (diving beetles, Dytiscidae, p. 192) (Humphreys 2006). The crustaceans include a number of highly unusual lineages and, as with other taxa present, have very restricted dispersal power or opportunity and are typically short range endemic species. Likewise, all Dytiscidae are endemics, but have radiated to become by far the richest stygofaunal group represented. However, they are not universally present, a substantial survey in the Pilbarra region (Eberhard et al. 2005) not yielding any beetles (or other insects) amongst the 78 species (in 21 families) recovered.

The importance of these unique habitats thereby rests on a variety of biota that depend on groundwater and have much to contribute to understanding Australia's relictual invertebrates. The beetles present are relatively recent (Tertiary) invaders of these habitats, but the broader elements of the stygofauna that merit conservation were listed by Humphreys (2006) as (1) lineages with marine origins and Tethyan affinities, and (2) ancient continental freshwater lineages, including many of the most unusual crustaceans, as well as (3) the Tertiary invaders. He considered that the three 'are of great scientific and conservation significance', and that many—perhaps most—species fulfil formal criteria for being 'vulnerable' or 'endangered' on grounds of the small areas inhabited or because they occur in only single, and restricted, sites.

## 3.4 Rock Pools

Small, mostly precipitation-dependent, rock pools are found throughout the world and, as with other small waterbodies such as phytotelmata, can support characteristic invertebrate communities. About 460 aquatic animal species have been reported from freshwater rock pools (Jocque et al. 2010). Because classical rock pools are not supplied by groundwater, that fauna is in general not water-transported but comprises immigrants through active dispersal and those able to persist through refuge use (p. 279) to survive unpredictable flooding cycles.

Rock pools in Australia, often termed 'gnammas' (from the Nyungar language) and many of them on granite outcrops, are each an isolated habitat island that over time may become filled with sediment and/or overrun by terrestrial plants as succession proceeds. Technically, gnammas are rock holes formed by chemical weathering and typically have rounded bottoms, but the term is often used more widely to encompass more irregularly-shaped pools formed by physical processes. Major groups of actively dispersing insects are Odonata, Hemiptera, Coleoptera and Diptera, while most crustaceans are 'passive dispersers' hatching from desiccation-resistant eggs in the pools. Records of diversity in Western Australia cited by Jocque et al. (2010) include (1) 66 species across 92 pools on two outcrops; (2) 88 species in 36 pools on 17 granite outcrops; and (3) 230 species from 90 pools on nine different rocks. Such richness renders old rock formations with pools in Western Australia possible 'biodiversity hot spots' as supporting specialised endemic taxa, largely of non-insect passive dispersers. Crustaceans are usually far more diverse than insects in these habitats. In the Great Victoria Desert, for example, only Culicidae were widespread, found in seven of 12 localities surveyed, Chironomidae and other Diptera were less frequent and the only other insects reported were Zygoptera, from a single locality (Bayley et al. 2011). Some other insects (Hemiptera, Coleoptera) noted as common in gnammas elsewhere were not found in that survey, possibly reflecting the absence in this arid environment of source areas such as farm dams.

## 3.5 Mound Springs

Springs fed from artesian water are a notable feature of Australia's extensive Great Artesian Basin, an area of about 1.76 million km$^2$—or about 22% of the Australian continent (Ponder 1986). Many are a permanent source of water, and many springs also support endemic fauna, including fish and invertebrates. Major radiations of hygrobiid snails occur, together with crustaceans and ostracods. However, although aquatic insect larvae can be varied and abundant, Ponder noted that 'They have not been studied but it is assumed that most are widespread species'. Other surveys revealed a considerable array of wolf spiders (Lycosidae), some endemic (Gotch 2001), and a variety of terrestrial insects in the immediate vicinity of individual springs (Greenslade 1985). However, more detailed investigations remain to be done.

Mound springs have considerable cultural significance, Degradation has occurred through trampling and fouling by stock, and some have been modified by digging or damming to increase water supply, steps noted by Ponder (1986, 2003) as leading to extinctions of some endemic invertebrates. Introduced mosquito fish occur in many springs. Proliferation of artesian bores has also led to extinctions through draw-down of water for pastoral use, which has led to water conservation measures by capping bores to restrict wastage. Losses of many springs, and degradation of many others would need considerable rehabilitation to redeem their former condition. However, as with subterranean aquifers (above) their scientific and evolutionary revelations are likely to be considerable.

Geographical knowledge is still very uneven. In contrast to the relatively well-documented South Australian springs, insects of those in Queensland are poorly documented; they include many on pastoral leases that are degraded and without any conservation measures in place (Ponder 2003). Molluscs are rather better known—but Ponder also commented that probably more than half the Queensland springs were then so heavily degraded that 'any fauna they may have contained is now extinct'. There is thus urgent need to conserve remaining springs with endemic fauna.

## 3.6 Lakes

As Timms (1985) commented, there are rather few large freshwater lakes in Australia, together with rather more medium-sized lakes, of 1–10 $km^2$. The variety present nevertheless makes faunal comparisons, such as of benthic invertebrates, difficult, with different depths, salinity and morphology all affecting community structure. Semantic differences to distinguish lakes from ponds abound, but water depth is a fundamental parameter. Lakes are deep enough for most or all of their bottom to be free of rooted vegetation (Bayley and Williams 1983), whilst ponds are sufficiently shallow to allow rooted vegetation to occur. The dynamics of Australia's inland lakes reflect that most of them are episodic, many being dry for much of the time, and saline when filled. Timms (2001) noted that the arid zone (Fig. 1.1, p. 3) contains some permanent or near-permanent lakes and wetlands, although many are small and isolated reaches (waterholes) of intermittent rivers. Invertebrate assemblages present in the variable environments of such lake systems are influenced by water permanence, salinity and turbidity, collectively providing a complex mosaic of conditions. Detailed invertebrate data are available for some large arid zone lakes, but many species are widespread—Timms commented that the majority of species are 'ubiquitous'—and rather few species seem to be catchment-level endemics. Levels of endemism may be higher in some smaller wetlands.

Four major categories of salt lakes are found in Australia (De Dekker 1988), and their restricted invertebrate fauna is related to widespread lack of ability to omosmoregulate in saline water and that many lakes are ephemeral. Salt lakes are also generally rather shallow. In general, invertebrate diversity is lower in salt lakes than in freshwater lakes, but the former assemblages are largely dominated by crustaceans (mostly endemic), with rather few insects, such as ephydrid flies, present. Chironomidae collectively span most salinity regimes present. Collectively, the insects present have received far less attention than the more diverse crustaceans. With few exceptions, Australian salt lakes have concentrations of sodium chloride similar to those of sea water. Salinity ranges over about 3–350% total dissolved solutes (TDS), where 'TDS' as saturated salt solution is ten times sea water. Many occur near the coast, and contrast with the large intermittently filled inland basins such as Lake Eyre (when in flood extending over up to about 10,000 $km^2$) in which environmental extremes are wide.

More broadly, high salinity may harm some invertebrates in rivers and elsewhere, but specific measures of tolerance levels and ranges are rather sparse (p. 95). Numerous human activities threaten salt lakes, and Williams (2002) considered many of those changes irreversible. He noted that impacts of increased salinisation extended well beyond the lakes themselves as this can also affect freshwater bodies and widely threaten water resources in arid and semi-arid regions. The major effects of secondary salinisation from human activities were listed as: (1) changes in the natural character of many water bodies in semi-arid regions, frequently including changes to natural hydrological patterns; (2) replacement of a less salt-tolerant biota by a more salt-tolerant one; and (3) decrease in biodiversity. Raising awareness of the values of salt lakes, and of the threats to them were seen as necessary preliminary steps to initiating informed management and conservation measures to minimise and mitigate human impacts.

Deliberate flooding to create reservoirs or to expand pre-existing lakes (such as the classic case of Lake Pedder, flooded for hydroelectricity supply in south western Tasmania) have led to modified substrates, with shallower areas now including drowned forest—some with characteristic dead trunks, stags, protruding above water and providing substrates and nutrition for a considerable variety of invertebrates. The shallow shores of Lake Pedder include rocky substrates that have allowed freshwater sponges to proliferate, and support a very large population of an endemic spongefly (Neuroptera, Sisyridae, p. 196), with the dead stags providing surfaces for larvae to leave the water and form cocoons.

## 3.7 Billabongs

Billabongs are permanent or semi-permanent waterholes in lowland flood plains that are often dry except during floods and the wet season. They include waterbodies such as oxbow lakes and meander cutoffs and vary greatly in form and features. As Hillman (1986) put it 'The 'standard' billabong, like the standard snowflake, does not exist'. Larger lakes also occur, also inundated only during periods of high river flow and spread of water across the floodplain. Floodplains in Australia are a characteristic of lowland rivers, and the areas with features that allow for water retention over any significant period are generally termed 'wetlands'. These tend to have high biodiversity and, more widely, lowland rivers comprise by far the most dominant river form in Australia. This reflects the largely flat nature of the continent, and Thoms and Sheldon (2000) defined these as river reaches below 300 m elevation for inland systems, and between 40 m and the tidal limit for coastal systems. On those criteria, lowland rivers comprise about 97% of the total Australian river length. Most (83%) are inland systems, with semi-arid to arid catchments, and may cease to flow seasonally or under more protracted drought.

Billabongs support numerous invertebrates, in some cases with Chironomidae predominant (Balcome et al. 2007). Macroinvertebrate communities of billabongs can differ markedly from those of their associated flowing waters and can show strong seasonal patterns. Comparing species from Ryans Billabong (Albury-Wodonga,

**Table 3.2** Insect species representation in Ryan's Billabong, Victoria, and three nearby river sites, 1978–1982 (Hillman 1986)

| | No. of species collected | | |
|---|---|---|---|
| Order | Billabong only | River sites only | Both environments |
| Ephemeroptera | 0 | 10 | 3 |
| Odonata | 9 | 7 | 5 |
| Plecoptera | 1 | 3 | 0 |
| Hemiptera | 9 | 11 | 8 |
| Coleoptera | 6 | 10 | 6 |
| Diptera | 13 | 28 | 16 |
| Trichoptera | 1 | 10 | 9 |
| Lepidoptera | 1 | 1 | 1 |

Victoria) with those from three sites on the River Murray and Kiewa River up to about 15 km from the billabong, only 66 of the 209 macroinvertebrate species occurred in both environments, so that characteristic floodplain or mainstream taxa were found (Table 3.2) (Hillman 1996).

## 3.8 Streams and Rivers

Running waters, by definition, do not exhibit the levels of isolation of many lakes and ponds—they are essentially continuous connected systems from headwaters to discharge, along pathways that are fundamentally linear, but may have many networking ramifications and linkages. Different patterns (exemplified in Fig. 3.2) imply different levels of functional connectivity and likely distributional associations between inhabitants. Headwater streams can support insects and others under relatively isolated conditions and can acquire considerable conservation importance for these roles. However, as Clarke et al. (2008) noted, the epithet 'important' so often used in this context is ambiguous, and it is often not clear whether it is intended to convey (1) that they contain restricted headwater species that do not occur elsewhere; (2) a number of undescribed species or other poorly known taxa; or (3) simply that they have high taxonomic richness. Clarke et al. found some evidence to support the first two of these conditions, but also pointed out the widespread lack of knowledge of the diversity patterns of macroinvertebrates in headwater streams. Those streams play significant roles in catchments as refuges, for example from introduced predators found in lower reaches, and for the extremes of temperature and flow that can become more evident downstream. The variety of studies reviewed by Clarke et al. (2008) confirmed considerable differences for invertebrate richness, and headwater streams do not necessarily support higher richness than downstream reaches—the variety of sampling methods used by different workers can preclude effective comparisons across some surveys.

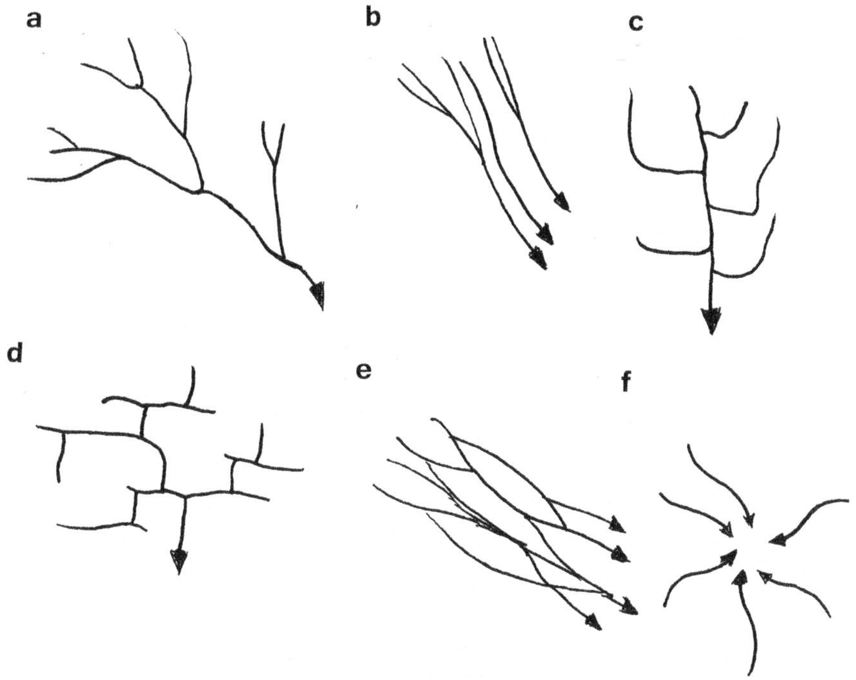


**Fig. 3.2** Selected patterns of stream flow, to illustrate variety and extent of possible connectivity: **a**, dendritic; **b**, parallel; **c**, trellis; **d**, rectangular; **e**, braided; **f**, centripetal (based on Bayley and Williams 1981)

The great range of headwaters (defined by Richardson 2019 as streams having no permanent tributaries, as the first flowing streams in a network) differ from other freshwater categories in being more isolated, with their biota largely depending on allochthonous resources and as potential refuges (p. 279) from larger predators found downstream. Their insect populations, limited in part by the small extent of their habitats, tend to be small and isolated—in themselves factors fostering taxonomic differentiation and also rendering them vulnerable to extirpation or extinction. Reasons for headwaters sustaining species not found downstream, tabulated by Richardson (2019), are (1) enemy-free space; (2) lack of competition; (3) unique physical and chemical environments, documented for some Odonata; (4) seasonal refuges, as from varying flow or temperature; (5) breeding and rearing sites, fostering higher survival; and (6) detritus-based food webs, as cited for some stoneflies. Hundreds to >1000 species, many of them characteristic of headwater environments, may occur in such streams.

Higher elevation headwaters fed by snowmelt can support restricted cold-adapted endemic aquatic insects in Europe (Brown et al. 2007), North America (Kubo et al. 2013) and New Zealand (Winterbourn et al. 2008), and Australian alpine streams are no less significant, as illustrated by Plecoptera (p. 188) and others. Increased documentation and protection based on knowledge of such key species, is needed to ensure the future of these unique and sensitive ecosystems.

The contrasts between 'linear' and 'network' surveys to detect supposed linear patterns of species richness are summarised in Fig. 3.3. The former, based on samples from single sites on headwater, mid-order and high-order streams, might reveal

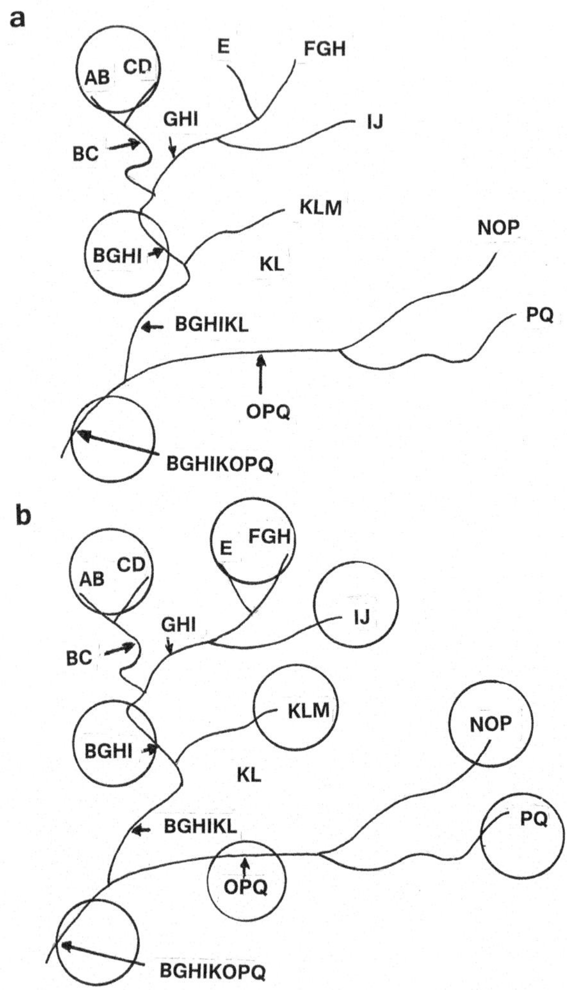


**Fig. 3.3** Two contrasting sampling designs for investigating longitudinal changes in macroinvertebrate richness: **a** traditional 'linear' approach, inferring that sampling from one part of a headwater network (circles represent sampling sites) can enable interpretation/inferences about diversity patterns (taxa shown as letters) across all headwaters in the network; **b** a 'network' sampling approach ensures that empirical data are collected from across the headwater network, and illustrates how taxonomic richness in headwaters may be severely underestimated by traditional linear sampling approaches (Clarke et al. 2008)

higher richness in the high-order stream, whilst a network sampling approach that comprises the total richness from all headwater streams in the catchment with that of all mid-order streams and with the high-order main stream might reveal greatest combined taxonomic richness in the headwaters. One 'explanation' for this would be greater diversity among individual headwater streams (high β-diversity). Clarke et al. suggested that the traditional linear sampling approach may dramatically underestimate the taxonomic richness of headwater networks in areas where very low connectivity among different headwater streams of a catchment occurs (Gomi et al. 2002, p. 231). Headwater biota may be especially vulnerable to disturbances in their nearby landscapes (Lowe and Likens 2005), and these influences are likely to be exacerbated by anticipated climate changes. Elevation may also be an important correlate of community composition in headwaters.

Attributing any changes in aquatic insect representation in running water to any specific cause is difficult. For stream and river taxa, effects of changing water flow, alien species, local sedimentation and water quality, nutrient availability and others can occur simultaneously, and partitioning observed changes amongst these can become very subjective and uncertain. However, Bunn and Arthington (2002) averred that 'flow regime' is a key driver of river and related ecosystems, with changes 'often claimed to be the most serious and continuing threat' to their wellbeing. They proposed four major principles about influences of flow regimes on aquatic biodiversity, with the major interacting features shown in Fig. 3.4. From their text, the principles are (1) flow is the major determinant of the physical habitats in streams, these in turn being a major determinant of biotic composition (with changes largely from floods or drought); (2) aquatic species have evolved life history strategies primarily in direct response to natural flow regimes; (3) the maintenance of natural patterns of both longitudinal and lateral connectivity (p. 233) is essential to the viability of populations of many riverine species; and (4) the invasion and success of alien and introduced species in rivers is facilitated by the alteration of flow regimes.

Influences of flow pattern on invertebrate diversity are reasonably well documented for benthic fauna, but less so for the hyporheic zone (p. 65). However, in the Selwyn River, New Zealand, the large array of hyporheic invertebrates—with 30 of the 56 taxa collected being insects—increased in diversity, density and stability of assemblages with increasing permanence of flow (Datry et al. 2007). That study examined invertebrates from stations along a 52 km stretch of river in which different regions have consistently different flow periods.

Intermittent streams, by definition, undergo periods when surface water is absent, often during a more-or-less prolonged dry season, and for some only under more protracted drought conditions. Aquatic macroinvertebrates must avoid desiccation during such times, adopting behavioural, physiological or phenological strategies to do so. If their resistance is effective, recovery can be rapid once water is again available. As Leigh et al. (2016) remarked, far less is known about invertebrate community responses to dryness than about responses to flooding. Both processes are intricate and functionally far-reaching (Chap. 5). Biotic responses to flow changes in intermittent streams are more difficult to monitor than those in perennial streams, reflecting the additional transitory conditions from wet to dry (England et al. 2019).

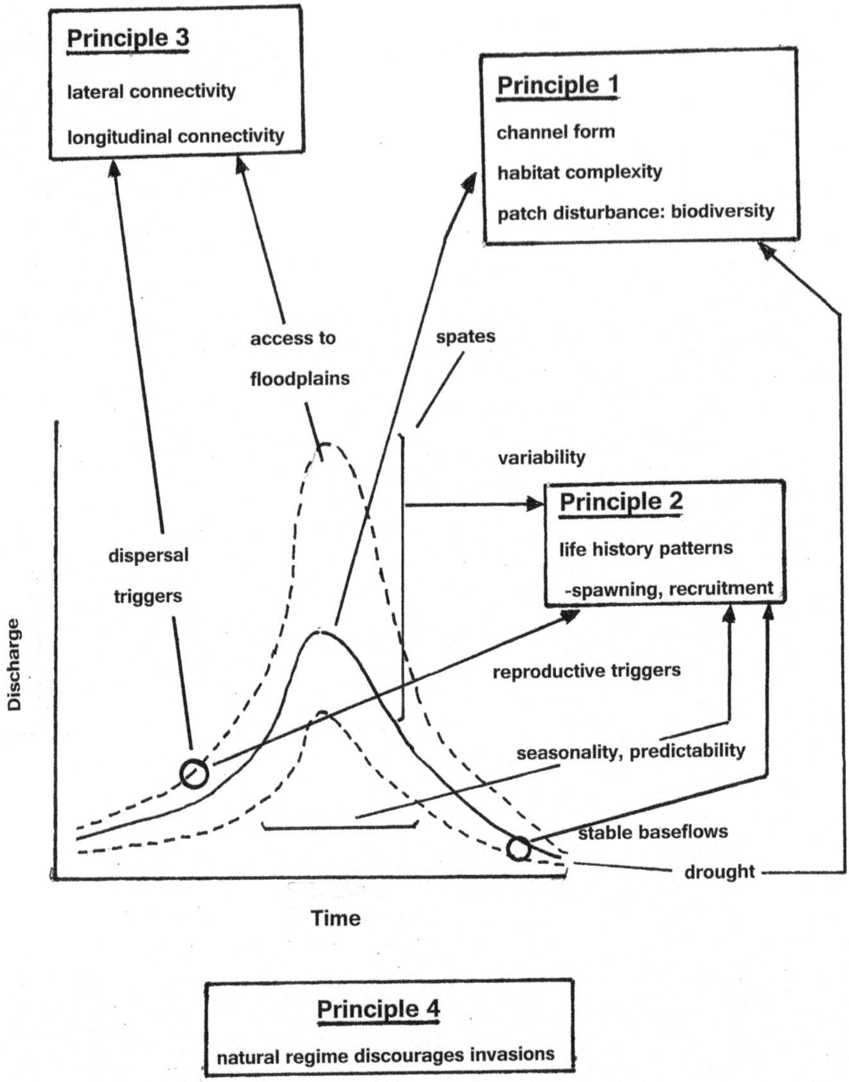


**Fig. 3.4** Relationships between natural flow regimes and aquatic biodiversity: four principles reflecting that native biota have evolved in response to the overall flow regime (solid line) which may increase or decrease (dashed lines). 1, the relationships may be driven by large events that influence channel form or shape; 2, timing, selectivity and predictability of particular flow events influence life-history patterns; 3, access to otherwise disconnected habitats can be triggered by flow regimes; and 4, incursions by alien species to the detriment of native biota can depend on their adaptation to modified flow regimes (Bunn and Arthington 2002)

Whilst indices such as LIFE (p. 121) can assess responses to changing flow regimes (Extence et al. 1999), England et al. incorporated wider considerations by including aquatic, semi-aquatic and terrestrial taxa to assess changes in habitat availability from hydrological variability. That 'MIS-index' (Monitoring Intermittent Stream-index) may have wide potential, but England et al. noted that further testing was necessary.

Two categories of streams that cease to flow have long been distinguished (Bayley and Williams 1973), as (1) intermittent, that flow only seasonally and mostly in Australia drain semi-arid regions, and (2) episodic, that drain only arid regions and flow only after unpredictable rainfall. A third category, ephemeral streams, flow only in direct response to snow melt or rainfall and receive no appreciable water from other sources. The unifying feature across all these categories is the repeated, not necessarily regular, onset and cessation of water flow. Whatever the category, many temporary streams occur in headwater regions, so supply both local invertebrate habitats and resources for downstream ecosystems (Buttle et al. 2012).

Levels of predictability of flow thereby differ greatly, and more than half the Australian continent is drained by intermittent rivers and streams (Boulton and Suter 1986). In the few Australian temporary streams (in Victoria and South Australia) for which data were then available, insects comprised more than three-quarters of the invertebrate fauna, with generally similar proportions of the major orders present. Diptera were the main component, followed by Coleoptera and Trichoptera. Permanent streams supported greater richness of Trichoptera and Ephemeroptera. However, the relatively little work that had then been undertaken of the wider ecology of temporary streams in Australia led Boulton and Suter to comment that 'it is virtually impossible to judge how vulnerable they are' to anthropogenic disturbance, and that most management and monitoring programmes have been developed from work on permanent streams alone.

## 3.9 Exposed Riverine Sediments

Variations in water level in lotic systems lead to areas of the bedding substrate becoming exposed in dry periods—in extreme cases forming more-or-less permanent ridges or banks within or bordering the main flow channel. These exposed riverine sediments (ERS) have received increased attention over recent decades as their ecological significance has been recognised. Bates et al. (2009) distinguished three major themes in assessing the importance of the arthropod assemblages supported by ERS as (1) the diversity of species and incidence of rare taxa; (2) transfer of energy and nutrients between aquatic and terrestrial systems; and (3) anthropogenic impacts on ERS. Although the term 'ERS' generally conveys the idea of relatively unvegetated alluvial areas within the river/stream channel and that comprise the 'aquatic-terrestrial ecotone' (Bates et al. 2007), the definition used by Bates et al. (2009) is more precise, as 'Exposed within channel, fluvially deposited sediments (gravels, sands and silts) that lack continuous vegetation cover, whose vertical distribution lies between the levels of bankfull and the typical base flow of the river'.

They are created and maintained by flood pulses—the latter because high flow disturbances impede vegetation succession and also thwart competitive exclusion of ERS specialist insects by less specialised taxa. ERS are characteristically warm and dry over much of the year when arthropods are most active, because of their freely draining substrates. Many ERS habitats are characterised by high temperatures and low moisture, combined with high rates of disturbance and inundation.

The fauna of ERS has been surveyed most thoroughly in parts of Europe and North America, and show a substantial number of rare beetles, together with rare spiders and other insect groups. Thus, in the United Kingdom 131 beetle species are regarded as ERS specialists, and 86 of these have been recognised more formally by inclusion in the Red Data Book or designation as 'Nationally Scarce' (Bates and Sadler 2005). However, determining the precise conservation status of many of these awaits further documentation of their distribution and population sizes and stability. In the meantime, and with a substantial array of threats to ERS habitats, their arthropod assemblages are 'an important conservation priority' (Bates et al. 2009). Threats to ERS can be either direct or indirect and many such systems have been altered considerably by unnatural disturbances. Many of the 'indirect' threats influence community change by changing flood regimes, as summarised in Fig. 3.5. Threats noted by Bates et al. (2009) include livestock trampling, extraction of gravel or aggregate from the substrates, invasive plant species, modified channelisation

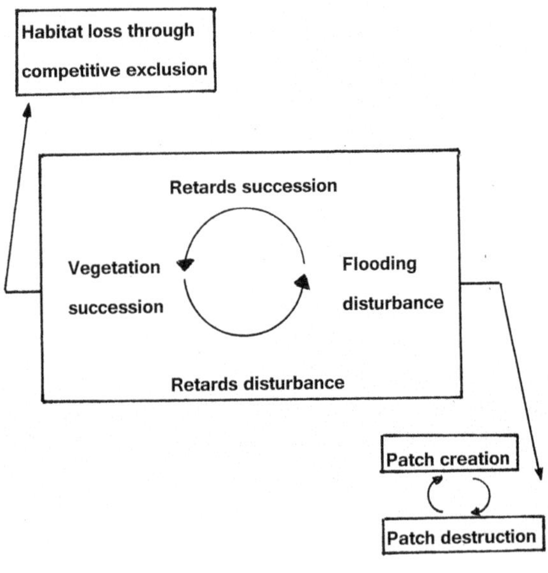


**Fig. 3.5** The dynamic equilibrium between succession and flood disturbance. Any shift upward in diagram (from enhanced rate of succession or reduced frequency/intensity of flood disturbance) will reduce available habitat as more generalist floodplain species competitively exclude exposed riverine sediment species that are adapted to unvegetated areas (Bates et al. 2009)

and river regulation, and climate change. That variety is itself strong imperative for further study of the vulnerabilities of ERS, exemplified by the statement that 'extensive channel modifications and engineering works have removed ERS from much of the European river network' (O'Callaghan et al. 2013). In some places they are essentially remnants of formerly much more widespread ecosystems. Losses of specialised species, exemplified by beetles, can reflect particular microhabitats as regimes of sediment, moisture, light and cover, as well as interspecific competition outcomes. Greater microhabitat heterogeneity allows greater diversity of ERS beetles to thrive (Bates et al. 2007), with lateral environmental gradients a strong influence.

Key adaptations of ERS specialists for survival in such exposed areas include, firstly, those that reduce mortality from river inundation—Bates et al. listed these as (1) avoidance, by moving elsewhere when flooding occurs or is likely; (2) escape, by swimming, flying or otherwise reaching safety; (3) phenology, so that vulnerable life stages are not present during normal seasons of high water flows; (4) surviving immersion; and (5) 'risk spreading', with extreme outliers allowing continued survival when flooding occurs. Secondly, they may show adaptations for inhabiting unvegetated substrates—features such as rapid running, concealment by camouflage, or hiding or burrowing to escape predators and also to help conserve water.

Drawing on the wider body of information on arthropods of ERS in Europe, Framenau et al. (2002) sought to compare these with an Australian scenario, alpine streams in north eastern Victoria. The two predominant riparian taxa in Australia were wolf spiders (Lycosidae, 12 species and representing >60% of the catches from time-limited direct searches in sites 0.5–1 m from the water edge) and groundbeetles (Carabidae, 27 species, of which five each reached at least 10% of the total catch in one of the two surveys). Collectively, the catches included five families of spiders (plus 'others') and three families of beetles in addition to small numbers of ants, bugs, grasshoppers, flies and Lepidoptera. The faunas thereby resembled those reported elsewhere in being dominated by active surface-hunting predators. However, the communities also differed from those in the northern hemisphere in that Lycosidae by far outnumbered Carabidae. The carabid dominance by species of *Bembidion* in the northern hemisphere is replaced by a far more diverse generic array in Victoria. Framenau et al. suggested that the dominance of lycosids in Victoria might reflect differences in larval mobility of these two groups and the less predictable flooding regimes. In alpine rivers in Italy and Switzerland, abundance of riparian spiders on gravel bars was associated negatively with inundation frequency (Paetzold et al. 2008), and combination of this with loss of interstitial habitats through increased 'filling' was associated with general reduction of riparian arthropods.

Comparison of upland and lowland sites showed Lycosidae to be more abundant on upland streams, whilst carabid densities were similar in the two zones.

In practice it may be difficult to maintain natural flooding regimes in the face of river use for hydroelectric power and consequent regulation measures, but this may be vital in order to conserve natural riparian communities such as the above. Paetzold et al. (2008) recommended that riparian arthropods should be included in river assessments, for several reasons: (1) they are sensitive to structural and hydrological changes; (2) are substantial components of riverine biodiversity; (3) are important in connecting aquatic and terrestrial food webs; and (4) can be prey for other taxa, including vertebrates of conservation interest.

## References

Balcome SR, Closs GP, Suter PJ (2007) Density and distribution of epiphytic invertebrates on emergent macrophytes in a floodplain billabong. River Res Appl 23:843–857

Bates AJ, Sadler JP (2005) The ecology and conservation of beetles associated with exposed riverine sediments. CCW Contract Science Report no 688, Bangor

Bates AJ, Sadler JP, Perry JN, Fowles AP (2007) The microspatial arrangement of beetles (Coleoptera) on exposed riverine sediments (ERS). Eur J Entomol 104:479–487

Bates AJ, Sadler JP, Henshall SE, Hannah DM (2009) Ecology and conservation of arthropods of exposed riverine sediments (ERS). Terr Arthr Revs 2:77–98

Bayley IAE, Williams WD (1973) Inland waters and their ecology. Longmans, Sydney

Bayley IAE, Halse SA, Timms BV (2011) Aquatic invertebrates of rockholes in the south-west of Western Australia. J R Soc Western Australia 94:549–555

Boulton AJ, Suter PJ (1986) Ecology of temporary streams—an Australian perspective. In: DeDekker P, Williams WD (eds) Limnology in Australia, CSIRO Melbourne/W. Junk, Dordrecht, pp 313–327

Brown LE, Hannah DM, Milner AM (2007) Vulnerability of alpine stream biodiversity to shrinking glaciers and snowpacks. Glob Change Biol 13:958–966

Bunn SE, Arthington AH (2002) Basic principles and ecological consequences of altered flow regimes for aquatic biodiversity. Environ Manage 30:492–507

Buttle JM, Boon S, Peters DL, Spence C, van Meerveld HJ, Whitfield PH (2012) An overview of temporary stream hydrology in Canada. Can Wat Res J 37:279–310

Clarke A, Mac Nally R, Bond N, Lake PS (2008) Macroinvertebrate diversity in headwater streams: a review. Freshw Biol 53:1707–1721

Datry T, Larned ST, Scarsbrook MR (2007) Responses of hyporheic invertebrate assemblages to large-scale variation in flow permanence and surface-subsurface exchange. Freshw Biol 52:1452–1462

Davis J (1997) Conservation of aquatic invertebrate communities in central Australia. Mem Mus Vict 56:491–503

Davis J, Sim L, Thompson RM, Pinder A, Brim Box J (and four other authors) (2016) Patterns and drivers of aquatic invertebrate diversity across an arid biome. Ecography 41: 375-387

Dean J, St Clair R (2006) *Dinotoperla walkeri* sp. nov. (Plecoptera: Gripopterygidae), a brachypterous stonefly from western Victoria. Zootaxa 1230:55–62

De Dekker P (1988) Biological and sedimentary facies of Australian salt lakes. Palaeogeogr Palaeoclim Palaeoecol 62:237–270

Eberhard SM, Halse SA, Humphreys WF (2005) Stygofauna in the Pilbara region, north-west Western Australia. J R Soc WA 88:167–176

England J, Chadd R, Dunbar MJ, Sarremajane R, Stubbington R, Westwood CG, Leeming D (2019) An invertebrate-based index to characterise ecological responses to flow intermittence in rivers. Fundam Appl Limnol 193:93–117

Extence CA, Balbi DM, Chadd RP (1999) River flow indexing using British benthic macroinvertebrates: a framework for setting hydroecological objectives. Reg Riv Res Manage 15:543–574

Fairweather PG, Napier GM (1998) Environmental indicators for national state of the environment reporting: inland waters. Environment Australia, Canberra

Framenau VW, Manderbach R, Baehr M (2002) Riparian gravel banks of upland and lowland rivers in Victoria (south-east Australia): arthropod community structure and life-history patterns along a longitudinal gradient. Aust J Zool 50:103–123

Gomi T, Sidle RC, Richardson JS (2002) Understanding processes and downstream linkages of headwater systems. Bioscience 52:905–916

Gotch TB (2001) Wolf spider assemblages in mound springs and drains of South Australia. In: Halliday L (ed) Proceedings of 4th Mound Spring researchers forum. Department of Environment and Heritage, Adelaide, pp 46–53

Greenslade P (1985) Terrestrial invertebrates of the mound springs bores, creek beds and other habitats. In: Greenslade J, Joseph L, Reeves A (eds) South Australia's mound springs. Nature Conservation Society of South Australia Inc., Adelaide, pp 64–77

Hillman TJ (1986) Billabongs. In: De Dekker P, Williams WD (eds) Limnology in Australia. CSIRO Melbourne, W Junk, Dordrecht, pp 457–470

Humphreys WF (2006) Aquifers: the ultimate groundwater-dependent ecosystems. Aust J Bot 54:115–132

Jocque M, Vanschoenwinkel B, Brendonck L (2010) Freshwater rock pools: a review of habitat characteristics, faunal diversity and conservation value. Freshw Biol 55:1587–1602

Kubo JS, Torgersen CE, Bolton SM, Weekes AA, Gara RI (2013) Aquatic insect assemblages associated with subalpine stream segment types in relict glaciated headwaters. Insect Conserv Divers 6:422–434

Leigh C, Bonada N, Boulton AJ, Hugueny B, Larned ST, Vorste RV, Datry T (2016) Invertebrate assemblage responses and the dual roles of resistance and resilience to drying in intermittent rivers. Aquat Sci 78:291–301

Lowe WH, Likens GE (2005) Moving headwater streams to the head of the class. BioScience 55:196–197

O'Callaghan MJ, Hannah DM, Williams M, Sadler JP (2013) Exposed riverine sediments (ERS) in England and Wales: distribution, controls and management. Aquat Conserv Mar Freshw Ecosyst 23:924–938

Paetzold A, Yoshimura C, Tockner K (2008) Riparian arthropod responses to flow regulation and river channelization. J Appl Ecol 45:894–903

Ponder WF (1986) Mound springs of the Great Artesian Basin. In: De Dekker P, Williams WD (eds) Limnology in Australia. CSIRO Melbourne/W, Junk, Dordrecht, pp 403–420

Ponder WF (2003) Endemic aquatic macroinvertebrates of artesian springs of the Great Artesian Basin—progress and future directions. Rec S Austr Mus Monograph Series 7:101–110

Rackemann SL, Robson BJ, Matthews TG (2013) Conservation value of waterfalls as habitat for lotic insects of western Victoria, Australia. Aquat Conserv Mar Freshw Ecosyst 23:171–178

Thoms MC, Sheldon F (2000) Lowland rivers: an Australian introduction. Reg Riv Res Manage 16:375–383

Timms BV (1985) The structure of macrobenthic communities of Australian lakes. Proc Ecol Soc Aust 14:51–59

Timms BV (2001) Large freshwater lakes in arid Australia: a review of their limnology and threats to their future. Lakes and Reservoirs 6:183–196

Williams WD (1983) Life in inland waters. Blackwell, Melbourne

Williams WD (2002) Environmental threats to salt lakes and the likely status of inland saline ecosystems in 2025. Environ Conserv 29:154–167

Winterbourn MJ, Cadbury CJ, Ilg C, Milner AM (2008) Mayfly production in a New Zealand glacial stream and the potential effect of climate change. Hydrobiologia 603:211–219

# Chapter 4
# Monitoring Freshwater Macroinvertebrates

## 4.1 Scope and Needs for Assessments

That invertebrate assemblages and individual species respond, often in predictable ways, to numerous changes their environments, gives them enormous value in assessing and predicting changes in water quality, impacts of disturbances, and the resources needed to sustain diversity and functions. However, detecting those changes in species incidence, richness and community composition and defining their causes is rarely easy—not least because many of the causes may not be evident, or interact in unexpected or undetected ways. The foundations discussed by Vinson and Hawkins (1998, based on ecological principles elucidated earlier by Thienemann) underpin much later discussion. Those principles are (1) richness increases with increased diversity of conditions, as 'heterogeneity'; (2) the more that conditions change, or are changed, from the normal range to which most resident species are suited, the smaller the number of those species that occur there and the greater the abundance of those species that do occur; and (3) the longer the water body remains in the same condition, the richer and more stable its biotic community will become. Applied initially to stream insects, the patterns suggested have been investigated extensively to reveal considerable variety, in part reflecting scale of consideration—from individual water bodies to catchment or regional studies. However, as affirmed by Stendera et al. (2012), the study of those effects itself has many restrictions—not least of cost and available experience—but also because the four main categories of 'stressors' to freshwater ecosysystems commonly act simultaneously and with varied additive, antagonistic or synergistic effects. Those categories were recognised by Vorosmarty et al. (2010) as catchment disturbance (such as land use or deforestation), pollution, water resource development (such as alteration of natural flow regimes), and biotic factors (such as invasive species), all of which have potential to become 'threats' to biodiversity (Chap. 6) either individually or in concert. In general, stressors tend to cause declines in species richness and diversity.

As predominant, universal, easily sampled, ecologically diverse and taxonomically rich components of inland water communities, insects and their relatives


T. R. New, *Insect Conservation and Australia's Inland Waters*,
https://doi.org/10.1007/978-3-030-57008-8_4

have formed the bases of numerous 'indices' and allied metrics to measure and compare condition, conservation status and needs of water bodies over space and time, and to define targets for management and monitor trajectories toward those goals. Benthic macroinvertebrates have played major roles in the development of freshwater monitoring, and the changes that occur from both natural processes and human interventions. Indeed 'Aquatic insects and other benthic macroinvertebrates are the most widely used organisms in freshwater biological monitoring of human impact' (Bonada et al. 2006), but the best methods by which to achieve this have for long been debated. Formally, 'macroinvertebrates' (see also p. xx) can be delimited as those taxa retained by net mesh sizes in the range of >/= 20–500 μm, and development of their use in biomonitoring is treated extensively in a seminal series of essays edited by Rosenberg and Resh (1993), and that cover all major topics of then current concern, and display the wide variety of values in assessing freshwater ecosystem conditions and change. Developments have been guided by several priorities: Bonada et al. (2006) listed these as (1) the kind of information that a given method can provide; (2) the priorities of individual freshwater ecologists; (3) the spatial and temporal complexity of the freshwater systems, likely to require adaptations of approaches for particular areas; and (4) differences in the precision of outcomes needed. Whatever the primary purpose, however, changes in the abundance, taxonomic variety and ecological roles of the invertebrates as indicators of the condition of host environments and the impacts of disturbances or recovery measures necessitate careful selection of the methods used, and equally careful interpretation of the information accumulated. Monitoring changes, through repeated standard evaluations, is often in the context of monitoring distribution, commonly against highly incomplete evaluation of a taxon's complete range. The difficulties of definitively surveying the running waters of Australia become evident from Figs. 4.1 and 4.2, which show the rivers and substantial density of smaller watercourses in East Gippsland, Victoria (after Raadik 1992). The region includes parts of three river basins, the Snowy River (west), East Gippsland (central) and a small part of the Towamba River system (far south east, but with about 90% of its area in New South Wales). Much of the area shown is remote, difficult to access and explore, and highly undersurveyed for invertebrates. Even for fish, Raadik (1992) noted only small numbers of sites surveyed over 1967–1991, with 87 of the 114 sites surveyed only once during that period. Even 'fish data' are thus highly incomplete, and Raadik commented that (other than for freshwater shrimps and crayfish) the fauna is poorly known 'so we cannot define macroinvertebrate communities in many regions'. That comment remains valid, but the region is recognised as biogeographically important as the southernmost distribution limit of some east coast taxa. Clearly, and perhaps especially in some remote headwaters, novelties may await discovery.

Two rather different criteria are used commonly as designating high conservation value of a waterbody or other habitat and both are, in principle, measurable for direct evaluation and comparison across sites. They are (1) the presence of rare or threatened species and (2) a high species richness, without regard to threatened or other notable species. More rarely, these are combined, so that high richness also includes high species rarity and with value enhanced by presence of any local endemic

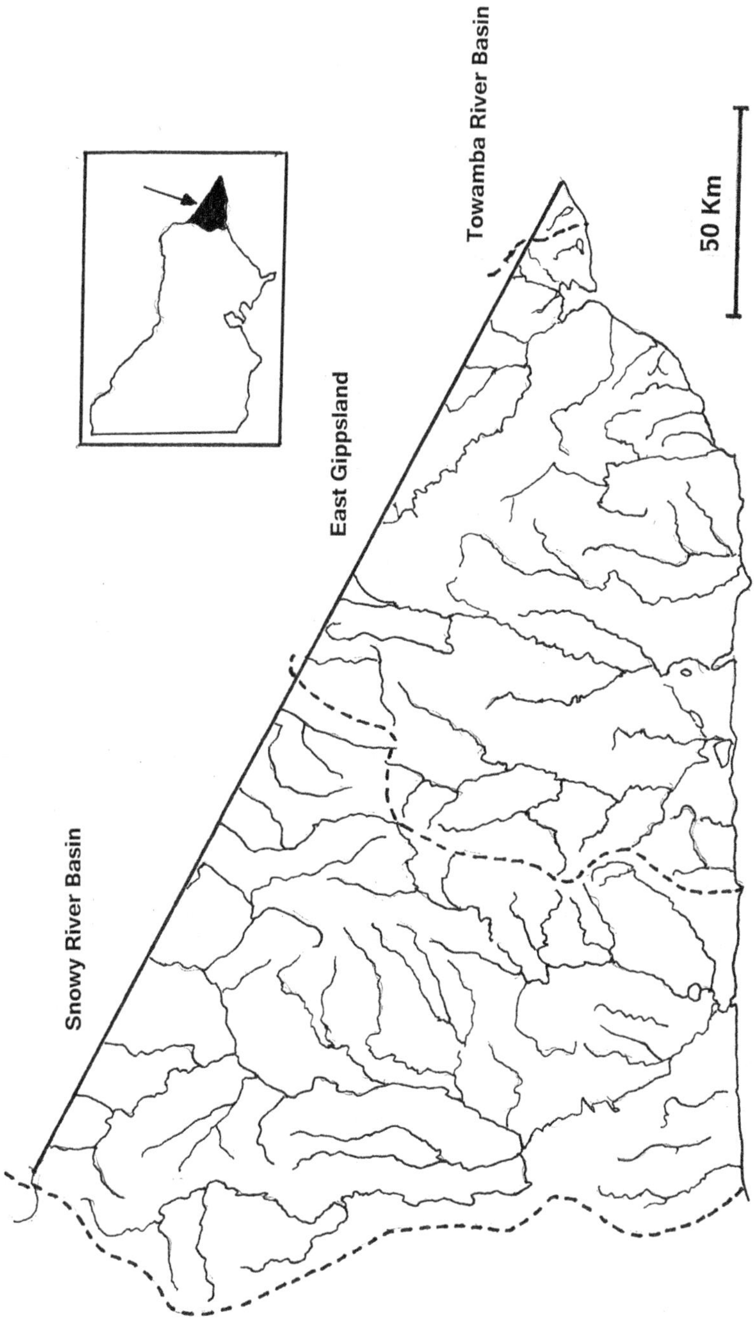


**Fig. 4.1** The extent of running waterways in East Gippsland, Victoria (locality map, black): the major rivers in the region (after Raadik 1992)

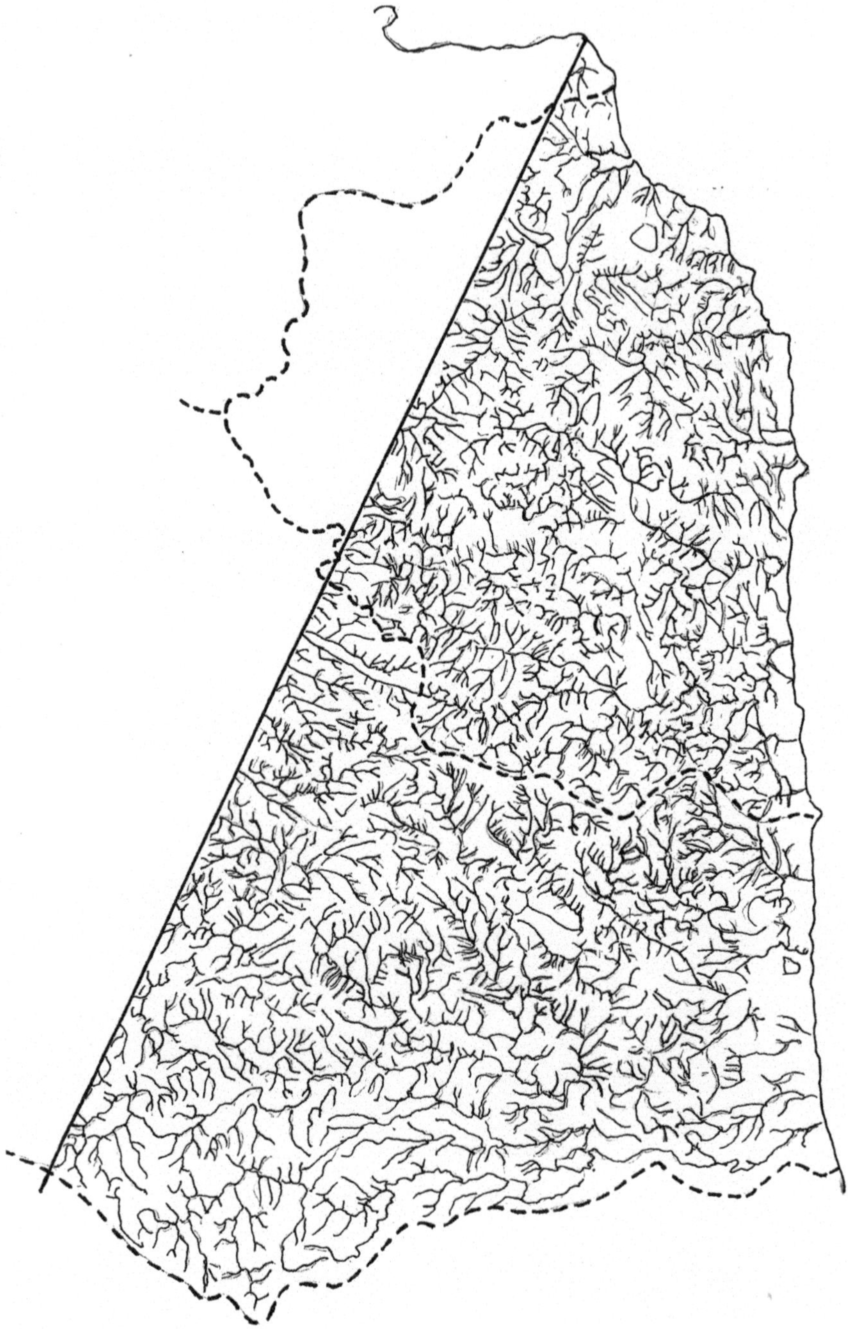

**Fig. 4.2** Running waterways in East Gippsland: the additional complexity and diversity implied from incorporation of streams (compare with Fig. 4.1) (after Raadik 1992)

taxa. Nevertheless, as Chadd and Extence (2004) noted, no then current index gave a combined general measure of these as a valid index of community conservation value. Any such index necessarily depends on sufficient knowledge being available to assess conservation status of all taxa present, and to identify them consistently and reliably to species level. Thus, for Britain, aquatic macroinvertebrates largely satisfy these needs. Much greater doubt occurs for Australia, with widespread acceptance that information is both 'patchy' and far from complete for most insect taxa. Any Australian parallel to the United Kingdom 'Community Conservation Index' derived for helping to set conservation priority needs for aquatic macroinvertebrates (Chadd and Extence 2004) is far off. The relatively sporadic information on distribution of many Australian aquatic insects can be contrasted with examples from elsewhere. The European Trichoptera, for example, are the focus of a detailed distributional survey in which more than 600,000 records georeferenced to particular sites (spanning more than 55,000 sites across 50 countries) provide a globally unique distributional profile for this order. Based on 1706 species and subspecies and contributions from more than 80 people working with caddisflies in the region (Schmidt-Kloiber et al. 2017), the compilation revealed several key centres of endemism, and also indicated the remaining geographical gaps in coverage, to constitute a sound template for conservation policy decisions based on evident need. Together with Odonata, also quite well documented, it gives aquatic insects a convincing profile in planning effective management. The far less complete information on most Australian taxa, although continuing to accumulate but from a small pool of contributors and with very restricted continuing support for such endeavours, seems likely to thwart such achievements over the forseeable future. That realisation demands development of other approaches, but also the need to be able to track condition, management progress, and changes realistically.

Needs for monitoring have generated numerous indices to express the meanings of changes in macroinvertebrate richness, abundance and assemblage composition through series of samples taken at intervals and interpreted to indicate 'improvement' or 'deterioration' in conditions over time or across sites. However, the biases flowing from different responses of insect species within larger taxonomic categories such as families are often considerable, and any simple index is likely to be less predictable and informative than anticipated. Brinkhurst's (1993) allusion to 'King Arthur's Index' remains salutary. He wrote 'The search for a Holy Grail Index that can be determined by the underfunded without exerting their mind persists and is as futile now as it was in King Arthur's Britain'. Without clear understanding of the taxa contributing to any index, including those used regularly and treated as definitive, inferences and conclusions can easily represent statistical treatment rather than biological reality. Resulting conservation measures may then be under-informed.

The twin themes that arise universally are (1) what taxonomic groups to use for monitoring or wider assessments, and (2) the level of identification needed for meaningful interpretations. As one example, both diatoms (Sonneman et al. 2001) and macroinvertebrates (Walsh et al. 2001) show distinct assemblages in streams near Melbourne, with strong compositional shifts between hinterland and metropolitan

**Table 4.1** Characteristics of 'key families' of aquatic insects that indicate good habitat and water quality within a region (Metzeling et al. 2004, 2006)

| | |
|---|---|
| 1. | Typically found in the types of streams that occur in the region |
| 2. | Representative of a particular habitat type—for example, of riffles, woody debris, fringing vegetation, macrophytes, pools |
| 3. | Are commonly collected when present, using a recommended method for sampling edges and riffles |
| 4. | Represent reasonable to good water quality, and tending to disappear as conditions deteriorate |
| 5. | Selection criteria: examine fidelity of each taxon to the region—used families occurring at >50% of sites sampled and with a fidelity score[a] of 1.5 or greater; expert judgement from experienced aquatic ecologists used to refine list of families expected in good condition streams in each region |

[a]see text for explanation

sites implying considerable effects from urbanisation (p. 126). However, key sensitivities apparently differ in these two taxa—diatom community composition reflected nutrient enrichment, whilst macroinvertebrates were better indicators of catchment distance. The particular impacts of concern could lead to one or other group, or both, being used for monitoring. Seeking key indicators that are relatively inexpensive and simple to use, and that link with rehabilitation goals and display a range of response times is a key need in rehabilitation of water bodies (Cottingham et al. 2005, on Australian rivers).

A somewhat different monitoring approach involves assessment of 'key invertebrate families', focusing on losses of taxa that are associated with (and indicate) good habitat and water quality. Metzeling et al. (2004, 2006) listed a rationale for selecting families (Table 4.1), drawing on the best available biological knowledge. Lists of suitable families, driven by these requirements, were generated for each of several regions of Victoria, allowing a wider perspective of monitoring based on the most characteristic insect/invertebrate taxa. The benthic invertebrate fauna enabled distinction of five such 'biological regions' (Fig. 4.3), excluding the dryer north west of the state, for which no specific biological objectives were nominated. The five regions are based on distributions of aquatic macroinvertebrates and reflect elevational and physiographic features (Wells et al. 2002), each with unifying features of water quality and vegetation that strengthen regional differences and variations that are useful in regional characterisation. Not least, they are useful in developing regional conservation objectives.

The designation of regions was retrospective, drawing on the extensive records of macroinvertebrates from almost all Victorian catchments, from rapid assessments (below) based on live-sorted samples from stream edge habitats, rocks and riffles of 199 reasonably unmodified reference sites, most of them sampled on four occasions over 1990–1996, and that yielded a total of 1025 macroinvertebrate taxa. That number was later increased slightly (to 1188 taxa up to 1997: Wells et al. 2002). Allocation to region was based on a 'fidelity score' (Newall and Wells 2000) in turn derived

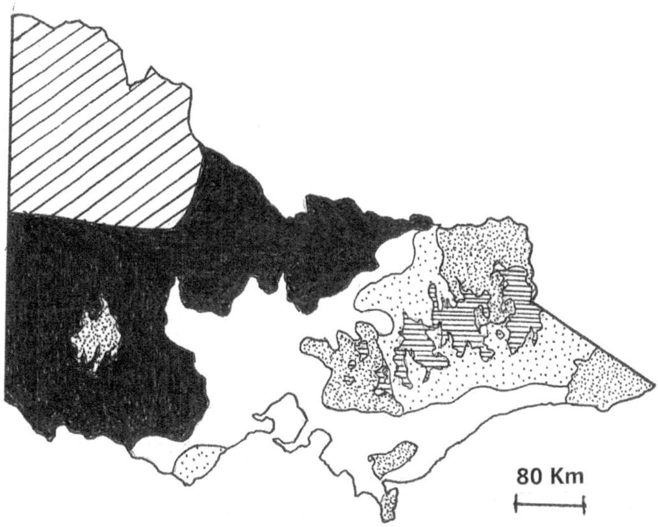


**Fig. 4.3** Biological regions across Victoria, based on aquatic macroinvertebrates. (regions are: Murray and Western Plains, black; cleared hills and coastal prairie, open; forests-A, dense dots; forests-B, sparse dots; highlands, horizontal hatching; unclassified, diagonal hatching) (Metzeling et al. 2004)

from 'constancy'. The latter is the proportion of sites at which the particular species is found—so that constancy of '1' means that it was present at all sites sampled. The fidelity score reflects a species' preference for a selected region, calculated as the ratio of the species constancy for that region to constancy across all sites in the data set. A fidelity score of '>1' indicates preference for the selected region, and of '<1' implies preference against the region. Either 'preference' increases with distance from '1', with '>2' suggesting strong preference.

Each region supported characteristic taxa. For example, Region 1 (the 'high country') was characterised by the mayfly *Nousia* spAV1, the caddisfly *Aphilorheithrus* spAV4, a blackfly (*Simulium victoriae*) and a midge (*Podonomopsis* sp.), all of which had strong constancy and fidelity to highland sites and occurred in medium- to fast-flowing streams at higher elevations. Similarly, three taxa characterised Region 2—the stoneflies *Riekoperla tuberculata* and *Acruroperla atra*, and the dragonfly *Austroaeschna* sp. Both the stoneflies are considered to be indicators of good water quality.

Monitoring of macroinvertebrates in streams and rivers has traditionally relied heavily on samples of benthic communities, obtained either directly (through easily replicable standard techniques such as use of Surber samplers or 'kick-net' sampling: Samways et al. 2010) or deploying artificial substrates on the stream bed and examined after periods sufficient for invertebrates to colonise them. Both approaches can provide valid comparisons across sites and time. Focus on benthic

**Table 4.2** Advantages of using benthic macroinvertebrates in rapid biodiversity assessments of streams and wadable rivers (after Barbour et al. 1999)

| | |
|---|---|
| 1. | Macroinvertebrate assemblages are good indicators of localised conditions. Many have limited migration patterns or have a sessile mode of life, so can show site-specific impacts (such as upstream/downstream comparisons) |
| 2. | Macroinvertebrates integrate effects of short-term environmental variations. Sensitive life stages will respond quickly to stress, overall community will respond more slowly |
| 3. | Assemblages are made up of species that constitute a broad range of trophic levels and pollution tolerances, so providing strong information for interpreting cumulative effects |
| 4. | Sampling is relatively easy, with minimal detrimental effects on the resident fauna |
| 5. | Invertebrates are a primary food source for fish |
| 6. | Benthic invertebrates are abundant in most streams; many small streams support diverse macroinvertebrate fauna but only limited fish fauna |
| 7. | Most water quality agencies include macroinvertebrates in routine biological survey data collection |

EPT (Ephemeroptera, Plecoptera, Trichoptera) larvae, as a commonly used limitation, allows reliable taxonomic interpretation to genus/species levels of these relatively well-documented taxa, whilst sweeping nearby emergent/riparian vegetation to obtain adult insects is far less certain, highly seasonal, and usually less representative of the assemblages present. However, more broadly, other taxa (such as algae or fish) may be selected in the interest of developing 'rapid biodiversity assessment' (RBA) measures (Barbour et al. 1999). Advantages of using macroinvertebrates are diverse (Table 4.2), and can be linked directly with habitat quality, characteristics and changes. Likewise, use of artificial substrates, or any alternative sampling method(s) must be considered carefully in relation to the information sought: each method has both positive and negative attributes (Table 4.3).

'Rapid biodiversity assessment' has been used extensively for macroinvertebrates of Australian rivers since the early 1990s and has drawn on (1) use of semi-quantitative sampling gear; (2) subsampling only a small proportion of the organisms collected; and (3) identifications only to higher taxonomic levels. Subsamples are commonly taken on site by 'hand-picking' (based on standard times or a standard number of individuals), and which has several practical advantages (Metzeling et al. 2003), as (1) speed of overall assessment; (2) avoiding retention of debris and unnecessary and unwanted specimens; (3) facilitated by animal movement rendering organisms more conspicuous; (4) more convenient than hand-collecting indviduals from habitats such as rocks or logs; and (5) is a cost-effective way of detecting human impacts. However, small or cryptic taxa may easily be overlooked, and the skills and observation powers of operators can vary considerably. In essence, the greater precision from more detailed examination of samples is traded off against demonstrable cost savings. Focus on individual species may still be possible by good direction to the operators, but evaluations of richness beyond family level are clearly restricted.

**Table 4.3** Advantages and disadvantages of using artificial substrate samplers in assessments of freshwater benthic invertebrates (after Barbour et al. 1999)

| |
|---|
| *Advantages* |
| 1. Allow sample collection in locations that are otherwise difficult to sample effectively |
| 2. As a 'passive' sample method, artificial substrates permit standardised sampling by eliminating subjectivity, with consistent setting and retrieval effort |
| 3. Confounding effects of habitat differences are minimised by providing a standardised microhabitat. This may promote selectivity for specific organisms if microhabitat different from that available naturally at a site |
| 4. Sample collection may require less skill and training than direct sampling of natural substrates |
| *Disadvantages* |
| 1. Artificial substrates require a return trip, and may be an important consideration if logistic resources limited or distances large |
| 2. Artificial substrates are prone to loss, natural damage or vandalism |
| 3. Material of the substrate will influence composition and structure of the community; solid substrates will favour attached forms over motile forms and compromise use of any siltation index |
| 4. Orientation and duration of exposure of the substrate will influence composition and structure of the community |

Consistency of sampling scale is a prerequisite for valid comparisons across sites or habitats. As Sheldon and Walker (1998) discussed, river invertebrates can be monitored or compared at three major scales: (1) macroscale, individual waterbodies, with macrohabitats as the major zones of rivers; (2) mesohabitats, the major subdivisions, such as backwaters, billabongs, channels and floodplain areas; and (3) microscale, the major components of each of these, such as submerged and emergent vegetation, submerged wood, and other defined substrates. Comparisons across these scales in the Murray-Darling River system showed insects to be the predominant taxon in both rivers, with Diptera, Coleoptera and Hemiptera the richest orders. Fifty taxa were found in both the rivers. Microhabitat diversity and associated structural complexity was the greatest influence on assemblages at the intermediate mesoscale, but hydrology and geomorphic features were more influential at the macroscale, with these essentially governing the diversity of habitats available at the finer scale(s).

Biological responses at the three scales approximated to a hierarchy, in parallel (Table 4.4, after Walker et al. 1995) and also reflect rather different time scales. The nestedness of macrohabitats including mesohabitats, and the latter including localised microhabitats, each affected by an equivalent scale of processes, encapsulates many of the needs to manage flow regimes and monitor their impacts.

Level of taxonomic interpretation for samples also needs careful thought—the balance between costs of sampling (far less with identifications only to level of family or genus levels, rather than species) and loss of ecological information (with the species in many families or genera very variable in their levels of ecological specialisation, resource needs and responses to change) is critical, and the need for species-level identifications is a major guiding factor to select the groups targeted,

**Table 4.4** Scale-dependent relationships of biological features to processes and responses, as summarised for dryland rivers (abbreviated from Walker et al. (1995)

| Feature | Process | Response | Scale | |
|---|---|---|---|---|
| | | | Space ($m^2$) | Time (y) |
| Ecosystem | Fluxes of nutrients and energy | Evolutionary: | >100,000 | >100 |
| Community, population | Competition, mortality, recruitment | Ecological: changes in community structure | 1000–$10^8$ | 1–100 |
| Organism | Life-history strategies | Physiology, behaviour: diapause, migration, reproduction | <10,000 | <1 |

and the values of different level interpretations a theme of continuing debate. Many invertebrate taxa cannot be reliably identified or separated to that level, whilst the more critical recognition to family category is reliable. Even such 'coarse-level' interpretations can reveal considerable diversity: invertebrate samples from stormwater management ponds and comparative non-managed ponds in Ottawa, Canada, yielded 55 families, with 37 of them insects (Hassall and Anderson 2015), with community structure and richness rather similar in the two pond categories. That diversity alone endorsed the functions of stormwater ponds as biodiversity reservoirs in urban areas and—despite considerable differences in water chemistry—supported their conservation values.

The logistic and financial attractiveness of coarse-level sorting of benthic invertebrates, most commonly to family level, has led to its widespread adoption in Australia, where claims have been made that family numbers may adequately reflect species richness (Growns and Growns 1997). Regression of number of families against total numbers of taxa (species or morphospecies) showed the number of families to 'explain' 91% of species richness, with numbers of species in each of Ephemeroptera, Coleoptera, Diptera and Trichoptera each 'explaining' 60–85% of variation in overall species richness. In some contexts, at least, family-level sorting may be sufficient to tentatively evaluate richness (Fig. 4.4)—but, of course, does not evidence the detailed content of that richness.

The greater information to be gained from species-level interpretations may involve far greater costs, and its worth must be considered carefully in relation to priority needs from a survey. Thus, Marshall et al. (2006) nominated four contexts in which this finer detail is beneficial or needed, as (1) the greater information content over less penetrating separations can result in more reliable site characterisation or description; (2) it can give greater resolution in detecting differences between 'test' and 'reference' sites; (3) it is necessary if detecting the presence of rare or threatened species is an objective; and (4) may be needed to identify or rank sites for protection or priority conservation management. If analyses to genus or family alone cannot satisfy themes such as these, the additional costs for species-level (whether as identified species, or consistently-recognised 'morphospecies') sorting may be the necessary foundation of a survey. However, in surveys across 29 study sites in

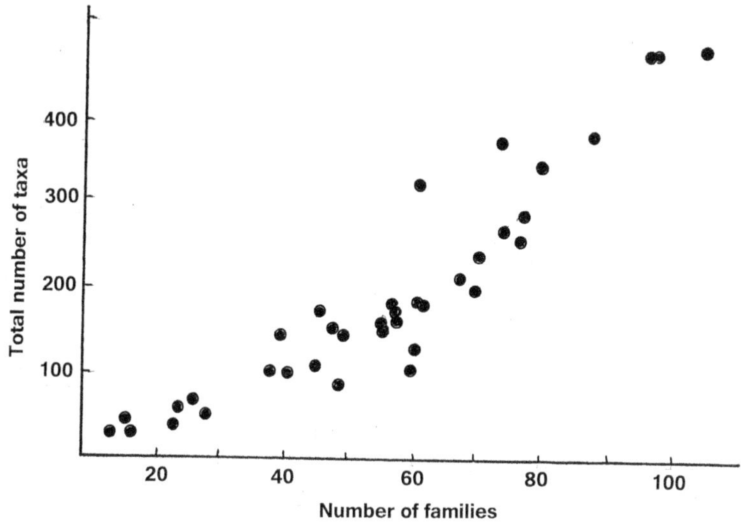


**Fig. 4.4** The total numbers of freshwater macroinvertebrate taxa plotted against the number of families across studies undertaken in Australia and Papua New Guinea (from Growns and Growns 1997, simplified by omission of locality information)

southern Queensland Marshall et al. found that little information was lost by the coarser analyses, and that quantifying the abundances of taxa was more useful for detecting patterns than presence/absence data alone. In that survey, the only insect orders that constituted the dominant invertebrate *order* in a stream were Diptera (in seven streams) and Ephemeroptera (two streams)—but in three of the 'Diptera streams' the predominant invertebrate *species* was either a copepod or a mollusc. One 'Ephemeroptera stream' was dominated by a mollusc, and conversely, a crustacean-dominated stream had a mayfly as the predominant species. The relative costs of sample analyses at the different taxonomic levels differed greatly (Table 4.5), as a major consideration transcending many studies of this nature.

Whilst analyses of samples to the finest possible level (species) allows for the greatest retrieval of biological information, as encapsulating the full taxonomic variety present in assemblages, the practical problems of achieving this for many

**Table 4.5** Estimates of the efforts and costs (AU$, in 2006) of identifying aquatic macroinvertebrates to different taxonomic levels, excluding time required to process the samples (Marshall et al. 2006)

| | Species | Genus | Family | Order |
|---|---|---|---|---|
| Skills and expertise | Very high | High | Moderate | Low |
| Operational time | 6–7 months | 4–6 months | 2–3 weeks | 3–5 days |
| Approximate cost | $20,000 | $15,000 | $3000 | $1000 |

aquatic insects in Australia remain formidable for reasons such as presence of undescribed taxa, incomplete and highly uneven documentation, and inability to associate larvae and adults correctly. It is not surprising that many workers have explored and to some extent justified the values of separations to family or other higher level alone—as well as their disadvantages and limitations. Indeed, the level of taxonomic resolution needed for benthic macroinvertebrates has been described as 'one of the most contentious issues' in interpreting samples (Bailey et al. 2001), with the real need sometimes dictated by the purpose of a study. Clearly, conservation of individual species necessitates unambiguous recognition of that/those species and separation from their close relatives, but wider monitoring may be needed for comparisons of assemblages across sites, regimes or treatments in which indices derived from the sample data reflect tolerance levels and broader responses among higher taxa. Many inferences on factors such as pollution tolerance, feeding habits and other attributes of species are in practice commonly extrapolated from a few known examples within a higher taxon. Family level information might suffice, and Bailey et al. noted the need to judge carefully whether the extra resources needed to obtain genus- or species-level identifications is justified.

Lenat and Resh (2001) emphasised the large numbers of aquatic insect species that can occur: their examples included >900 species from a creek in South Carolina and >700 species from a river in Germany, so that any attempt to identify all to species level can become a protracted and expensive task. Nevertheless, species-level interpretations of responses to pollutants or other disturbance may differ from genus-level indications. The survey by Resh and Unzicker (1975) is a classic example of this. Amongst the 89 genera of Trichoptera for which water quality preferences were known for more than one species, 61 genera have different responses when each species was assessed as 'tolerant', 'facultative' or 'intolerant'. The largest group of genera contained species with tolerance values in all three categories. Use of family-level interpretations in essence gives far less certainty in the outcomes—and use of family-level data to compute biotic indices of water/habitat quality leads to declines in ability to detect extremes or small differences in water quality or degradation levels.

Comparative analyses, most notably ordinations as the approach used most frequently for comparisons of aquatic invertebrate assemblages, have been investigated for consistency at different scales of taxonomic resolution. In ordinations of macroinvertebrates from nine rivers in Victoria, five data sets were compared by Marchant et al. (1995), in addition to the original quantitative species-level interpretations. These further sets were (1) binary (presence/absence); (2) quantitative or (3) binary data on families; and quantitative data on EPT (4) families and (5) species, each seeking to reduce the level of quantification and taxonomic resolution from the original information and so reduce the costs of valid surveys. The three main gradients shown by the invertebrates (namely, elevational, substrate and combined water pH and conductivity) were evident in all the ordinations. This analysis was based in a collective 417 species across 114 families, of which EPT comprised 201 species in 31 families. Full reasons for the correspondence are not clear, but Marchant et al. postulated that related species have similar ecological requirements, so that these gradients remain evident at the family level. They also suggested redundancy in the data between less closely related taxa, citing that EPT—clearly a subset of the entire range—had ordination patterns largely indistinguishable from those from the complete data set. Marchant et al. considered that the level of identification need might also depend on the spatial scale of a survey, so that family level separations might suffice for ordinations of spatial patterns if from over a wide area such as several different catchments. The study strongly endorsed that concentration on EPT taxa alone might indeed be sufficient in rapid comparative surveys.

For use of benthic EPT larvae as indicators, Lenat and Resh (2001) gave recommendations for the two contexts of (1) family-level identifications and (2) more precise identifications, and those suggestions (Table 4.6) led to conclusions that biological monitoring studies had greater benefits if the finer levels could be undertaken or afforded. The EPT Index is based on the premise that loss of any families within these orders indicates disturbance. However, whilst used very widely in stream systems, some natural biogeographical trends among the insects can prelude its worth. Thus, many Plecoptera and some families of Ephemeroptera are largely restricted to cool, well-oxygenated streams, and seldom occur elsewhere, so that the number of families occurring in some slow-flowing warmer waters in Victoria

**Table 4.6** Recommendations for appropriate taxonomic levels for sorting aquatic invertebrates: situations in which less precise or more precise identifications are needed (from Lenat and Resh 2001)

| |
|---|
| *Situations where family level identifications are appropriate*: |
| 1. Low level of taxonomic expertise available, such as volunteers; high degree of uncertainty in the data obtained |
| 2. Very large numbers of sites must be sampled, with logistic inadequacy preventing thorough sampling, and necessitating shortcuts; loss of precision inevitable |
| 3. Goal is to determine relatively large between-site differences, such as treatment versus reference sites; monitoring for water quality, for example, may need few groupings, such as 'good', 'intermediate' and 'bad' |
| 4. Sampling in areas known to have low taxon richness: low genus: family or species: family ratios may give similar results at lower and higher taxonomic sorting levels; depends on prior knowledge of the above relationships |
| *Situations where more precise identifications are required*: |
| 1. Conclusions need to have high degree of confidence, such as in guiding management decisions |
| 2. Small between-site or between-date differences must be detected, such as for precise allocation among a series of quality grades |
| 3. Survey results will be used to deduce both magnitude and type of water quality problem; indicator assemblages may be useful only at genus or species levels of discrimination |
| 4. Selection of areas for high levels of protection, such as by supporting rare species and/or high biodiversity |
| 5. In multispecies surveys, where invertebrates are sampled together with other biota and physical/chemical features, to give more holistic information |

is low (6–10), leading to possible misinformation simply from sampling error, and thwarting selection of clear EPT objectives (Metzeling et al. 2004).

Whilst larval EPT are recognised widely as responding to changed conditions in freshwater environments, adult Odonata are more easily sampled and identified to species level and are also a widely acknowledged surrogate or indicator group for such changes. The use of any such surrogate to reflect wider changes across sites and time, and conservation need, may need to be appraised carefully in any specific study. The concept of an 'umbrella index' discussed by Fleishman et al. (2000) extends the values of indicators of diversity to using surrogates whose conservation leads also to protection of other species not assessed directly. A comparison of those values for EPT and Odonata in the Cape Floristic Province, South Africa, suggested that conservation managers could choose either group for adequate planning (Kietzka et al. 2019), with the selection influenced by specific questions and the resources and knowledge available. Clearly, availability of species-level information on responses to the relevant change(s) is beneficial—but is unlikely to be available for most such taxa in Australia. However, as Kietzka et al. pointed out, Odonata is often the most managable taxon for study, largely because of the very conspicuous adult stage (p. 176) that can often be identified to species level by citizen scientists and other non-specialist observers.

In discussing development of biotic indices for river macroinvertebrates, Chessman (2003) noted that family-level distinctions were most commonly used by government agencies in Australia, whilst surveys by community groups focused mainly on only the order level. The uses of 'reference sites' for standard background information levels and comparison with sites undergoing varying levels of disturbance or pollution are also becoming more difficult—Chessman implied that for many river types, 'pristine' sites are no longer available, and 'least disturbed sites' that have undergone impacts of little known effects may be the best available for comparison, but perhaps showing species-level changes not revealed by less penetrating analysis of the samples. Many kinds of disturbance, however, cannot be excluded reliably in selecting such sites, but may be minimised by approaches such as assuring the presence of uninterrupted riparian zones and the absence of obvious point-sources of pollution.

The concept of 'reference site' or 'reference condition' is central to monitoring, in which repeated measures to determine condition or change from some 'norm' are the key activity. A reference site is the standard for comparison with condition or divergence in test or study sites, and workers such as Stoddard et al. (2006) have argued convincingly that the concept should be standardised or rendered as consistent as possible. Levels of disturbance can differ greatly, so that in many local areas, the 'best available' reference condition can also differ greatly as a putative standard, as implied in Fig. 4.5, which also demonstrates the dilemmas arising from having multiple definitions of this state: any reference site may be far from pristine or 'natural'. The several categories of reference condition recognised by Stoddard et al. are noted here to show that variability, which can easily be overlooked in practice. They differentiated (1) minimally disturbed condition, the condition in the absence of significant human interference and regarded as the best approximation to

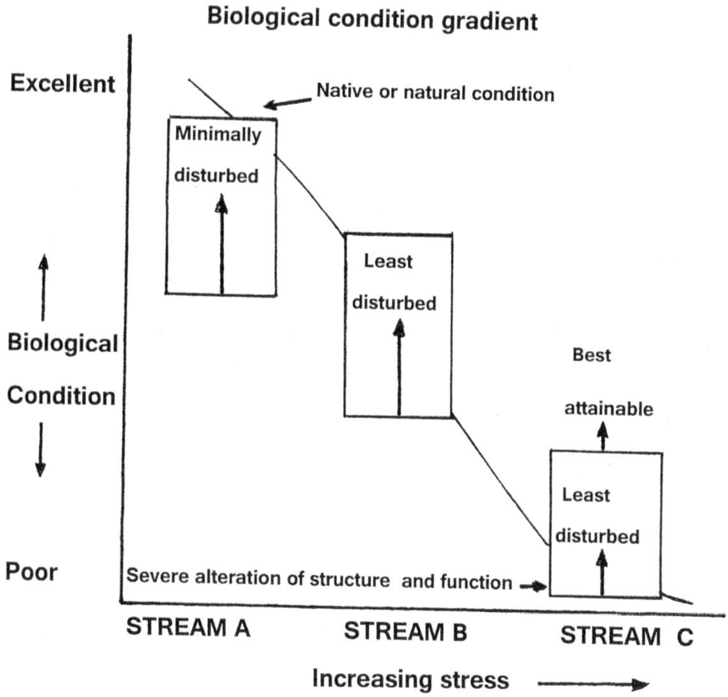


**Fig. 4.5** Differing levels of human disturbance of landscape in different ecological regions (or stream categories) create situations where the least-disturbed sites in each region describe very different 'reference conditions'. Here, the level of degradation is greater for 'C' than for 'B', and a reasonable 'reference condition' for 'C' might not exist but could perhaps be achieved (as 'best attainable', shown); in contrast, the 'least disturbed' sites for 'B' might represent reasonable reference conditions for those streams (Stoddard et al. 2006)

biological integrity available for comparison with current conditions; (2) historical condition, representing condition at some defined point in site history, ideally at a point preceding human interference but often used in relation to more recent nominated baseline events such as settlement or start of agriculture or industry; (3) least disturbed condition, in conjunction with the best available physical, chemical and biological habitat conditions in the contemporary landscape, with election of explicit local criteria to define 'best', so reflecting characteristics and use of the landscape being evaluated; and (4) best attainable condition, equivalent to the expected condition of least-disturbed sites if the best possible management has been deployed for some time. The terms apply variably, to specify different states and, in order to reduce or minimise confusion, Stoddard et al. (2006) recommended using reference condition (expressed more fully as 'reference condition for biological integrity') in its originally applied sense of naturalness or biological integrity, implying absence of significant changes or disturbance from human activities.

In a rather different approach, the functional composition of benthic invertebrates in running waters may provide a useful reference condition assessment across larger regions in which the detailed taxonomic composition of the assemblages differs. Statzner et al. (2001) developed this approach to biological monitoring through examining a wide array of species' traits, incorporating feeding habits, reproductive potential, dispersal potential and others for a total of 63 categories across 11 major biological traits. Undertaken in France, and the premise expanded to include wider comparisons across Europe, led to suggestion that the defined 'functional composition' of natural communities was very stable, so that differences between natural and impacted streams may be a valid index of naturalness. According to Stazner et al., its stability across different categories of running waters in Europe (1) indicated human impacts and (2) could potentially indicate specific kinds of human disturbance.

However, also as Chessman (2003) pointed out, development of indices such as 'SIGNAL' (Stream Invertebrate Grade Number Average Level) and their incorporation into wider assessment systems for stream and river condition evaluations, however initially valuable they appear, must be used as part of a wider portfolio of measures. SIGNAL was developed initially for use in eastern Australia (Chessman 2003) and was later broadened (as SIGNAL 2) to cover a wider variety of family and higher level groups (most non-insect groups were not taken beyond order or similar level) across 210 taxa and the whole of Australia. The score for any macroinvertebrate sample is calculated as an average of the pollution sensitivity grade levels (ranging from 10 [most sensitive] to 1 [most tolerant]), and this can be associated with indicative levels of water quality. Chessman (1995) used levels of clean water, possible mild pollution, moderate pollution, and probable severe pollution in his initial survey in New South Wales. SIGNAL scores reflect those levels as: excellent (score > 7), clean water (6–7), mild pollution (5–6), moderate pollution (4–5), and severe pollution (<4), but each score is also a compromise when (1) species within a family respond in different ways to a pollutant, or (2) where the family responds differently to different pollutants. Taxonomic superficialities are overcome by counting each of the groups not identified to that level (commonly, groups such as Acarina, as well as Oligochaeta and many other non-arthropod taxa) as '1', and ecological complexity acknowledged by distinguishing the diverse Chironomidae to subfamily level.

The SIGNAL index and the number of families of EPT were the most sensitive indicators across 16 stream sites, representing a gradient of stormwater impacts in eastern Melbourne (Walsh 2006). The streams were selected as those in which urban stormwater was the predominant human impact, essentially satisfying the important criterion for indicator assessments of measuring a range of responses to a single defined disturbance. Samples from riffles and the stream edge in two seasons were analysed by three protocols, and the patterns from each were similar (Fig. 4.6).

Broad river monitoring programmes, such as the national Australian Rivers Assessment Scheme (AusRivAs) incorporate macroinvertebrates identified to family level in assessing river 'health' or ecological condition (Smith et al. 1999). These values are amplified in the next chapter, in the most frequent practical context of assessing threat impacts to freshwater ecosystems.

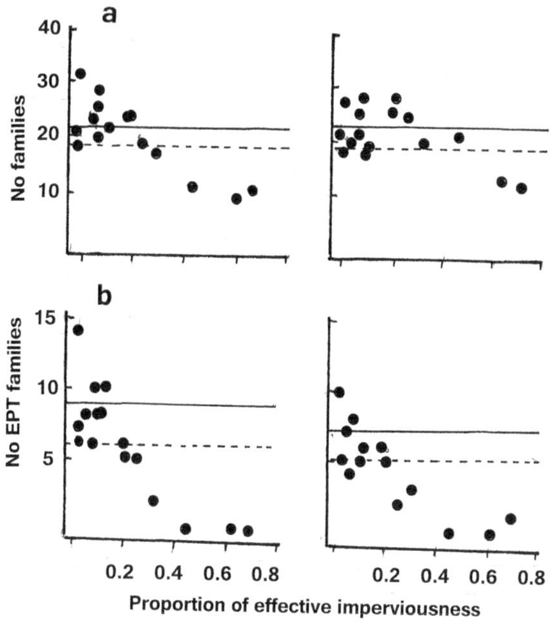


**Fig. 4.6** Relationships of macroinvertebrate biological indicators with extent of effective catchment imperviousness in riffle (left) and edge (right) sites: solid horizontal line indicates the regional objective for ecosystem protection, and dashed line indicates the 'urban objective' for the Melbourne region; top **a** total number of invertebrate families; bottom **b** number of EPT families (Walsh 2006)

## References

Bailey RC, Norris RH, Reynoldson TB (2001) Taxonomic resolution of benthic macroinvertebrate communities in bioassessments. J N Am Benthol Soc 20:28–286

Barbour MT, Gerritsen J, Snyder BD, Stribling JB (1999) Rapid bioassessment protocols for use in streams and wadeable rivers: periphyton, benthic macroinvertebrates, and fish. Second edn. EPA 841-B-99-002, United States Environment Protection Agency, Office of Water, Washington DC

Bonada N, Prat N, Resh VH, Statzner B (2006) Developments in aquatic insect biomonitoring: a comparative analysis of recent approaches. Annu Rev Entomol 51:495–523

Brinkhurst RO (1993) Future directions in freshwater monitoring using benthic macroinvertebrates. In: Rosenberg DM, Resh VH (eds) Freshwater biomonitoring and benthic macroinvertebrates. Chapman and Hall, New York, pp 442–460

Chadd R, Extence C (2004) The conservation of freshwater macroinvertebrate populations; a community-based classification scheme. Aquat Conserv Mar Freshw Ecosyst 14:597–624

Chessman BC (1995) Rapid assessment of rivers using macroinvertebrates: a procedure based on habitat-specific sampling, family-level identification and a biotic index. Aust J Ecol 20:122–129

Chessman BC (2003) New sensitivity grades for Australian river macroinvertebrates. Mar Freshw Res 54:95–103

Cottingham P, Bond N, Lake PS, Outhet D (2005) Recent lessons on river rehabilitation in eastern Australia. Technical Report, CRC for Freshwater Ecology, Canberra

Fleischman E, Murphy DD, Brussard PF (2000) A new method for selection of umbrella species for conservation planning. Ecol Appl 10:569–579

Growns JE, Growns IO (1997) Predicting species richness for Australian freshwater macroinvertebrates: do we want to know? Mem Mus Vict 56:483–490

Hassall C, Anderson S (2015) Stormwater ponds can contain comparable biodiversity to unmanaged wetlands in urban areas. Hydrobiologia 745:137–149

Kietzka GJ, Pryke JS, Gaigher R, Samways MJ (2019) Applying the umbrella index across aquatic insect taxon sets for freshwater assessment. Ecol Ind 107:10565. https://doi.org/10.1016/j.ecolint.2019.105655

Lenat DR, Resh VH (2001) Taxonomy and stream ecology—the benefits of genus- and species-level identifications. J N Am Benthol Soc 20:287–298

Marchant R, Barmuta LA, Chessman BC (1995) Influence of sample quantification and taxonomic resolution on the ordination of macroinvertebrate communities from running waters in Victoria, Australia. Mar Freshw Res 46:501–506

Marshall JC, Steward AL, Harch BD (2006) Taxonomic resolution and quantification of freshwater macroinvertebrate samples from an Australian dryland river: the benefits and costs of using species abundance data. Hydrobiologia 572:171–194

Metzeling L, Chessman B, Hardwick R, Wong V (2003) Rapid assessment of rivers using macroinvertebrates: the role of experience, and comparisons with quantitative methods. Hydrobiologia 510:39–52

Metzeling L, Wells F, Newall P, Tiller D, Reed J (2004) Biological objectives for rivers and streams – ecosystem protection. Information Bulletin, Environment Protection Authority Victoria, Melbourne

Metzeling L, Tiller D, Newall P, Wells F, Reed J (2006) Biological objectives for the protection of rivers and streams in Victoria, Australia. Hydrobiologia 572:287–299

Newall P, Wells F (2000) Potential for delineating indicator-defined regions for streams in Victoria, Australia. J N Am Benthol Soc 19:557–571

Raadik TA (1992) Aquatic fauna of East Gippsland: fish and macroinvertebrates. VSP Tech Report no.16, Department of Conservation and Natural Resources, Melbourne

Resh VH, Unzicker JD (1975) Water quality monitoring and aquatic organisms: the importance of specific versus generic identifications. J Water Poll Control Fed 47:9–19

Samways MJ, McGeoch MA, New TR (2010) Insect conservation. A handbook of approaches and methods. Oxford University Press, Oxford

Schmidt-Kloiber A, Neu PJ, Malicky M, Pletterbauer F, Malicky M, Graf W (2017) Aquatic biodiversity in Europe: a unique dataset on the distribution of Trichoptera species with important implications for conservation. Hydrobiologia 797:11–27

Sheldon F, Walker KF (1998) Spatial distribution of littoral invertebrates in the lower Murray-Darling River system, Australia. Mar Freshw Res 49:171–182

Smith MJ, Kay WR, Edward DHD, Papas PJ, Richardson KStJ (and seven other authors) (1999) AusRivAS: using macroinvertebrates to assess ecological condition of rivers in Western Australia. Freshw Biol 41:269–282

Sonneman JA, Walsh CJ, Breen PF, Sharpe AK (2001) Effects of urbanization on streams of the Melbourne region, Victoria, Australia II. Benthic diatom communities. Freshw Biol 46:553–565

Statzner B, Bis B, Doledec S, Usseglio-Polatery P (2001) Perspectives for biomonitoring at large spatial scales: a unified measure for the functional composition of invertebrate communities in European running waters. Basic Appl Ecol 2:73–85

Stendera S, Adrian R, Bonada N, Canedo-Arguelles M, Hugueny B (and three other authors) (2012) Drivers and stressors of freshwater biodiversity patterns across different ecosystems and scales: a review. Hydrobiologia 696:1–28

Stoddard JL, Larsen DP, Hawkins CP, Johnson RK, Norris RH (2006) Setting expectations for the ecological condition of streams: the concept of reference condition. Ecol Appl 16:1267–1276

Vinson MR, Hawkins CP (1998) Biodiversity of stream insects: variation at local, basin and regional scales. Annu Rev Entomol 43:271–293

Vorosmarty CJ, McIntyre PB, Gessner MO, Dudgeon D, Prusevich A (and six other authors) (2010) Global threats to human water security and river biodiversity. Nature 467:555-561

Walker KF, Sheldon F, Puckridge JT (1995) A perspective on dryland river ecosystems. Reg Riv Res Manage 11:85–104

Walsh CJ (2006) Biological indicators of stream health using macroinvertebrate assemblage composition: a comparison of sensitivity to an urban gradient. Mar Freshw Res 57:37–47

Walsh CJ, Sharpe AK, Breen PF, Sonneman JA (2001) Effects of urbanization on streams of the Melbourne region, Victoria, Australia. I. Benthic macroinvertebrate communities. Freshw Biol 46:535–551

Wells F, Metzeling l, Newall P (2002) Macroinvertebrate regionalisation for use in the management of aquatic ecosystems in Victoria, Australia. Environ Monit Assess 74: 271-294

# Chapter 5
# Threats: The Background Variations in Condition

## 5.1 Introduction

Threats to inland water insects can only be appraised realistically in the wider perspective of threats to freshwater biodiversity and are dominated by changes to their habitats from human activities. The major threats to freshwater ecosystems are driven by human population growth associated with anthropogenic climate change which itself exacerbates previously less intensive threats such as fire and vegetation removals. Many commentators have listed major categories of threats, either generally or in more specific contexts. Thus, in appraising the reasons for listing wetland insects in the British Red Data Book (Shirt 1987), Foster (1991) listed the following groups, each with more particular reasons for listing subsets of species: agricultural development (67 references), natural changes (14), recreational (18), urban and industrial development (26), and with direct causes of loss as drainage (52) and pollution (29). These broad groups express the collective variety of concerns, although many of the literature comments were difficult to verify—Foster noted, in particular, allusions to 'disturbance' with its great variety of interpretations, so that several references to water beetles were 'nebulous and perhaps false'. Dudgeon et al. (2006) distinguished five major and complex categories of threats, all universal, and all of which interact in a variety of ways (Table 5.1). Many studies on invertebrates recognise these broad groups, and most emphasise one or more as predominant in their influences in any particular context. Their local impacts are augmented by the superimposition of broader environmental challenges that occur on wide or global scale. Dudgeon et al. thus noted nitrogen deposition and climate change through global warming and consequent changes in precipitation and run-off patterns. Biodiversity declines from these multiple influences in inland waters seem to be faster than those in marine and terrestrial systems. Very broadly, collective anthropogenic changes and the responses of freshwater ecosystems to those pressures involve alteration of physical structure, water chemistry and temperature, biodiversity representation (as composition, diversity and distribution), and ecosystem processes, with measurement of outcomes ranging from relatively straightforward to highly uncertain. Estimates


T. R. New, *Insect Conservation and Australia's Inland Waters*,
https://doi.org/10.1007/978-3-030-57008-8_5

**Table 5.1** The five major threats to global freshwater biodiversity, each interacting with all others (Dudgeon et al. 2006)

| | |
|---|---|
| 1. | Overexploitation—primarily affects vertebrates |
| 2. | Water pollution—pandemic and, despite increasing controls and regulations, also increasing in variety of constituents and impacts |
| 3. | Flow modification—ubiquitous in running waters and can become highly variable from human interventions |
| 4. | Destruction or degradation of habitat—of universal concern |
| 5. | Invasion by exotic species—includes deliberate introductions, often exacerbating changes from other threats |

of overall levels of threat are inevitably imprecise, but often alarmingly high. In the United States, 43% of Plecoptera and 18% of Odonata have been designated as 'species at risk'. Morse (2009) noted that, because Plecoptera are more characteristic of cold headwater streams and Odonata more typical of warmer waters, fauna of headwater streams may be at the higher risk.

Whilst, largely for clarity, the various threats are discussed individually here and in the next chapter, this approach is clearly overly simplistic. Many of the threats are synergistic, interacting and compounding impacts in many ways so that the collective outcome for any individual context becomes very difficult to predict in detail.

Much more attention has been paid to threats to security of human water supply than to inland water biodiversity conservation, so that Vorosmarty et al. (2010) could not produce any globally meaningful estimate of biodiversity threat—but nevertheless noted that perhaps 10000–20000 freshwater species are extinct or at risk, with those risks likely to persist well into the future. Conservation of critical habitats and ecosystem processes was seen as the best strategy to contain risk, but how this might be achieved remains highly uncertain, despite parallel risks to humanity as water supply is also eroded.

Dudgeon et al.'s (2006) review of the major threats to freshwater environments and their biota has formed the foundation for numerous more recent investigations, which continue to proliferate. However, in addition to these broad and varied categories, Reid et al. (2019) identified a suite of 12 emerging threats to freshwater biodiversity that have either appeared or intensified since 2006. These (Table 5.2) collectively raise many additional concerns, with impacts having been documented or inferred strongly for all major aquatic animal taxa, with potential for major ecosystem-level changes to flow from these. Ramifications from several of these are poorly understood. As examples (1) global electronic commerce is linked with internet sales of novel invasive species, increasing chances of deliberate or accidental releases of aquarium fish stocks by increased numbers of hobbyists or traders; (2) 'emerging contaminants' include novel antimicrobial agents and pesticides such as endocrine-disruptors; and (3) nanomaterials affect sediment accumulation, with proliferation of particle surface coatings for particular agricultural, pharmaceutical or other purposes. Table 5.2 also includes possible mitigation options for each of these emerging threats, as listed by Reid et al. Vertebrate considerations largely dominate many of these options—but

**Table 5.2** List of the twelve emerging threats to freshwater ecosystems, to imply major concerns that have arisen or intensified since the review by Dudgeon et al. a decade beforehand (see Table 5.1) (from Reid et al. 2019)

| Threat | Possible mitigation measures |
|---|---|
| Changing climates | Global commitment; expand protected areas; restore thermal refugia |
| e-commerce and invasions | Online container accountability tools; awareness campaign |
| Infectious diseases | Improve surveillance; manage to favour ecosystem controls |
| Harmful algal blooms | Improve surveillance; manage to favour ecosystem controls |
| Expanding hydropower | Ameliorate passage infrastructure; assess all project inputs |
| Emerging contaminants | Improve medication disposal; advance wastewater treatments |
| Engineered nanomaterials | Improve detection and characterisation; create targeted formulations |
| Microplastic pollution | Reduce plastic usage; enact legislation to curb use of specific products |
| Light and noise | Identify less harmful types; reduce usage; educate users |
| Freshwater salinisation | Control point sources; strategic use of freshening flow |
| Declining calcium | Further reduce acid precipitation; replenish calcium in watersheds |
| Cumulative stressors | Identify multipurpose solutions that protect biodiversity hotspots |

concerns for fish and others transfer easily to many invertebrate conservation needs, through measures such as habitat protection, arresting declines in extent and condition of waterbodies, and appreciating needs for heterogeneity, connectivity, freedom from pollution and alien species, and other shared but fundamental needs. All of these, in Reid et al.'s words, 'continue to be persistent and ubiquitous threats to freshwater biodiversity'.

Any threat or 'stressor' to a waterbody can vary from 'low' to 'high', with a parallel gradient of increased degradation from the more natural conditions. The scheme summarised in Fig. 5.1 notes six major biological changes that can occur in streams as threat levels increase. That model (Davies and Jackson 2006) was presented to help provide a more consistent assessment of aquatic ecosystem status. This Biological Condition Gradient had four objectives: (1) describing the complete range of conditions from natural to severely changed; (2) to be capable of synthesising existing field observations and the generally accepted interpretations of patterns of biological degradation within a common framework; (3) based on measurable ecologically significant attributes (or those likely to be measurable in the future) that help to assess the extent to which a system may have departed from natural conditions; and (4) consistent with empirical evidence documenting the trajectories of ecological attributes across stressor gradients (Davies and Jackson 2006). Anthropogenic changes can be to physical, chemical and/or biological components of several major features, summing to produce an 'Index of Biotic Integrity' (IBI) to evaluate changes

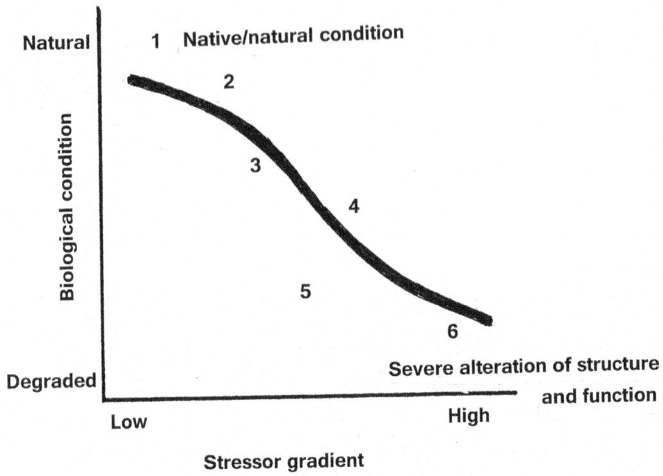


**Fig. 5.1** A gradient of biological conditions: increased 'stress' links with lowering of environmental condition (Davies and Jackson 2006). Intermediate numbers reflect the following (this text on the original figure); 2, minimal loss of species (some density changes may occur); 3, some replacement of sensitive/rare species, functions fully maintained; 4, some sensitive species maintained but notable replacement by more-tolerant taxa, altered distributions, functions largely maintained; 5, tolerant species show increasing dominance, sensitive species rare, functions altered

in stream conditions. IBI is based on a series of measurements, including species richness and composition, local indicator taxa, trophic structure and others, each sensitive to environmental stress, and has been adapted for use with macroinvertebrates (Kerans and Karr 1994).

The values for benthic invertebrates were evident from species richness of each of the EPT orders in the original description of IBI in Tennessee among the attributes related to water quality (Kerans and Karr 1994). IBI and another widely advocated index, RIVPACS (the River Invertebrate Prediction And Classification System) have rather different applications. RIVPACS is focused primarily on chemical contamination, so that 'healthy' sites can be selected, and given conservation priority on being largely free from pollution. IBI incorporates a wider set of biological attributes in a broader index of river condition or health, rather then evaluating species incidence and abundance, as does RIVPACS. Karr and Chu (2000) noted the many practical benefits of IBI as having 'proven more sensitive to human impacts than many conventional models of those impacts', and as adaptable for use in numerous different contexts whilst placing biological attributes (rather than chemical parameters) at the core of water management.

Combinations of multiple stressors and threats are perhaps most severe in lowland waters, because the surrounding environments are often severely influenced and changed by people, and the riparian, terrestrial and atmospheric environments all provide some forms of disturbance to the waters embedded in the local landscapes (Reid et al. 2019). In their initial 'Action Plan for Australian Freshwater Fishes',

Table 5.3 The threats to Australian freshwater fish, as listed by Wager and Jackson (1993)

| |
|---|
| *Regulation or modification of water flow*: |
| Altered flow volumes; altered seasonality of flow; reduced frequency of floods; altered river levels; increased rates of change of water level |
| *Geomorphic alteration*: |
| Catchment deforestation and overgrazing; removal of riparian vegetation; erosion and sedimentation; desnagging; barriers to fish movement; river engineering; alteration of habitat |
| *Water quality*: |
| Toxic substances; temperature; dissolved oxygen; suspended sediments; eutrophication; salinity |
| *Invasive exotic and native species*: |
| Invasive exotic species; invasive native species; diseases and parasites<br>Overfishing<br>Lack of knowledge |

Wager and Jackson (1993) remarked on the difficulties of attributing species declines to any single factor, because broad groupings of 'threatening processes' can interact strongly, and sometimes unpredictably. In their example, they noted that water quality problems can be exacerbated by factors such as reduced flow rates, changes in water temperatures, and sediment load. The same confounding effects apply to invertebrates, and the broad threat categories noted for fish are equally relevant (Table 5.3). An important addition to the lists noted above is acknowledgment that 'lack of knowledge' is a key concern for evaluating threats to fish; it is perhaps even more so for most invertebrates. It is perhaps realistic to note that, should the threat alleviation measures and management recommendations proposed for the 24 fish species documented by Wager and Jackson be accomplished, much benefit would flow to macroinvertebrates in Australian waters. For many of those fish, comments on then current status included poor knowledge of distribution and very general appraisals of the main causes of decline (or of suspected declines), with objectives for recovery including protection of known populations and occupied sites, additional surveys, and needs for monitoring, with a variety of more-or-less focused management activities listed for each species. As for many invertebrates, the releases and spread of alien predatory fish, notably trout and mosquito fish (p. 114), have been associated with many declines of native fish.

Flow-related disturbances, including droughts and floods, impose massive impacts on lotic environments and communities. In essence, they disrupt the biota through changes in resources, habitat availability and suitability and a variety of environmental conditions. Van Looy et al. (2018) distinguished three ways in which flood or drought lead to fundamental disturbances to river ecosystems (Fig. 5.2), as (1) shifts in productivity influencing resource availability and the food webs and community dynamics; (2) increasing or interrupting physical continuity, with ramifications for availability and metacommunity structure; and (3) changing environmental conditions affecting habitat quality and heterogeneity, and in turn community diversity.

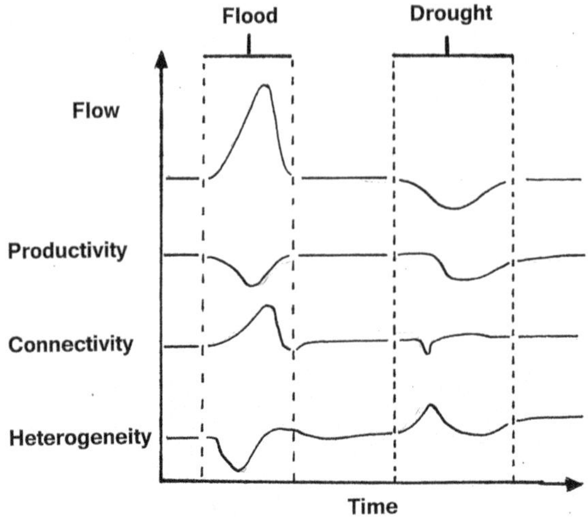


**Fig. 5.2** River disturbances produced by high ('flood') and low ('drought') flows and the potential effects on three major ecosystem properties over time (Van Looy et al. 2018)

Many of the themes interpreted as 'threats' in particular contexts result directly from changes in water flow and water quality.

However, changes in these parameters, and related changes such as sediment depositions, are commonly natural, and sometimes predictable features of Australian inland waters. It follows that native aquatic biota have a long history of association with such variable conditions—incorporating morphological, behavioural and ecological adaptations to resist or otherwise counter their impacts. In many cases, anthropogenic influences have changed 'more natural' regimes, leading to highly uncertain outcomes. Natural fires occur, for example, but more deliberate burning can accelerate and increase post-fire impacts on aquatic fauna from physical and chemical changes—destruction of riparian vegetation can increase water temperatures, sediment loads, ash-induced chemical changes, and reduced dissolved oxygen levels. The last alone can induce mass mortality of macroinvertebrates, but the most commonly documented impacts from fires include increased turbidity, suspended solids and nutrients. 'True threats' must therefore be distinguished carefully from the background normalities, but these can clearly include anthropogenic changes to those long-evolved systems. Many anthropogenic impacts on freshwater ecosystems are not novel and may have operated over decades to centuries (Gangloff et al. 2016). Many, however, continue to increase in scale and variety, with potentially catastrophic implications for biota and environments that have withstood—at least to some extent—the earlier impositions. Both droughts and floods may be changed in season and intensity, for example, and water quality and chemical content affected. Both events, beyond 'normal incidence levels', are understandably publicised most for their often catastrophic impacts on human welfare and economy, and impacts on aquatic biota are signalled most dramatically by large-scale fish kills in rivers with diminished flow or oxygen level, or influxes of vast numbers of waterbirds that can breed in these previously uninhabitable inland areas. Irregular events, such as flooding of the normally dry vast area of Lake Eyre, are significant attractions for tourists.

The extent of some postulated threats is usually unclear but may be substantial. For example, Harrison et al. (2007) reported that about 30,000 km of streambed is degraded by fine sediment in Australia. If management is to be undertaken, this scale emphasises needs to identify sources of sediment in any site and also to evaluate various practical options. Figure 5.3 summarises a management framework devised by Wood and Armitage (1997), with the major contributions to management including mitigation to protect key areas from fine sediments and allowing degraded areas to recover.

As in other contexts, isolated island freshwater insect faunas are especially susceptible to waterside development and change, as demonstrated for the endemic Diptera of Hawai'i (Benbow et al. 2005), where these insects dominate cascades and related torrential habitats. Notable endemic taxa occur in several families: Chironomidae (*Telmatogeton*), Canacidae (*Procanace*), and Ephydridae (*Scatella*). All are susceptible to flow reductions caused by damming streams and diverting water for human use. Impacts differed on the different taxa, with an inverse relationship between the densities and biomass of *Telmatogeton* and *Procanace*. This suggested a possible competitive interaction, with reduced flow increasing suitable habitat (exposed rocks) for *Procanace* and reducing larval colonisation by *Telmatogeton*. In turn, this may reflect that *Procanace* is a relative generalist and that the more specialised *Telmatogeton* can thrive only in high water velocity habitats. Such interactions can be complex to interpret, especially when other species or threats—such as native and alien fish—are also present.

The over-riding universal concern is the loss, degradation and fragmentation of habitats, so that knowledge of the resource needs and living conditions needed by any

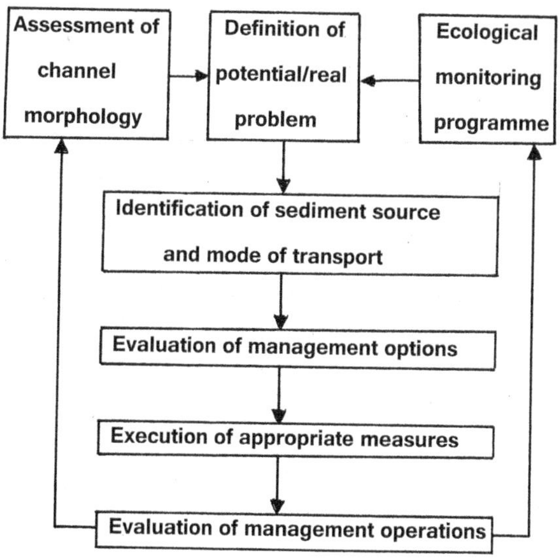


**Fig. 5.3** A holistic management framework for evaluating the effects of sediment accumulation in rivers (taken from Harrison et al. 2007)

focal taxon is necessary for practical conservation management to be superimposed on 'habitat protection'. That knowledge also allows for ranking and assessment of vulnerability as changes occur. Thus, for waterbeetles in Spain, Abellan et al. (2005) distinguished four main groups of impacts ('threats') as (1) infrastructure—dredging and stream channelisation, drainage, urbanisation, and other human developments; (2) agriculture—impacts of agricultural activity on the aquatic systems, embracing land clearing for crops, water extraction for irrigation, non-point pollution, and others; (3) pollution—from waste water and industrial discharges; and (4) other impacts—such as tourism and recreational activities, alien species, grazing, and other human disturbances. These groupings were used to construct a 'habitat vulnerability value', from '0' (no impact) through increasing levels of impact (1–4), and thence a 'habitat loss value' for each species as the sum of the above scores divided by the number of habitats occupied, so that highest scores were for species in habitat types suffering the greatest impacts. In the two Spanish provinces assessed, scores were given also for distribution (general and level of endemism), rarity, persistence, habitat rarity ('0' very common to '3' extremely rare), and habitat loss. At local level, all 209 invertebrate species species from Albacete and 145 from Murcia were ranked in decreasing order of conservation priority. None was regarded as having 'very high' vulnerability, but 10 and 12 species respectively were ranked with 'high vulnerability', and 44 and 35 species with 'moderate vulnerability'. Abellan et al. noted that their system is 'certainly arbitrary', but that the criteria may validly indicate the level of threat. All the high priority species from Albacete are rare at national and global scales but, in contrast, most of those from Murcia are common nationally despite their local rarity. This was especially so for Elmidae (Coleoptera), and perhaps reflected their dependence on headwater or middle reach streams which have suffered greatly from human interventions in the region.

This chapter introduces the background to understanding the expected and long-coevolved impacts of droughts and floods on Australia's inland aquatic invertebrates, as a prelude to assessing the wide range of additional threats imposed by people and which are the most frequent stimuli for conservation.

## 5.2 Drought

The wide impacts of drought in Australia, discussed in the context of the prolonged 'millenium drought' regarded as one of the worst droughts since European settlement, showed responses from scientists and managers to be 'both haphazard and uncoordinated' (Bond et al. 2008). Concerns were expressed that human disturbances to landscapes had markedly lessened the long-assumed capacity for natural survival and recovery to occur—a popular expectation for once droughts break, but not soundly backed up by scientific evidence. Bond et al. also noted that management responses to drought are typically reactive, and that development of sound water management strategies is vital. Call for effectively coordinated and well-resourced research programmes on droughts and their varied ecological, hydrological, social and economic impacts reveal that impacts of droughts are far wider than for conservation alone and apply across both running and standing waters, and across almost all aspects of human life.

There is no single unifying definition of 'drought', but many workers have differentiated between 'seasonal drought' (essentially predictable from normal rainfall patterns) and 'supra-seasonal drought' (far more unpredictable, usually for far longer periods, and typifying what many people consider as 'drought'). Boulton (2003) noted the stepwise process of drought progressing through the threshold levels of discharge or water levels as drying continues (Fig. 5.4), as a gradient reflected also in decreasing numbers of invertebrate species. The categories of more-or-less predictable cyclic seasonally intermittent flows and more 'episodic' and unpredictable flows are both widespread (Bayley and Williams 1973) as descriptors of 'temporary waters'.

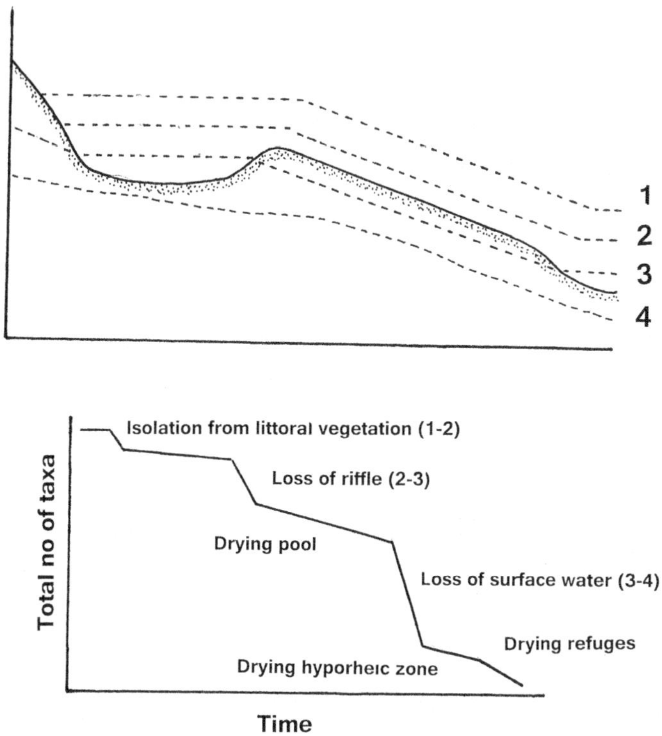


**Fig. 5.4** Changes in the composition of macroinvertebrate assemblages as a 'stepped' process during transitions across threshold discharges or water level during drying. Total numbers of taxa are implied to decline sharply during isolation of submerged or trailing littoral vegetation as it becomes separated from free water (stages 1, 2), then as flow ceases in riffles (2, 3) and when surface water disappears (3, 4). Further declines might occur in the hyporheic zone if subsurface water levels fall (Boulton 2003)

Although numerous definitions have been proposed, Kallis (2008) conceived drought as 'a temporary lack of water which is, at least partly, caused by abnormal climate conditions and is damaging to an activity, group, or the environment', and as a state that affects more people than any other climatic hazard, through economic, social, health, and a great variety of environmental and socioeconomic factors. The adaptive capacities of people and insects are both tested severely as availability of water decreases, and the true impacts of any drought are immensely difficult to conjecture. Droughts are thus difficult to both define and study, but are unified by recognition that prolonged declines in water availability have severe consequences for freshwater ecosystems and their associated biota. As Boulton and Lake (2008) noted, the progressions are difficult to appraise—replication of studies is largely impossible, and any drought may not be acknowledged or even detected until it is well in train; studies on drought impacts are often serendipitous, and most lack the sound pre-flood data needed to define a baseline condition. Boulton and Lake considered droughts a 'ramp disturbance', in which severity of disturbance increases steadily in time and space, intensifying until adequate rainfall resumes. The progression of changes in stream habitats and associated biological responses shown in Fig. 5.5 exemplifies the broad sequence of events as drought progresses, with some of the habitat changes leading directly from changed water quality rather than quantity alone. Particular families of insects may typify the fauna occurring in each stage (Table 5.4)—but collectively mark a transition from full water supply to dry, with the latter extreme supporting only rather few desiccation-resistant taxa able to take refuge in more-or-less moist stream beds. These conditions parallel those for ponds and lakes as they dry out, with progressive reductions to pools with poor water quality, and in which remnant populations of a few insects and others that can tolerate that warm stagnant water may persist. Some insects may, in course, be able to fly to alternative refuges—and several authors have noted that some of these tolerant insects are highly mobile (Velasco and Millan 1998). Boulton and Lake (2008) also noted the transitions of remnant pools to 'predator soup' as densities of remaining insects increased. In Australia, many predatory species invade stream pools from other waterbodies. In the final stages of drought, stream insects may adopt a variety of potential refuges and microhabitats in which moisture might be retained or its presence prolonged.

The ecological consequences of changes in insect assemblages during drought—even when those changes are reasonably predictable—are usually highly uncertain. However, Boulton and Lake (2008) suggested that management should heed the reality that drought is a natural component of all stream ecosystems. For many insects, refuge use is critical in preventing losses, so that management could usefully preserve and enhance refuge availability and variety, and couple this with patterns of water allocation for environmental purposes—invariably involving complex and controversial political decisions at times when water is scarce, and when multiple conflicting pressures on its use for economic purpose and welfare are greatest. Changes in assemblages from drought can extend well beyond the period of the drought itself. Indeed, the complex of influences noted in Fig. 5.4 are closely related, with confounding effects varying over time, habitats and specific aquatic environments (Boulton 2003). A 'dry phase' can be viewed as a constraint on lentic invertebrate diversity simply

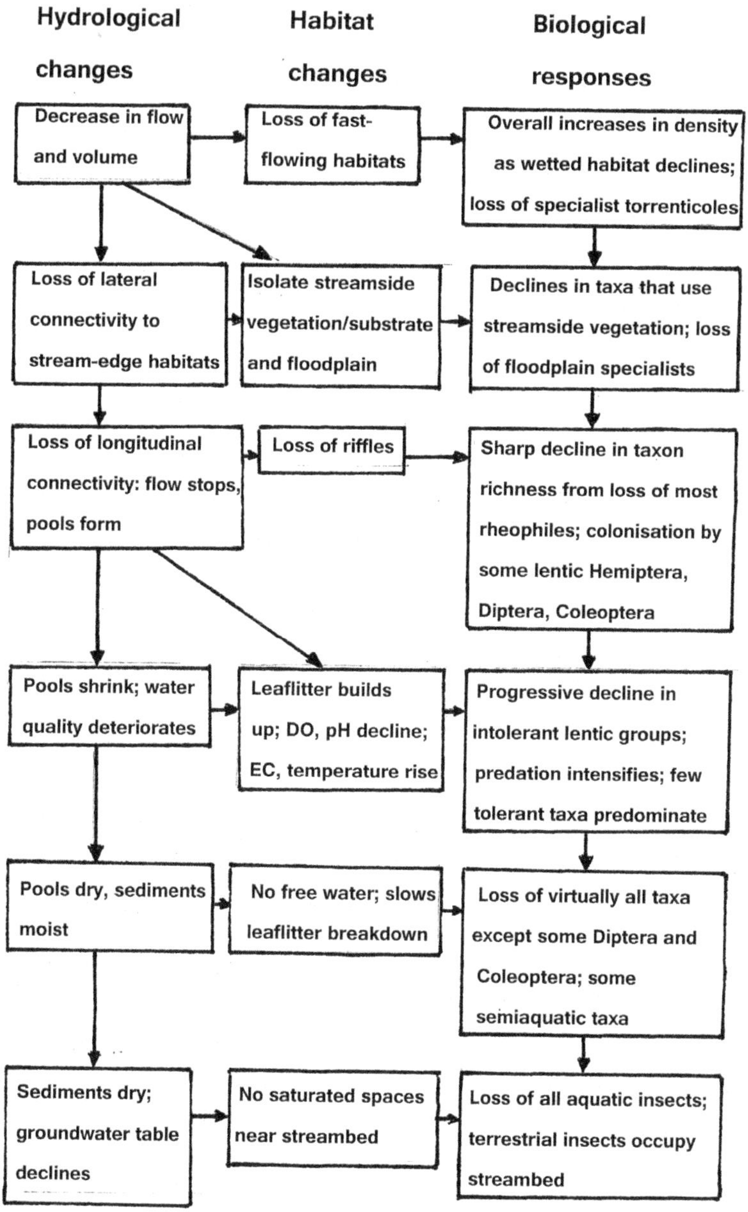


**Fig. 5.5** Changes in stream habitat and biological responses during the progressive hydrological stages of drought. (DO: dissolved oxygen; EC: electrical conductivity) (Boulton and Lake 2008)

**Table 5.4** The occurrence of many families/subfamilies of Australian aquatic insects with sequential stages of incidence as habitat changes with progress of drought, from initial decrease of water flow/volume through stages of pool loss to dry (from table by Boulton and Lake 2008)

| Habitat or habitat change; ecological groups | Families/Subfamilies |
|---|---|
| Decrease in flow/volume; specialist torrenticoles or need fast-flowing oxygenated water | Blephariceridae, Amelotopsidae, Coloburiscidae, Siphloneuridae Eustheniidae, Notonemouridae, Hydrobiosidae, Philorheithridae |
| Loss of lateral connectivity; uses stream-edge habitats or vegetation | Hydraenidae, Thaumaleidae, Pyralidae, Lestidae, Synlestidae, Hydroptilidae |
| Loss of longitudinal connectivity, flow ceases; rheophiles | Psephenidae, Ptilodactylidae, Empididae, Simuliidae, Baetidae, Leptophlebiidae, Corydalidae, Nevrorthidae, Austroperlidae, Gripopterygidae, Notonemouridae, Calocidae, Coenosucidae, Glossosomatidae, Helicopsychidae, Hydrobiosidae, Hydropsychidae, Leptoceridae, Philopotamidae, Philorheithridae, Polycentropodidae. Tasimiidae |
| Early stages of pool shrinking; moderately tolerant (or extremely versatile) lentic taxa | Elmidae, Hydraenidae, Hydrochidae, Athericeridae, Tanypodinae, Orthocladinae, Tipulidae, Tabanidae, Dolichopodidae, Psychodidae, Baetidae, Caenidae, Leptophlebiidae, Naucoridae, Sialidae, Sisyridae, Coenagrionidae, Isostictidae, Lestidae, Protoneuridae, Synlestidae, Hemicorduliidae, Libellulidae, Gomphidae, Gripopterygidae, Atriplectidae, Calocidae, Coenosucidae, Ecnomidae, Hydroptilidae, Leptoceridae, Odontoceridae |
| Final stages of pool drying; highly tolerant lentic taxa | Dytiscidae, Gyrinidae, Hydrophilidae, Scirtidae, Ceratopogonidae, Chironominae, Culicidae, Ephydridae, Muscidae, Stratiomyidae, Leptophlebiidae, Belostomatidae, Corixidae, Gerridae, Hydrometridae, Nepidae, Notonectidae, Veliidae, Mesoveliidae, Calamoceridae, Leptoceridae |
| Pools dry: taxa with desiccation Resistant stages or able to survive in moist stream beds | Dytiscidae, Ceratopogonidae, Tabanidae, Simuliidae, Gelastocoridae, Gripopterygidae |

because a normally suitable habitat becomes seasonally or sporadically unavailable (Anton-Pardo et al. 2019). Both temporal and spatial factors affect species' dynamics and, with considerations of dispersal (p. 231) affect community richness and composition. Many studies imply that permanent lentic habitats support greater invertebrate diversity than temporary waters, but in some cases the differences decreased as conditions change over time and are revealed only by longer sampling periods (Anton-Pardo et al. 2019).

Seasonally intermittent waters, notably streams (p. 277), have given important lessons for understanding how macroinvertebrates can respond to such imposed changes. Comparisons of the assemblages in intermittent and permanent (perennial) streams, together with 'before and after' samples from the former have suggested that (1) species that persist at intermittent sites were dominated by those resistant to desiccation, and (2) taxa persisting at perennial sites had fewer traits associated with resistance but more that conferred resilience (Leigh et al. 2016). The intermittent site faunas were dominated by Diptera, with families such as Tipulidae and Ceratopogonidae considered to be highly resistant to stream drying. Perennial sites are important sources of resilient taxa that can colonise (or specialise in) intermittent sites. In this way, both resistant (that is, with ability to persist) and resilient (with ability to recover quickly) taxa, in combination with both intermittent and perennial reaches of a stream, enable rich invertebrate communities to persist (Leigh et al. 2016). Successions of species across seasons can lead to marked differences in community structure and function from a running water season, a 'pool stage' as water flow ceases and a (most commonly summer) dry season. Functional changes found in Victoria (Boulton et al. 1992) included seasonal balance of functional feeding groups. However, year-to-year variation within a single stream can parallel the differences between streams with different levels of water permanece in a given year, and the assemblages of normal intermittent streams may not differ greatly from those of nearby permanent streams (Feminella 1996, for Alabama). That high similarity endorses that they cannot be overlooked in planning for conservation of regional freshwater communities.

Conservatively, 'intermittent rivers' (embracing 'all temporary, ephemeral, seasonal and episodic streams and rivers in defined channels': Datry et al. 2014) have been estimated to comprise >30% of the global river network. In Australia, about 70% of the 3.5 million Km of river channels are intermittent. In addition, water abstraction and impoundments, as well as interceptions of overland flows for agriculture and mining (Sheldon et al. 2010) over the last few decades have led to many formerly perennial rivers becoming intermittent. The ecological roles of perennial rivers may thus also change considerably, with loss of hydrological connectivity affecting virtually all their ecological processes (Datry et al. 2014). Lotic biodiversity often decreases in response to increasing intermittence of flow, whilst lentic and terrestrial biodiversity may increase and compensate for lotic losses. All three faunal groups should be considered within the scenario that lotic communities are replaced gradually by lentic (pond-like) communities and semiaquatic communities, with diverse terrestrial insect communities found in dry riverbeds. In arid regions, riverbeds and disconnected pools are valuable refuges for semiaquatic and aquatic taxa (Boulton et al. 2008).

Pitfall trap catches of invertebrates in the dry beds of intermittently flooded streams in South Africa revealed considerable diversity. The 327 taxa, representing 19 orders, were dominated by beetles (5568 of the 9850 individuals trapped) but 10 other insect orders were also recovered (Wishart 2000). In contrast to some temporary Australian streams in which amphipods and hydraenid beetles were the most abundant terrestrial invertebrates represented (Boulton and Suter 1986), carabids and scarabaeids were more abundant in South Africa. Inputs of organic material from terrestrial insects were significant in both countries. Fires on dry river beds can then also severely affect hatching of eggs of more truly aquatic invertebrates, as found from comparative samples of eggs from soil in burned and unburned sites in South Africa (Blanckenberg et al. 2019), possibly through mortality from raised soil temperatures.

Dry river beds differ from riparian habitats in many ways and, although many are devoid of vegetation they can also be where the greatest local vegetation variety and density occurs. Steward et al. (2002) noted the latter in and around areas such as Lake Eyre in suggesting that dry riverbeds are significant landscape features with considerable ecological values. They are often overlooked in river management, so also in monitoring and assessment programmes for river biodiversity. Steward et al. suggested that this widespread neglect by ecologists might reflect that dry river beds are outside the traditional domains of both aquatic and terrestrial ecosystems and not conventionally included in the primary interests of either.

In many intermittent river systems, river waterholes represent the only permanent aquatic habitats during flowless periods. Many such waterholes in the Lake Eyre and Murray-Darling Basins are <2–3 m deep and can persist for several years in the absence of local rain or surface flows. They function as 'aquatic islands' or refuges in the then essentially terrestrial landscape. However, the variety of invertebrate needs and capabilities during drought implies that no single type of waterhole will be a 'universal refuge'. Rather, a mosaic of refugia that provide different conditions and levels of isolation may be necessary for different components of a regional insect fauna to survive dry periods. Land-use practices that change the features or presence of waterholes could markedly affect their roles and, in turn, the resistance and diversity of the dryland river systems they may otherwise support.

Nevertheless, the spectacular 'booms' of Lake Eyre and similar systems during rare periods of flooding are in large part due to presence of refugia that are effective over extended periods. The variety of 'refuges' available for insects over waterless periods can thus collectively increase opportunities for resistance. Thus, in Ontario, Canada, eight distinct categories of refuges for a temporary stream over the dry summer were distinguished (Fig. 5.6), and over-summering in particular stages of the life cycle (Williams and Hynes 1977 exemplified species that pass the period as eggs or adults, rather than more vulnerable aquatic stages) may also be a significant refuge strategy. From invertebrate surveys in two temporary streams, three main groups were recognised (Williams and Hynes 1976, 1977), and categorised as follows: (1) permanent species are not especially adjusted to living in temporary waters, but have wide tolerances that enable them to survive in streams that undergo a short dry period; (2) facultative species occur in both lotic and lentic ecosystems; and (3) specialised species that are highly adapted and often restricted to temporary waters. The 'dry period' (in Canada, typically in summer) represents a critical time associated with a variety of strategies adopted by insects to withstand the changes imposed.

In two intermittent streams northwest of Melbourne (the Lerderderg and Werribee Rivers), both draining catchments with tall open eucalypt woodland and heavily shaded (Boulton et al. 1992), those refuges included the hyporheic zone below the stream bed (used by eight taxa), under dried litter or rocks (11 taxa), in crayfish burrows (six taxa), and dry sediments in riffles (18 taxa). Permanent pools along

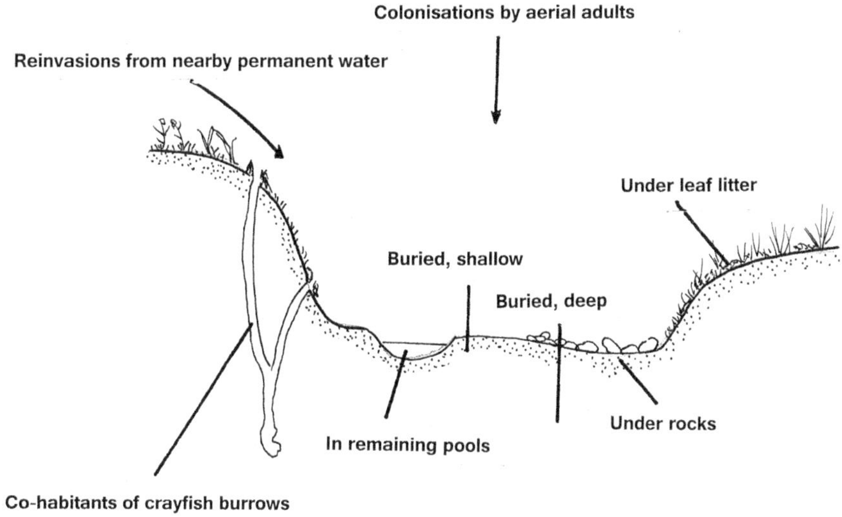


**Fig. 5.6** Summary of the oversummering methods used by fauna of a temporary stream in Canada (Williams and Hynes 1977)

the streams, as a further refuge category, were not appraised. In that survey, only nematodes and oligochaetes were common in the hyporheic zone. The latter zone has been defined as 'the saturated interstices below the stream bed and adjacent banks that contains some proportion of channel water' and, following Boulton et al. (2010), can be regarded as a dynamic ecotone between the stream (above) and groundwater (below). Such subsurface systems (as for underground aquifers, p. 22) can support diverse assemblages of invertebrates, some highly characteristic of these restricted environments, and both merit and need conservation as part of broader planning to conserve freshwater ecosystems.

However, the hyporheic zone is an important potential refuge for benthic invertebrates (Stubbington 2012), with studies variously revealing increased abundance in extremes of either drought or flood. Stubbington's review demonstrated that refuge potential was affected by four categories of influences, as (1) use varies in space in relation to physical parameters such as sediment porosity; (2) use is variable in time and reflects disturbance characteristics; (3) use is taxon-specific, reflecting a range of structural and behavioural traits; and (4) the behaviours governing use vary, with both migration and passive habitat use contributing to community persistence. These all interact to create considerable variation in refuge suitability to any given taxon.

The hyporheic zone can vary greatly in thickness, depending on the substrate. In one example, from China, it is only a few cm deep in sandy substrates, but up to 60 cm among cobble stones, with far more occupants in the latter (Xu et al. 2012). The key resource determinants of habitat accessibility were living space, food, and oxygen level. Strong relationships were found between taxon richness and the size of interstitial spaces, as evaluated through particle size (Fig. 5.7). Whilst greatest

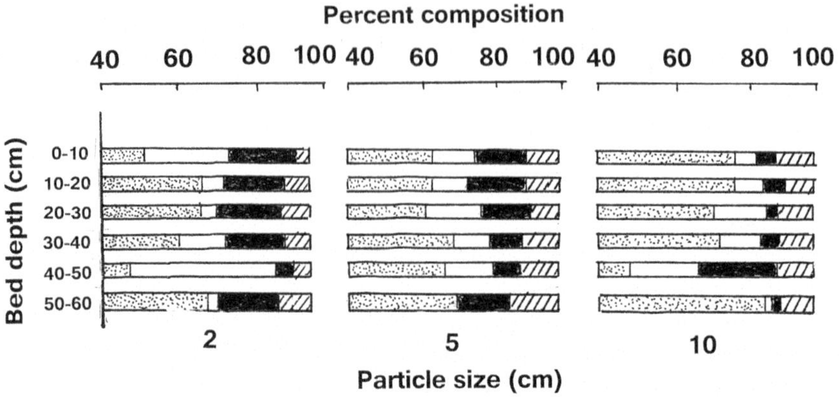


**Fig. 5.7** Relative abundance (%) of some predominant macroinvertebrate taxa at different bed depths (10 cm vertical intervals) and bed particle sizes (2, 5,10 cm diameter): dots, Ephemeroptera; open, Turbellaria; black, Trichoptera; diagonal hatching, others—including Chironomidae, Odonata, Coleoptera, Gastropoda and others) (simplified from Xu et al. 2012)

macroinvertebrate biomass occurred in the deeper levels (40–50 cm), highest richness and abundance were in shallow zones (0–30 cm). In the Thomson River, Victoria, most of the benthic invertebrate fauna was in the 0–10 cm range, so was easily accessible to standard sampling techniques (Marchant 1988). In China, however, some taxa—notably, larvae of Corydalidae—were more numerous in the deeper levels (Xu et al. 2012).

In addition, the fauna typical of hyporheic zones are not always refugial, but characteristic of the habitat, as groundwater specialists or moving regularly between benthos and the hyporheos. Nevertheless, the 'hyporheic refuge hypothesis' essentially encompasses (1) that benthic invertebrates can increase their chances of surviving adverse conditions in the stream, and (2) when conditions improve, the hyporheic zone is a source of colonists that can facilitate community recovery.

Use of the hyporheic zone as a refuge reflects three main criteria that must all be met (Fig. 5.8) for active movements into the refuge to take place. Stubbington (2012) exemplified these from particular taxa of Ephemeroptera, Plecoptera, Trichoptera and Diptera, each of which also contains taxa that do not migrate actively, and Coleoptera (Elmidae) larvae. Traits of taxa that do migrate actively in this context are very varied (Table 5.5).

The deeper subsurface sediments, largely overlooked as a habitat for aquatic arthropods, support substantial numbers down to a depth of at least 1.1 m (Langhans and Tockner 2014). Richness in subsurface and surface sediments in the Tagliamanto River (Italian alps) was similar, and the deeper sediments function as both refuge and primary habitat. Loss of interstitial spaces from deposition of fine sediment (p. 87) or livestock trampling may prevent insects using this apparently important region of gravelbed rivers.

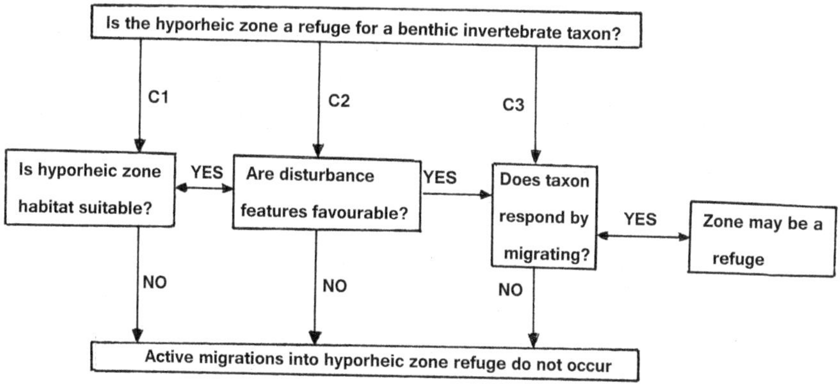


**Fig. 5.8** A flow chart to predict use of the hyporheic zone as a refuge. The three criteria (C1, C2, C3) must be met in order for active refuge use to occur (Stubbington 2012)

**Table 5.5** Evidence of proposed traits of insect taxa identified as active migrants amongst benthic invertebrates associated with refuges in the hyporheic zone; trait allocation reflects that at least some members of the taxon fulfill the state (from Stubbington 2012) ('+', trait present; '–', trait absent)

| | Proposed trait of active migrants[a] | | | | | | |
|---|---|---|---|---|---|---|---|
| Taxon | 1 | 2 | 3 | 4 | 5 | 6 | 7 |
| Chironomidae | + | + | + | – | + | + | + |
| Elmidae, larvae | + | + | + | + | + | – | + |
| Chloroperlidae | + | + | – | + | – | – | – |
| Leuctridae | + | + | + | + | – | – | – |
| Leptoceridae | + | – | + | + | – | – | – |
| Sericostomatidae | + | + | + | + | – | – | – |
| Hydropsychidae | + | + | – | + | – | – | – |
| Leptophlebiidae | + | + | + | – | – | – | – |
| Polycentropodidae | + | – | – | + | – | – | – |

[a]1, no attachment (perhaps a prerequisite for migration); 2, burrowing habit (predisposition for interstitial environments); 3, depositional rheophily (implied adaptation to sediment-depositing environments); 4, high crawling rate (ability to respond rapidly at onset of disturbance); 5, small (< 9 mm) at maturity (facilitates use of interstitial habitats); 6, tegument respiration (seems not to influence refuge use); 7, vermiform shape (implies 'morphologically suited' to use refuges)

## 5.3 Flood

Increased water discharges to rivers, whether seasonal or episodic and more unpredictable from cyclones, storage dam releases, or similar events, vary immensely in extent, duration and frequency. They are, however, viewed widely as one of the main influences on structure of stream invertebrate assemblages and communities. That belief has led to repeated calls for 'natural flood regimes' (where these can be defined, for example by seasonal patterns) to be incorporated into regulated river management. Whatever their origin, floods involve large volumes of water moving, usually rapidly, downstream and leading to structural changes in substrates, sediment loads, and deposition of debris and pollution—including in many cases that from nearby land inundated by increased water levels, and ranging from small items of debris to large branches and trees, and soils treated with agricultural and industrial chemicals. Invertebrates are influenced accordingly, by the physical and biotic alterations imposed—scouring of substrates, for example, may remove them and their food supplies, and the refuges that might otherwise aid their protection. Most attention has been paid to reductions in benthic diversity and biomass, but Death (2008) and others have noted the paucity of knowledge of the processes involved in this.

Invertebrate assemblages can differ considerably between stable and flood-susceptible streams, with differences greater in open than in forested streams. Death (2008) summarised differences as related to taxa in flood-prone streams often being small, the adults highly mobile, with generalist habits, 'clingers' with flattened or streamlined bodies (so able to resist being swept away), and two or more life stages outside the streams.

As for drought, debate continues over whether such floods are really a tolerable disturbance or a severe threat, because of the long-term historical co-occurrence of floods and aquatic invertebrates, with many insect species having features (presumed, adaptations) that normally enable them to withstand impacts, and persist (Lake 2008; Resh et al. 1988). However, the major changes to substrates, with removal of materials and organisms to far greater than normal extents, almost inevitably cause local changes—but with the longer-term outcomes commonly uncertain and unpredictable. Defining the 'threshold flows' at which permanent changes occur is difficult, and in many instances recovery from floods appears to be rapid.

Many severe disturbances from floods clearly cannot be predicted reliably, but others—such as those from seasonal weather conditions—may be more certain and allow organisms to develop strategies that in some way buffer the most severe impacts. Lytle et al. (2008) discussed the impacts of flash flooding in the arid climate of Arizona and Mexico, where monsoon deluges can lead to sudden floods. There, a flightless bug, *Abedus herbertii* (Belostomatidae) is able to respond to rainfall and move out of streams to refuges in riparian vegetation where it is safe from flood—but also more vulnerable to desiccation and predation. After floods abate, the bugs return to the same stream. The behaviour can become quite complex, and different populations can have a characteristic response time (as the duration of rainfall needed to trigger the response) and response rate (the proportion of individuals that show that response). The wide differences across populations suggest that populations can evolve to respond to disturbance regimes over time, if sufficient genetic variability is present. The behaviour sequence for *A. herbertii* comprised eight elements, observed both in the field and in simulations undertaken in the laboratory, in which sprayed water simulated rain (Lytle and Smith 2004). Avoidance behaviour (1) was initiated only with torrential rainfall; (2) with reasonably consistent duration or regime; when (3) individuals began crawling out of a stream, (4) always in a negatively geotactic manner so that the steepest available routes (even nearly vertical walls) were used. Further, (5) startle behaviour, which normally causes the bugs to return rapidly to water when they are disturbed, was inhibited, and (6) after the behaviour sequence was initiated, only light rain is needed for it to continue. Removal of the rainfall (7) caused individuals to stop moving, but movement is resumed if rainfall recommences; and (8) individuals returned to the water within 24 h of better conditions occurring. Behavioural responses to rainfall enabling aquatic insects to avoid flash floods have been reported also for water striders (*Aquarius remigis*, Gerridae) and water scorpions (*Curicta pronotata*, Nepidae), but were not found in several non-hemipterans examined in Arizona (Lytle and White 2007) and for which alternative strategies such as seeking in site refuges might be needed.

Inundation zones, including some riparian areas, are the preferred—or only—habitats for many invertebrates and, as above, some of these animals have well-defined behavioural strategies that facilitate survival during flood. Another example involves some small ground-dwelling grasshoppers, Tetrigidae. Two related European species (*Tetrix subulata*, *T. tenuicornis*) are essentially semi-aquatic, and are easily transported by even moderate water flows. In response to lowered atmospheric pressure (as a harbinger of rain) *T. tenuicornis* moves to dense vegetation, where both species are more resistant to such disturbance and removal, in contrast to staying in their more usual and susceptible habitat of bare gravel or sparsely vegetated areas (Musiolek and Kokarek 2017). Deliberate seeking of dense vegetation may help to stabilise the tetrigid populations by reducing direct losses of individuals to flood.

Anthropogenic changes to flow regimes may alter levels of predictability from naturally evolved regimes, and influence insect survival. Lytle et al. noted that some other flood-avoidance measures do not rely on rainfall cues. As examples, they cited (1) that many smaller desert stream insect taxa are multivoltine, with fast development and rapid recolonisation compensating for mortality from flood; (2) life history simplification with monsoons—their example was the Sycamore caddisfly (*Phylloicus mexicanus*) in which most larvae develop to adults before flash-flooding risks become high; and (3) that some insects, such as Hemiptera, in which adults remain aquatic may need in some way to anticipate and avoid individual floods. Most are at risk, but the above–mentioned belostomatid is one example of how this can occur. Another is the Giant waterbug (*Lethocerus medius*) that migrates from flood-prone streams to non-flooding rain pools during the monsoon season.

Severe drought-breaking rains can inundate large areas. One consequence may be 'blackwater', with high levels of dissolved organic carbon which, through consumption by microbes, can lead to hypoxic conditions and severe consequences for aquatic organisms. King et al. (2012) assessed the impacts of one such case on crayfish and fish, in the Barmah-Millewa Forest (Murray-Darling Basin, Victoria) in 2010. Large numbers of dead or dying crustaceans were found, together with Murray crayfish (*Euastacus armatus*) leaving the water and becoming more vulnerable to predation and desiccation. Native fish were also reduced substantially, and it is highly likely that parallel effects would occur on other organisms, although these were not investigated in this study, and that similar impacts would occur on other floodplain rivers.

As King et al. put it ‘At the very least, managers should be suitably cautious that floods during warmer weather after “drought-breaking rains” increase the risk of fish kills, crayfish emergence or other detrimental outcomes for aquatic fauna’.

## References

Abellan P, Sanchez-Fernadez D, Velasco J, Millan A (2005) Assessing conservation priorities for insects: status of water beetles in southeast Spain. Biol Conserv 121:79–90

Anton-Pardo M, Ortega JCG, Mel0o AS, Bini LM (2019) Global meta-analysis reveals that invertebrate diversity is higher in permanent than in temporary lentic water bodies. Freshw Biol 64:2234–2246

Bayley IAE, Williams WD (1973) Inland waters and their ecology. Longmans, Sydney

Benbow ME, McIntosh MD, Burky AJ, Way CM (2005) The influence of stream flow reduction on the energetics of endemic Hawaiian torrenticolous aquatic insects, *Telmatogeton* Schiner and *Procanace* Hendel. J Insect Conserv 9:175–185

Blanckenberg M, Mlambo MC, Parker D, Reed C (2019) The negative impacts of fire on the resurrection ecology of invertebrates from temporary wetlands in Cape Flats sand fynbos in the Western Cape, South Africa. Austral Ecol 44:1225–1235

Bond NR, Lake PS, Arthington AH (2008) The impacts of drought on freshwater ecosystems: an Australian perspective. Hydrobiologia 600:3–16

Boulton AJ (2003) Parallels and contrasts in the effects of drought on stream macroinvertebrate assemblages. Freshw Biol 48:1173–1185

Boulton AJ, Lake PS (2008) Effects of droughts on stream insects and its ecological consequences. In: Lancaster J, Briers RA (eds) Aquatic insects. Challenges to populations, CAB International, Wallingford, pp 81–102

Boulton AJ, Suter PJ (1986) Ecology of temporary streams—an Australian perspective. In: DeDekker P, Williams WD (eds) Limnology in Australia, CSIRO Melbourne/ W. Junk, Dordrecht, pp 313–327

Boulton AJ, Stanley EH, Fisher SG, Lake PS (1992) Over-summering strategies of macroinvertebrates in intermittent streams in Australia and Arizona. In: Robarts RD, Bothwell ML (eds) Aquatic ecosystems in semi-arid regions: implications for resource management. Environment Canada, Saskatoon, pp 227–236

Boulton AJ, Fenwick GD, Hancock PJ, Harvey MS (2008) Biodiversity, functional roles and ecosystem services of groundwater invertebrates. Invertebr Syst 22:103–116

Boulton AJ, Datry T, Kasahara T, Mutz M, Stanford JA (2010) Ecology and management of the hyporheic zone: stream-groundwater interactions of running waters and their floodplains. J N Am Benthol Soc 29:26–40

Datry T, Larned ST, Tockner K (2014) Intermittent rivers: a challenge for freshwater ecology. Bioscience 64:229–235

Davies SP, Jackson SK (2006) The biological condition gradient: a descriptive model for interpreting change in aquatic ecosystems. Ecol Appl 16:1251–1266

Death RG (2008) The effect of floods on aquatic invertebrate communities. In: Lancaster J, Briers RA (eds). Aquatic insects. Challenges to populations CAB International, Wallingford, pp 103–121

Dudgeon D, Arthington AH, Gessner MO, Kawabara Z-I, Knowler DJ, Leveque C (and five other authors) (2006) Freshwater biodiversity: importance, threats, status and conservation challenges. Biol Revs 81: 163-182

Feminella JW (1996) Comparison of benthic macroinvertebrate assemblages in small streams along a gradient of flow permanence. J N Am Benthol Soc 15:651–669

Foster GN (1991) Conserving insects of aquatic and wetland habitats, with special reference to beetles. In: Collins NM, Thomas JA (eds) The conservation of insects and their habitats. Academic Press, London, pp 237–262
Gangloff MM, Edgar GJ, Wilson B (2016) Imperilled species in aquatic ecosystems: emerging threats, management and future prognoses. Aquat Conserv Mar Freshw Ecosyst 26:858–871
Harrison ET, Norris RH, Wilkinson SN (2007) The impact of fine sediment accumulation on benthic macroinvertebrates: implications for river management. In Wilson AL, Dehaan RL, Watts RJ, Page KJ, Bowmer KH, Curtis A (eds) Australian rivers: making a difference. Proc 5th Aust Stream Management Conf, Thurgoona, New South Wales
Kallis G (2008) Droughts. Ann Rev Environ Resour 33:85–118
Karr JR, Chu EW (2000) Sustaining living rivers. Hydrobiologia 422(423):1–14
Kerans BL, Karr JR (1994) A benthic index of biotic integrity (B-IBI) for rivers of the Tennessee Valley. Ecol Appl 4:768–785
King AJ, Tonkin Z, Leishcke J (2012) Short-term effects of a prolonged blackwater event on aquatic fauna in the Murray River, Australia: considerations for future events. Mar Freshw Res 63:576–586
Lake PS (2008) Flow-generated disturbances and ecological responses: floods and droughts. In: Wood PJ, Hannah DM, Sadler JP (eds) Hydroecology and Ecohydrology: past, present and future. John Wiley, Oxford, pp 75–92
Langhans SD, Tockner K (2014) Is the unsaturated sediment a neglected habitat for riparian arthropods? Evidence from a large gravel-bed river. Glob Ecol Conserv 2:129–137
Leigh C, Bonada N, Boulton AJ, Hugueny B, Larned ST, Vorste RV, Datry T (2016) Invertebrate assemblage responses and the dual roles of resistance and resilience to drying in intermittent rivers. Aquat Sci 78:291–301
Lytle DA, Smith RL (2004) Exaptation and flash flood escape in the giant waterbugs. J Insect Behav 17:169–178
Lytle DA, White NJ (2007) Rainfall cues and flash-flood escape in desert stream insects. J Insect Behav 20:413–423
Lytle DA, Bogan MT, Finn DS (2008) Evolution of aquatic insect behaviours across a gradient of disturbance predictability. Proc R Soc B 275:453–462
Marchant R (1988) Vertical distribution of benthic invertebrates in the bed of the Thomson River, Victoria. Aust J Mar Freshw Res 39:775-784
Morse JC (2009) Biodiversity of aquatic insects. In Foottit RG, Adler PH (eds) Insect biodiversity. Science and Society (1st edn) Wiley-Blackwell, Oxford, pp 165–184
Musiolek D, Kocarek P (2017) Effect of substrate on the risk of being washed away by floods for the groundhoppers *Tetrix subulata* and *Tetrix tenuicornis* (Orthoptera: Tetrigidae). River Res Appl 33:1071–1078
Reid AJ, Carlson AK, Creed IF, Eliason EJ, Gell PA (and 11 other authors) (2019) Emerging threats and persistent conservation challenges for freshwater biodiversity. Biol Revs 94:849–873
Resh VH, Brown AV, Covich AP, Gurtz ME, Li HW (and five other authors) (1988) The role of disturbance in stream ecology. J N Am Benthol Soc 7:433–455
Sheldon F, Bunn SE, Hughes JM, Arthington AH, Balcombe SR, Fellows CS (2010) Ecological roles and threats to aquatic refugia in arid landscapes dryland river waterholes. Mar Freshw Res 61:885–895
Shirt DB (1987) British Red Data Books. 2. Insects. Nature Conservancy Council, Peterborough
Steward AL, von Schiller D, Tockner K, Marshall JC, Bunn SE (2002) When the river runs dry: human and ecological values of dry riverbeds. Front Ecol Environ 10:202–209
Stubbington R (2012) The hyporheic zone as an invertebrate refuge: a review of variability in space, time, taxa, and behaviour. Mar Freshw Res 63:293–311
Van Looy K, Tonkin JD, Floury M, Leigh C, Soininen J (and 12 other authors) (2019) The three Rs of river ecosystem resilience: resources, recruitment, and refugia. River Res Appl 35:107–120
Velasco J, Millan VH (1998) Insect dispersal in a drying desert stream: effects of temperature and water loss. Southw Nat 43:80–87

Vorosmarty CJ, McIntyre PB, Gessner MO, Dudgeon D, Prusevich A (and six other authors) (2010) Global threats to human water security and river biodiversity. Nature 467: 555-561
Wager R, Jackson P (1993) The Action Plan for Australian freshwater fishes. Australian Nature Conservation Agency, Canberra
Williams DD, Hynes HBN (1976) The ecology of temporary streams I. The faunas of two Canadian streams. Int Rev Ges Hydrobiol 61:761–787
Williams DD, Hynes HBN (1977) The ecology of temporary streams II. General remarks on temporary streams. Int Rev Ges Hydrobiol 62:53–61
Wishart MJ (2000) The terrestrial invertebrate fauna of a temporary stream in southern Africa. Afr Zool 35:193–200
Wood PJ, Armitage PD (1997) Biological effects of fine sediment in the lotic environment. Environ Manage 21:203–217
Xu M-Z, Wang Z-Y, Pan B-Z, Zhao N (2012) Distribution and species composition of macroinvertebrates in the hyporheic zone of bed sediment. Int J Sed Research 27:129–140

# Chapter 6
# Major Imposed Threats

## 6.1 Introduction

Anthropogenic changes to inland waters have significantly affected an estimated >83% of land surface surrounding aquatic systems (Vorosmarty et al. 2010) and have led to suggestions that 10,000–20,000 freshwater species are extinct or at serious risk. Arthington (2012) asserted that impoundments and depletion of river flows are the clearest sources of threat to biodiversity because they directly degrade and reduce river and floodplain habitats. Species losses, from any cause, can reduce all other provisionary services (including water supply and storage, food production, wood and fuel production, and the array of more conventionally cited 'ecosystem services') for humanity. The complex synergistic and compounding interactions between different threats and their associated ecological changes pose equally complex problems for those attempting to counter them. Craig et al. (2017) identified three areas that need greater attention in this endeavour, as (1) linking observed changes to threats; (2) understanding when and where threats overlap; and (3) choosing metrics that best quantify the impacts of multiple threats. Such efforts require multidisciplinary teams to pursue and coordinate the complex investigations needed, not least to meet the challenge of avoiding unintended consequences and maximising conservation benefits.

The considerable variety of impacts from human activities pose numerous 'threats' to inland water environments and biota, and fall into several broad categories, each with numerous constituents. Of these major threats (Table 5.1, p. 58), overexploitation (both directly and by inclusions as 'bycatch' accompanying commercially desirable species) is perhaps the least important for most insects but has far greater relevance to more commercially desirable edible crustaceans and molluscs that are harvested extensively. Even those concerns are usually far less than for the widespread 'overfishing' of many freshwater fish, and take of dolphins, crocodilians and turtles for food and other income production. All those activities can, of course, influence local invertebrate communities. However, especially for Odonata, local overharvesting of dragonfly larvae ('mudeyes') for fishing bait in Australia may occasionally raise concerns.



T. R. New, *Insect Conservation and Australia's Inland Waters*,
https://doi.org/10.1007/978-3-030-57008-8_6

And, although small in relation to pressures on some butterflies and beetles highly desired by collectors and valuable in trade, unscrupulous collecting for commercial purposes of rare (and, consequently, desirable) dragonflies is occasionally suggested. Although likely to be infrequent, impacts cannot be discounted entirely for some rare species. Well-intentioned protective legislation for many insects (Chap. 8) includes some form of 'prohibition of take', but—however valid the need—is usually very difficult to monitor or enforce. Collector zeal can drive 'black market operations' in which the more usual principle of 'commercial extinction' as costs of obtaining the specimens exceed return no longer applies, as prices paid continue to drive desires for specimens. Hawking and Theischinger (2004) quoted a statement that 'there is considerably more material of tropical *Petalura* in overseas collections than in Australian collections', commenting that export control for *Petalura* (p. 221) is not effective and should be improved. *P. gigantea* in New South Wales has been the target of formal searches in the east of the state, but this step is unusual, and was fostered through public and commercial interest in this as a 'giant dragonfly' sought by collectors.

'Prohibition of take' is a common outcome of listing significant insects under conservation legislations but can actually impede progress through promoting anti-collector sentiments and thwart cases in which the capture and detailed examination of voucher specimens is needed to confirm identity—as is the case for many aquatic insects. For some, field identification from living individuals that are subsequently released still leaves considerable doubt over the correct identity of the insect reported.

Although referring to marine fisheries, the contention of Jackson et al. (2001) that overfishing was often the first disturbance in a historical progression, followed by others such as pollution and eutrophication, habitat destruction, introduced species and most recently climate change, transfers easily to many inland waters where the complex impacts of increased fishing intensity can become very intricate. Not least, many of the fish species targeted are apex predators, whose excessive removal may have cascade effects on the local previously stable food webs. Allan et al. (2005) noted that fishery decline, the most commonly observed syndrome, 'took place within a complex of other pressures'.

The other four threat categories in Table 5.1 are all highly relevant in Australia. Human need for assured and predictable water supply and electrical power have led to widespread modifications of running waters, with profound effects on those ecosystems and their inhabitants. In addition to direct loss of habitat, rapid transitions from undisturbed to anthropogenic or human-influenced landscapes have (1) altered water flows and reduced the overall variety of aquatic environments; (2) increased the inputs of sediments, pollutants, nutrients and other organic materials; and (3) increased clearing of riparian vegetation, in some places replacing this with alien plants, or leaving denuded waterside environments. Especially in lotic systems, composition of communities has been influenced strongly, often from the influences of sedimentation, deforestation or agricultural developments. Changes in water temperature from greater insolation or heated or cold water discharges can have severe impacts. The major modifications are thus to discharge (such as by impoundments, through which the amount, timing and rate of water release can be regulated), with this linking closely to flow patterns. Four major categories of change in water flow regimes are sometimes differentiated, as (1) reduced flow; (2) increased flow; (3) short-term fluctuations in flow; and (4) seasonal flow constancy, and since these were distinguished (Ward 1998) they have been used extensively to describe the variety of changes that can occur. As Brittain and Saltveit (1989) noted, discharge and flow pattern changes are inevitable once river regulation occurs, these linking with changes in water temperature, community composition and structure, and processes such as predation rates. Those changes can become the basis for conservation concern.

The general scheme of impacts of river discharges changed by river regulation illustrated for mayflies by Brittain and Saltveit (1989) is likely to apply widely to other taxa as well. They distinguished three major contexts: (1) intermittent water flows, leading to increased growth of algae and macrophytes that provide increased food for grazing/periphyton feeders; (2) lower flows can increase representation of lentic species; and (3) higher flows can foster increase of lotic species. Life history traits of mayflies also affected their fate—Brittain and Saltveit noted that unpredictable harsh conditions may favour opportunistic species with flexible, asynchronous life cycles or an active short larval stage that can exploit short-term favourable conditions. Their contrasts (Table 6.1) exemplify a wide range of traits, most of them relevant far more widely.

Extreme variation in water flow is a feature of many Australian rivers (Chap. 5), and flow may cease entirely at times of drought or low rainfall. In many dryland rivers of the arid/semi-arid zones, extended periods of dryness are not uncommon—but the extensive floodplains can also be inundated following periods of substantial rain. Major flooding across the Lake Eyre Basin, for example, results from incursions of moist tropical air from the north, and occurs on average about every six years. It is associated also with increased monsoonal activity. Rivers and their biota are thus subjected to extreme variation, and extensive fragmentation coupled with conditions that appear highly unsuitable to sustain invertebrates and other biota.

Individual sites may become more or less connected with others, as explored for channels and floodplains of Cooper Creek, a major river of Queensland's channel country, by Sheldon et al. (2002). Soon after protracted flooding, aquatic macroinvertebrates were sampled from the principal microhabitats in a series of channels

**Table 6.1** Summary of mayfly life history features that may be advantageous or disadvantageous in regulated rivers (Brittain and Saltveit 1989)

| Feature | Advantageous | Disadavantageous |
|---|---|---|
| Life cycle | Flexible | Fixed |
| | Bivoltine/multivoltine | Univoltine |
| Temperature relationships | Eurytherm | Stenotherm |
| Larval development | Asynchronous | Synchronous |
| Egg development | Long | Short |
| Feeding | Scraper | Collector/shredder |
| Larval size | Small | Large |

**Table 6.2** The composition of the invertebrate fauna in samples from the Coongie Lakes wetland complex, central Australia, in December 1991: abundance of individuals and no. of taxa (except for Oligochaeta) are shown (Sheldon et al. 2002)

| | % Individuals | % Taxa |
|---|---|---|
| Bivalvia | 0.7 | 4.5 |
| Gastropoda | 8.5 | 13.6 |
| Oligochaeta | 24.6 | – |
| Crustacea | 10.6 | 3.1 |
| Insecta | 55.5 | 78.8 |
| Ephemeroptera | 7.1 | 3.8 |
| Odonata | 0.4 | 5.8 |
| Hemiptera | 55.7 | 13.5 |
| Coleoptera | 8.2 | 26.9 |
| Diptera | 26.3 | 44.2 |
| Trichoptera | 2.3 | 5.8 |

and floodplain sites in the Coongie lakes series, which mostly contained water. Altogether, insects predominated among the 70 taxa obtained, with 79% of taxa and 55% of the more than 25,000 individuals retrieved (Table 6.2). Many appeared to reflect particular conditions. Lakes and their connecting channel habitats were dominated by Chironomidae, beetles, hemipterans and predatory caddisflies (*Ecnomus*); in contrast, larger channels harboured the caddisfly *Triplectides australis* and a variety of non-insect species. Nevertheless, representatives of six insect orders were retrieved, and overall richness resembled that reported earlier from several other dryland rivers.

Sheldon et al. (2002) noted the likelihood that after prolonged flooding, normally dry water bodies may progressively resemble more permanent ones, raising the need to distinguish between short-term 'hydrological history' and the relevant long-term 'hydrological regime', in conjunction with connectivity. Widespread lack of understanding of how dryland rivers are sustained emphasises the importance of sustaining water supplies for human needs without reducing the ecological integrity of the water sources and their floodplains. Hydrological changes are likely to affect both diversity and abundance of their biota in both space and time, through alterations in connectivity and water availability beyond any natural (however irregular) patterns.

This chapter is a brief overview of some major factors implicated as threats to Australia's freshwater and related ecosystems and their inhabitants.

## 6.2 Water Temperature

Water temperature is an important determinant of habitat quality for many invertebrates, and imposed increases or decreases from a 'normal' range are sometimes termed 'Thermal Water Pollution'. Whilst projected longer-term impacts of climate change include warmer water temperatures (p. 141), the more immediate effects from human activities have aroused considerable attention, and these are noted here. In general, these changes may result from heated industrial discharges, release of cooling water from power stations, river regulation, irrigation water returns, insolation changes, and other causes. Their collective importance in Australia is acknowledged formally by listing of ‘Alterations to the natural temperature regimes of rivers and streams’ as a potentially threatening process under Victoria’s Flora and Fauna Guarantee Act 1988, with much of the reasoning summarised in the ensuing Action Statement (Doeg and Heron 2003a, b). In the absence of appropriate management, recommendation of that listing (1992) reflected that imposed temperature changes pose or have the potential to pose a significant threat to (1) the survival of a range of flora and fauna; (2) survival of two or more taxa; and (3) the evolutionary development of two or more taxa. Priority areas were easier to define for cold water, largely from releases of water from storage dams: 24 of 49 such dams in Victoria drew water from more than 10 m deep, 11 from 6 to 10 m, and the remaining 14 were of lower concern. The highest priority dams (Ryan et al. 2001) were distributed across several areas of the state. At extremes from major impoundments, Dartmouth Dam releases water from more than 60 m deep, reducing downstream water temperatures by up to 15 °C, Lake Eildon from nearly 50 m deep with temperature reductions in the Goulburn River of around 8 °C, and Lake Hume from more than 30 m to reduce downstream temperatures by about 6 °C. Particularly during summer, substantial stratification of temperatures occurs in such large dams, but the cooler temperature from release can persist over considerable distances—in the above three examples, Ryan et al. (2001) reported noticeable impacts at 70, about 100, and 200 km downstream, respectively. Both species richness and representation of mayflies can be changed dramatically from deep-water releases, with reduced richness sometimes compensated by increased densities of the few tolerant species (Brittain and Saltveit 1989).

Natural thermal clines influence many freshwater habitats—examples include changes with water depth in lakes and from source to mouth in streams and rivers. The examples noted above indicate how human interventions can both disrupt those natural patterns and provide temperature ranges outside any natural occurrence—and which species may or may not be able to tolerate. Many of Australia’s southern aquatic insects are regarded as ‘cold water specialists’ from their long evolutionary history of association with these habitats and are considered likely to be intolerant of elevated water temperatures. Ever since Ward and Stanford (1982) discussed the idea of a ‘cool water ancestry’ for such taxa, measures of thermal tolerance for aquatic species have progressively sought to establish the extent of changes that limit viability, through both mortality and a range of sublethal effects affecting all aspects of their lifestyles. Thus, laboratory trials on larvae of four species from Western Australia demonstrated considerable differences between the species (Stewart et al. 2013) but nevertheless suggested that 21 °C was a likely upper water temperature threshold for a range of taxa. Stewart et al. also predicted that future structure of assemblages will be determined largely by taxonomic differences in thermal tolerances, with progressive loss of sensitive taxa and their replacement by taxa more tolerant of higher temperatures.

Concerns over raised water temperatures arise from more general evidence of losses of riparian vegetation, for which relatively undisturbed conditions are now largely confined to upper forested catchments, with most lowland and agricultural watersides severely degraded. Discharges of cooling water from electricity generating stations can also be much warmer than in the receiving waterbodies.

Several surveys have linked changes in river invertebrate communities or species with temperature changes. Thus, for the Mitta Mitta River downstream from the Dartmouth Dam (above), some species that need warm summer conditions for their normal seasonal development were eliminated completely during cold water storage release (Doeg 1984), and many other taxa became less abundant, leading to a situation in which a few previously uncommon cold-adapted species predominated. Increased water temperatures can occasionally induce adult insects (aquatic Heteroptera, Coleoptera: Velasco et al. 1998) to disperse, and a variety of less direct effects (such as warmer waters fostering algal blooms that reduce dissolved oxygen levels) can also be highly detrimental to aquatic fauna.

The subtle sensitivities of many aquatic insects to water temperatures, at any life stage, are a key component of determining development and distribution of those species. Many authors (see Conti et al. 2014) have suggested that those responses may contribute to understanding how aquatic insects are affected (or likely to be affected) by climate changes. Conti et al. examined the potential vulnerability of three predominant and well-documented insect orders in Europe, based on the ecological preferences and biological characteristics of the EPT orders, Ephemeroptera (344 taxa), Plecoptera (461) and Trichoptera (1137), across the 23 ecoregions recognised formally in Europe. They noted that in the past the ecological sensitivities of many of the constituent taxa had been used widely as equivalent to sensitivity to wider environmental changes. Thus, sensitive Trichoptera had been selected on criteria such as being endemic to any ecoregion, ecological specialists, having short emergence periods, preferences for cold water temperatures or springs (Hering et al. 2009), with species potentially susceptible to climate change being endemics with at least one other of these sensitivity parameters. Conti et al. analysed 23 traits (Table 6.3), reflecting those most widely available across a range of scales, and each with a number of states (modalities) recognised. The exercise demonstrated the relatively comprehensive evaluation possible for a well-documented fauna in well-characterised habitats, in which all species of EPT could yield contributary data—although the lower percentages coded for many ecological conditions are salutary in implying substantial remaining gaps in knowledge even for this best-known aquatic biota. At a European scale, general patterns of vulnerability were evident in all three orders. Plecoptera were the most vulnerable, because many species had very restricted ranges of temperature tolerance and elevational distributions, with strong preference for upstream zones. Conversely, Trichoptera showed the widest range of traits and were considered highly suitable for assessing sensitivity of changes. Mayflies were the least vulnerable of these orders.

Effects of water temperature on development of individual species are exemplified by egg development of a mayfly (*Coloburiscoides* sp., Coloburiscidae) (Brittain and Campbell 1991). Laboratory studies to define the ranges of temperature tolerances

**Table 6.3** The traits used by Conti et al. (2014) for analyses of ecological preferences and distributions of Ephemeroptera (E), Plecoptera (P) and Trichoptera (T) in Europe (percentages refer to proportion of richness; 'modalities' = character states)

| Trait | No. modalities | %E | %P | %T |
|---|---|---|---|---|
| Endemism | 2 | 100 | 100 | 100 |
| Micro-endemism | 2 | 100 | 100 | 100 |
| Stream zonation preference | 10 | 60 | 75 | 71 |
| Altitude preference (WFD) | 3 | 78 | 96 | 61 |
| Altitude preference | 8 | 53 | 94 | 59 |
| Microhabitat/substrate preference | 13 | 58 | 22 | 67 |
| Hydrologic preference | 5 | 78 | 0 | 18 |
| Current preference | 7 | 66 | 56 | 84 |
| Temperature range preference | 3 | 37 | 56 | 31 |
| Temperature preference | 5 | 36 | 40 | 7 |
| pH preference | 3 | 28 | 16 | 11 |
| Feeding type | 8 | 44 | 99 | 29 |
| Locomotion type | 6 | 17 | 63 | 9 |
| Respiration | 2 | 100 | 100 | 92 |
| Resistance/resilience to drought | 5 | 6 | 8 | 8 |
| Life duration | 2 | 48 | 86 | 9 |
| Aquatic stages | 4 | 11 | 8 | 14 |
| Larval development cycle | 5 | 58 | 13 | 7 |
| Emergence/flight period | 4 | 72 | 96 | 53 |
| Duration of emergence period | 2 | 70 | 83 | 60 |
| Reproductive cycles/year | 6 | 49 | 27 | 11 |
| Reproduction | 6 | 14 | 8 | 12 |
| Feeding specialism | 2 | 100 | 100 | 100 |

WFD, Water Framework Directive

and optima of individual species help define the limits of where they may occur in nature. Eggs of a probably undescribed species of this genus were incubated at constant temperatures of 5 °C intervals from 5 to 30 °C. No eggs hatched at either extreme, but >80% hatch occurred over 10–25 °C, where there was also strong relationship between water temperature and duration of the egg stage, so that day-degree accumulation for egg development rose from 325 (25 °C) to 550 (10 °C), and with egg hatching synchronous at all temperatures. These results paralleled those from Suter and Bishop (1990) on a wider range of South Australian Leptophlebiidae and Baetidae.

Many mayflies undergo several generations each year. For these, increased water temperatures as an accelerant of development rates can in principle lead to later emergence of adults of the latest generation—either from a direct phenological effect from changed development rates or by affecting a summer diapause. The latter can occur, for example, by diapause being prolonged by absence of the trigger for renewed development caused by lowered temperature (Knispel et al. 2006). This scenario was widespread for the European *Baetis liebenavae* in the River Gwda, Poland, where prolonged summer diapause shifted the mayfly's life cycle so that the last generation developed considerably later than in two unregulated rivers in which temperatures were unchanged (Glazaczow et al. 2016). Elevated water temperatures from discharges disrupted the cues by which the mayfly regulated seasonal development, providing an ecological trap (p. 138) from (1) increased summer temperatures delaying autumn emergences until October/November, when (2) the cold winter prevented the emerging subimagos from taking off from the water surface to moult and mate. These outcomes supported the 'lost generation hypothesis' posed by Van Dyck et al. (2015), who pondered more generally whether colder climates could lead to ecological traps for ectotherms.

In contrast, raised temperature treatments imposed over a year for the European *Cloeon dipterum* influenced mayfly size or abundance only slightly, leading McKee and Atkinson (2000) to suggest that direct outcomes of small temperature changes may have little significance in comparison with indirect effects involving predation and nutrient inputs.

Changed water temperatures as climate changes occur (p. 141) are largely inevitable, and their impacts on aquatic insects difficult to forecast. Changes in insect size at maturity, growth rate, fecundity, sex ratios and seasonal appearance within individual species have all been suggested, but only rarely subjected to critical experimental evaluations in the field. In Ontario, Canada, Hogg and Williams (1996) undertook a complex field manipulation in which a small (1 m wide, 3.5 cm deep, 60 m long) permanent first-order stream was divided longitudinally and one side (half) subjected to temperature increase of 2–3.5 °C by using a propane water heater with the heated water released from a holding tank. The non-heated stream half constituted a control, and the heating treatment continued for two years (that is, over two generations of univoltine insects) following a pre-treatment year that established a baseline against which to appraise any changes. Two insects were studied in detail, within a total of about 50 invertebrate species present: a stonefly (*Nemoura trispinosa*, Nemouridae) and a caddisfly (*Lepidostoma vernale*, Lepidostomatidae) are both largely restricted to cool stream habitats and so were anticipated to respond negatively to increased temperatures. Benthic larvae were sampled monthly (Surber sampler) in both stream sides and emergence traps traced adult emergences at intervals of 2–10 days. Adults of both species commenced emergence about two weeks earlier in the heated stream, and both 50% and peak emergence were similarly earlier. Sex ratio of *N. trispinosa* shifted from male-biased to female-biased earlier, and body size (dry mass) decreased in the second year of manipulation. The last did not occur for *L. vernale*. Changes in life history and development parameters such as these may be more sensitive in indicating gradual shifts in response to global warming than are the more usually considered changes in community composition and richness or densitiy of individual taxa—but are also correspondingly more difficult and laborious to evaluate in any realistic conditions. One need advanced by Hogg and

Williams (1996) was to maintain a natural thermal diversity of regional habitats—for example including hot/cold springs and temporary/permanent streams and ponds—to facilitate insect dispersal and a variety of conditions that can collectively support most species in the area.

## 6.3 Sedimentation

Increased inputs of solid materials into waterbodies, and their presence in features such as stormwater run-off from urban areas can lead to substantial changes in both substrate and water column characters. The 'top three' most important contaminants of rivers in Australia are nutrients, salinity and sediments (Kefford et al. 2010). Sediments are generated by erosion and many anthropogenic activities, and effects on invertebrates can include increasing the amount of drift and reducing capacity to re-attach from the drift, decreasing feeding efficiency (especially of grazers and filter-feeders), and burying habitats such as by filling interstitial spaces and covering rocks. Fast- and slow-flowing streams may suffer different consequences. In Australia, fast-flowing streams are relatively rare, and many lowland streams have high suspended sediment content. In slow flow conditions, benthic invertebrates may drift for only rather short distances.

Any land-based activities that lead to sedimentation of running waters are regarded as potential threats, and many studies have explored the impacts of physical and related chemical changes on fauna, in some cases by comparisons of the fauna above and below the sediment entry points. At sites in the Murrumbidgee River, the mayfly *Tasmanocoenis tillyardi* (Caenidae) was abundant. At densities of up to 13600/m$^{-2}$, it was by far the most abundant invertebrate and comprised up to 85% of all animals captured (Hogg and Norris 1991). However, numbers decreased considerably downstream from the entry of Tuggeranong Creek, and several other benthic insect groups (Odonata, Coleoptera, Trichoptera) were also rare in those lower reaches. The creek flowed through urban areas, and run-offs after storms contained considerable amounts of suspended solids (Fig. 6.1), with their concentrations inversely correlated with mayfly abundance. Losses were attributed to deposition of fine inorganic sediments changing the substrate, but Hogg and Norris anticipated that recovery was likely once high flow events occurred and flushed out the sediments. Should that occur, sedimentation might constitute a short-term, although severe, 'pulse' disturbance. Unlike some other studies that have focused on impacts in riffle areas, the Murrumbidgee survey was on pools, which might be more susceptible to sedimentation because they are sites in which such materials may be mainly deposited.

The varied impacts of sediment discharges were exemplified by study of inputs of clay from alluvial gold-mining activities in streams in New Zealand, an activity paralleled in parts of Australia (p. 91). Davies-Colley et al. (1992) and Quinn et al. (1992) showed that the benthic invertebrates had far lower densities downstream from mining activity, with a variety of possible effects from the operations (Table 6.4).

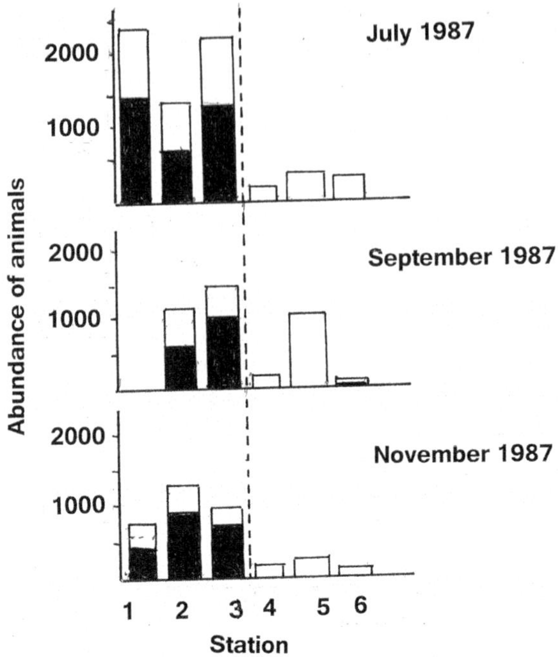


**Fig. 6.1** Distribution of invertebrates above and below the entry point (dashed vertical line) of Tuggeranong Creek to the Murrumbidgee River, ACT, at three stations (1–3) upstream and three stations (4–6) downstream, at three times during 1987 (no sample from station 1 in September) (black bars, Ephemeroptera; open bars, all others (Hogg and Norris 1991)

**Table 6.4** The mechanisms of impacts of fine inorganic sediments in stream ecosystems, as deduced from studies of alluvial gold-mining in New Zealand (Davies-Colley et al. 1992)

| | |
|---|---|
| 1 | Reduced light penetration into turbid water—reduced primary productivity |
| 2 | Reduction in quality of epilithon as food for invertebrates |
| 3 | Clogging/infilling of streambed gravels |
| 4 | Avoidance reactions by invertebrates, fish and aquatic birds |
| 5 | Accumulation of particles on body surfaces and respiratory structures |

Constructions of dams and similar disruptions to water bodies are almost inevitably accompanied by increased sediment disturbances from sand and silt, and many benthic insects are very sensitive to the changes that occur. Thus, two previously widespread mayflies (*Atalophlebioides* sp., *Coloburiscoides* sp.) declined markedly once dam constructions commenced on the Mitta Mitta and Thomson rivers in Victoria (Blyth et al. 1984; Doeg et al. 1987). Conversely, in another Victorian example, in the Tanjil River a burrowing leptophlebiid mayfly (*Jappa* sp.) increased with greater sedimentation, whilst free-living *Baetis* decreased in abundance (Chessman et al. 1987).

Accumulation of sediment is recognised widely as harmful to benthic invertebrates, and to degrade their living conditions in running waters. It can, for example, kill flora and fill the normal interstitial spaces between larger substrate particles and, in so decreasing suitable habitat, can lead to increased drift of displaced animals. Fine sediments (defined variously as of particle sizes of <4 mm or <2 mm) can thus markedly change invertebrate assemblages. Harrison et al. (2007), following Wood and Armitage 1997) distinguished four ways in which changes in the substrate can occur, as (1) changing substrate composition and its suitability for some taxa; (2) increasing drift, as above; (3) deposition of sediment directly on respiratory structures such as gills, so that oxygen uptake is affected; and (4) affecting food supply and feeding activities, such as by reducing prey density and periphyton. Most relevant studies have emphasised assemblage changes, predominantly within larvae of the 'EPT group' (that is, Ephemeroptera, Plecoptera and Trichoptera together), and any of the above four contexts can be present. The more limited array of studies on single species reveal species-specific responses. Attempts to evaluate impacts of fine sediments on benthic community change also include associated changes to algal and plant communities and are exemplified by the PSI Index ('Proportion of Sediment-sensitive Invertebrates') discussed by Extence et al. (2013).

Factors found by Wood et al. (2005) in their comparative observations on four benthic species buried by fine sediments in laboratory trials included (1) larvae of the mayfly *Baetis rhodani* could not extricate from any of a range of sediment sizes tested; (2) larvae of the caddisfly *Hydropsyche pellucidula* were trapped in sediments of < 500 μm in diameter; and (3) an isopod, *Asellus aquaticus*, was trapped in sediments of >2 mm. They noted also that greater deposition of sediment particles of >2 mm in diameter on the larvae of *H. pellucidula* led to increased effort for larvae to extricate and perhaps increased their vulnerability to predators at that time.

Ability to move vertically in changing stream sediments is relevant in assessing the impacts of sediment and its role as a refuge (p. 279) during dry periods and maintaining stream bed 'porosity' to facilitate movements can become important for some species. Vadher et al. (2017) used artificial acrylic pipes containing transparent particles of different sizes (and, so, different interstitial spaces) to compare vertical movements of several macroinvertebrate taxa, including single species of EPT, in England. All three were able to penetrate the hyporheic zone, but whilst larvae of the stonefly *Nemoura cambrica* were 'stranded' in any particle regime as water level was decreased, others showed more varied responses. The mayfly *Heptagenia sulphurea* moved only 37.4 (± 5) mm and the caddisfly *Hydropsyche siltalai* rather more, to 62.8 ± 6.3 mm. Feeding habits of both the latter (filter-feeder, grazer, respectively) imply need for sediment surface or near-surface regimes, and that food supply might be reduced considerably with greater sediment depth. However, about half the caddis larvae tested reached the base layer of the coarsest (with largest interstitial spaces) sediments tested, so that vertical movement may still enable access to hyporheic refuges. However, the inter-taxon differences implied from these comparative tests confirm the difficulties of detecting more general patterns and predicting outcomes within the complex mosaics of sediment changes in space and time.

A much wider comparative survey of fine sediment impacts on invertebrate communities in the north western United States involved 1394 streams or stream segments, from which a total of 707 invertebrate taxa were recorded (Relyea et al. 2012). Omitting rare taxa and those for which only coarse levels of identification were possible, 206 taxa were suitable for comparative evaluations. Of these, 93 showed some sensitivity to fine sediments, and four levels of sensitivity were distinguished among these: extremely sensitive (11 species, including Plecoptera [5], Trichoptera [3], Ephemeroptera [2] and Coleoptera [1]), very sensitive (22 taxa), moderately sensitive (30) and slightly sensitive (30). The variation in response, whereby species differed across a gradient of 'severely affected' to 'unaffected' by fine sediments enabled construction of an index specific to fine sediment impacts by which stream 'health' might be assessed. This was based on modelling the relative abundance (as a percentage of the total sampled taxa abundance in each stream) for each of the 206 amenable taxa against percent of fine sediment, in that study defined as <2 mm particle diameter. Each taxon was allocated to one of six categories (Table 6.5), expanding from the above levels of sensitivity, and the outcome (termed the Fine Sediment Biotic Index: FSBI) targets only fine sediment impacts on the stream organisms. In essence, the index uses the presence of common taxa to imply varying extents of degradation, with the considerable practical advantage that 'difficult' groups (such as Chironomidae, p. 201) are avoided. Nevertheless, reasons for the sensitivity of individual taxa are not always clear, and considerable interspecific differences within any group occur. The advantages of the FSBI were listed as (1) a single pollutant, fine sediment, is targeted; (2) it is easy to use; (3) lists of taxa can draw on previous studies; (4) only a selection of taxa need to be identified; and (5) no identification of insects is necessary. Although developed for a region of North America, the approach appears to be applicable to other areas in which sufficient data are available to appraise macroinvertebrate-substrate relationships.

Sedimentation of lowland streams is a frequent outcome from extensive disturbance such as forest clearing and grazing in upland areas, and large amounts of fine sediments, notably sand, can be deposited. The phenomenon of 'sand slugs',

**Table 6.5** The six categories of macroinvertebrate sediment-tolerance levels, based on proportions of fine sediment (particles <2 mm diameter including sand, silt and clay, measures given at increasing 10% intervals) in streams in the north-western United States, as used to construct a 'Fine Sediment Biotic Index' (FSBI) to assess impacts (Relyea et al. 2012). Values are calculated for each taxon, and the FSBI score is the sum of values of all taxa from the four most sediment-sensitive categories

| Fine sediment content (%) | Classification | FSBI taxon value |
|---|---|---|
| 0–20 | Extremely sensitive | 20 |
| 21–30 | Very sensitive | 15 |
| 31–40 | Moderately sensitive | 10 |
| 41–50 | Slightly sensitive | 5 |
| 51–70 | Moderately resistant | 0 |
| >70 | Extremely resistant | 0 |

represents distinct volumes of sand that are transported only very slowly downstream and, as Downes et al. (2006) implied, these could cause changes and losses of fauna and aquatic habitats. Recovery is generally slow, and some sand slugs may even become fixed—so that recovery cannot occur. Comparisons of benthic invertebrates (mostly non-insects) in upstream and downstream areas of three streams in Victoria gave rather inconsistent implications of slug impacts. Very few of the 40 'abundant taxa' were of value as indicators of sand sedimentation impacts, but the analysis at family level only necessarily amalgamates taxa with differing biologies and sensitivities—the presence of 44 species of Chironomidae, nine of Dytiscidae, and eight of Ceratopogonidae, each aggregated in this analysis, may clearly mask more individualistic species effects.

The ecological impacts of sand slugs can be very complex (Lind et al. 2009). They may form 'plugs' across tributary junctions in rivers, for example, and lead to development of wetland along those tributaries, with consequent changes to invertebrate assemblages associated with those macrophytes. A model proposed for the Glenelg River (western Victoria) demonstrated the consequences of such large and discrete amounts of sediments in that context. In that region, more than 75% of natural wetlands had been lost in the agricultural landscapes, and the slugs led to development of new wetland habitats that in part replaced those lost elsewhere in the local landscape. Such anthropogenic habitats may have important conservation roles, and some assessment of these may be needed before restoration is undertaken (Lind et al. 2009).

### *6.3.1 Mining*

Historical gold-mining, especially in the south east of Australia has had profound effects on stream and river ecosystems. Use of educator dredges to recover alluvial gold from stream beds has been particularly widespread in Victoria, where the effects on the stream environments have been adjudged 'impossible to assess' (Hall 1988). The practice became popular for both professional prospectors and hobbyists, with annual licences issued since 1979—but the real number of operators greatly exceeded the number of licences issued: Hall quoted 1200 more dredges than those licenced! Sediment may be dredged at depths of 50–200 cm, with preliminary preparation of the substrate including, where necessary, removing rocks or other large objects likely to block the intake pipe. Studies on outcomes elsewhere showed reduced abundance of benthic and drifting invertebrates, with detectable impacts both at the dredge site and the downstream areas into which sediment is discharged. Additional pollution effects can occur through the disturbance of chemicals from earlier operations. For example, mercury was used widely in historical gold extractions, and may be extensively redistributed through dredge use. Dredging followed earlier phases of gold extraction based largely on sluicing.

The history of gold mining in Victoria (Grove et al. 2019) revealed dramatic effects on the state's rivers, where a quarter of the length of the main streams was damaged, with 'catastrophic' effect on river systems. Rivers were used as a source of water for winnowing the fine sediments from terrestrial diggings by processes such as cradling and sluicing, and the volumes of sediments introduced to rivers caused massive changes in loads, with losses of waterholes, channels and vegetation. With increased high-volume mining, hydraulic sluicing, depending on large volumes of water, largely replaced earlier sluicing methods.

Impacts of these processes on stream/river biota in Australia is not known in detail, but the legacy of substrate contaminants that may be mobilised in the future is a salutary warning of dangers that could emerge.

## 6.4 Pollution

The condition of 'pollution' is easy to comprehend, but far more difficult to quantify or rank and compare across different water bodies and different constituent causes. Many attempts have been made to devise a satisfactory 'index' to assess likely impacts and needs to redress causes, many of them involving benthic macroinvertebrates. Hilzenhoff (1977, 1987) introduced and, later, modified a biotic index for stream pollution levels based on 'pollution tolerance values' for each arthropod species found in the North American streams evaluated. Every species (or genus if species were not recognisable easily) was allocated a tolerance value from '0' (the most intolerant of pollution) to '5' (the most tolerant, with a range of intermediate values. The later (1987) modification expanded the range from '0 to 10' provide greater precision and reflecting greater accumulated knowledge. That later appraisal is based on reassessed tolerance values for 359 taxa, with comparisons facilitated by a standardised sampling procedure. The resulting biotic index (referred to by some workers as the 'Hilzenhoff index') is an average for all the taxa collected from a site. The range of pollution values and index correlates is summarised in Table 6.6.

Precipitation and drainage/runoff directly convey many of the effects and changes made on 'dry' land by people to water bodies. Whilst some of those materials may, at some level, be beneficial, many others are not, and are deemed 'pollutants'. Their

**Table 6.6** Scheme for evaluation of water quality using biotic index values (after Hilzenhoff 1987)

| Biotic index | Water quality | Degree of organic pollution |
|---|---|---|
| 0.00–3.50 | Excellent | No apparent organic pollution |
| 3.51–4.50 | Very good | Possible slight organic pollution |
| 4.51–5.50 | Good | Some organic pollution |
| 5.51–6.50 | Fair | Fairly significant organic pollution |
| 6.51–7.50 | Fairly poor | Significant organic pollution |
| 7.51–8.50 | Poor | Very significant organic pollution |
| 8.51–10.00 | Very poor | Severe organic pollution |

impacts are exacerbated by the direct uses of water bodies by local people, both sporadically (such as for recreation) or as part of their every-day life. The latter can lead to substantial ecological changes.

As in many island communities, the human population of Mayotte (one of the main islands of the Comoro archipelago, between Africa and Madagascar) is concentrated around the lowland perimeter. Impacts of people on watercourses there are substantial and, as for Hawaii (p. 117) and elsewhere, those impacts move increasingly upstream to affect the less generalised biota of the more isolated and less disturbed streams. Some Comoro dragonflies are endemic and are largely stream inhabitants susceptible to human disturbances. Use of streams for laundry leaves many of them 'white with detergent' (Samways 2003), and Samways also noted Odonata with white-powdered wings from presences of the detergents. Together with removal of native vegetation and its replacement by alien plants, this pollution is inducing declines of endemic specialist dragonflies, whilst leaving several eurytopic generalist species little affected.

The concept (and reality) of water pollution is universal, but formal definition far more difficult in view of the wide variety of components that contribute to it. Bayley and Williams (1981), for discussion purposes, defined water pollution as 'a significant and deleterious change in the natural character of a water resulting from the addition of material or heat by man' but noted also that legal and other official definitions are generally just as imprecise! Roback (1974) emphasised that 'pollution' is a subjective and value-charged term: 'one person's pollution can be another person's necessary price for progress' and noted also that in discussing pollution 'it is easy to sink in a morass of undecipherable verbiage'. Nevertheless, there is abundant evidence of increased anthropogenic chemical and other inputs to waterbodies and that these can result in substantially changed aquatic environments with lessening of biodiversity and habitat quality for aquatic organisms. Referring especially to Odonata in Europe, van Tol and Verdonk (1988) noted that several natural aquatic biotopes had become so heavily contaminated that hardly any invertebrates can survive, from consequences such as eutrophication from sewage and fertiliser inputs, acid precipitation, mass mortality from industrial chemicals, and (rivers) as drainage systems for numerous effluents. Losses of some burrowing mayflies in Europe have been attributed directly to pollution; in particular the previously abundant *Palingenia longicauda* disappeared from many parts of Europe, notably within the Danube River system, during the twentieth century, and Polhemus (1993) also attributed losses of similar burrowing species in North America to pollution effects, illustrating the more general scenario that such species in large rivers subject to urban and industrial waste materials may be more vulnerable than species restricted to small headwater streams. Many categories of pollutants are also recognised and several may co-occur and create synergistic or other confounding impacts leading to more drastic and often less predictable effects. Bayley and Williams noted the categories listed in Table 6.7, with an array of impacts ranging from very general to far more specific.

Collectively, each of these categories continues to diversify—but as each is recognised it may become subject to controls and regulations. Thus, detergents (above) are one of the major components of 'domestic refuse'. Many are not broken down

**Table 6.7** Categories of water pollutants noted by Bayley and Williams (1983) to indicate the variety of contributing materials and sources

| | |
|---|---|
| Domestic refuse | Notably sewage and detergents: large amounts of organic material and can also contain pathogenic bacteria and other disease-producing organisms |
| Organic trade residues | Many kinds of waste effluents from food industries and others; oils and other chemicals also significant |
| Inert material | More-or-less chemically inert wastes that are insoluble: examples are coal dust, wood pulp and fibres, and heavy oils |
| Inorganic poisons | Includes mining wastes and industrial discards |
| Organic poisons | Compounds such as pesticides, from domestic to industrial scales, including many agricultural chemicals; many now regulated and uses diminished from major historical concerns |
| Hot water | 'Thermal pollution' as discharges from power stations and industry |
| Non-poisonous salts | Brine wastes and others can elevate water salinity |
| Radioactive wastes | Significant pollutants in some industrial regions of the world |

easily, but traditional high reliance on these has gradually given way in many places to use of 'softer', less stable, detergents that break down rapidly to nominally harmless residues. The other major domestic component, sewage, is also progressively subject to more advanced treatments rather than release of 'raw' sewage, which is a prime contributor to deoxygenation of receiving waterbodies. Deoxygenation (1) leads to many species that cannot tolerate lowered oxygen tensions being killed, effectively eliminating much of the natural biota; and (2) leads to their replacement by an impoverished biota of species that can tolerate the new conditions. In streams and rivers, such changes may extend far downstream. Adult and larval Odonata decreased markedly just below a sewage outfall into Bulimba Creek, Queensland (Watson et al. 1982). The creek's rich dragonfly fauna (36 species, 26 of them regarded as common) was sampled at sites at various distances below the outfall and compared with upstream unpolluted sites. Declines were more persistent in larvae than adults but in both stages were more severe closer to the outfall. The three sites up to 4.3 Km downstream were all severely affected and even the lowermost site (at 7 Km) had far fewer insect species than the control sites. Especially at the first downstream site (at 0.8 Km), toxic contamination was associated with drastic reductions in richness and abundance. Two kinds of impacts might occur—direct mortality and the elimination of prey organisms for the dragonfly larvae. Watson et al. believed the former to be more likely, in part because some food items (pollution-tolerant fly larvae) persisted at the most polluted sites.

Eutrophication, essentially an outcome of increased nutrient loads, is a major problem in lakes in many parts of the world and poses severe problems for managers seeking to preserve and restore water quality. Restoration efforts are diverse, but many can succeed only if external nutrient loading is stopped or reduced substantially. Key processes commonly focus on removal of fish, especially plankton- and benthos-feeders, both directly and by introduction of fish-eating taxa (such as with Pike,

*Esox lucius*, in Europe: Sondergaard et al. 2007). Both measures have yielded mixed or inconsistent outcomes.

Concerns over nutrient inputs and excessive deposits that lead to threats of eutrophication have led to many measures to reduce or counter these. However, trends toward increased fires (p. 124), such as to decrease ground fuel loads in forest, may indeed give significant increases in such deposition, both aerially (smoke, ash) and in runoff. Summer fires, for example, could lead to heightened nitrogen and phosphorus inputs, stimulating phytoplankton growth and extending well beyond the immediately burned areas. Large forest fires may have correspondingly large effects, but in evaluating the impacts of fires on freshwater eosystems 'there remains a scarcity of data collected in the midst of large forest fires' (Spencer et al. 2003).

## 6.5 Salinisation

Increased salinisation is recognised widely as a significant form of aquatic pollution. The term encompasses broadly the processes by which concentrations of dissolved salts (salinity) increase in inland waters, or the results of these processes (Williams 1987). 'Secondary salinisation', that caused by human activities, arises from two major causes (Halse et al. 2003), as (1) irrigation salinity from continued addition of water to poorly drained soils, leading to the water table rising to very near the soil surface, where concentration of salts in the subsurface layers increases soil salinity; and (2) dryland salinity, from the clearing of deep-rooted perennial vegetation, and its replacement with shallow-rooted annual crops that do not so severely dry out at the soil profile—again, water tables rise and groundwater salinity is increased, leading to death of vegetation. The trend toward increased salinity may be linked also to increased removal of water for human uses diverting it from flowing into, and diluting salt concentrations in, water bodies. In principle this may be countered by restoring water flow that offsets continued evaporation and reduction of standing water volume. Receding water levels may affect marginal vegetation and allied invertebrate habitats. Mayflies may be especially susceptible to even minor increases in salinity (Kefford 2018), rendering them amongst the most salinity-sensitive of all stream macroinvertebrate groups.

Environmental and economic costs of both irrigation and dryland salinity are severe, and dryland salinity is especially severe in the south west of Western Australia, encompassing a 'biodiversity hotspot'. Halse et al. cited surveys recording 957 species of aquatic invertebrates in the wheatbelt area—more than in any other region then surveyed in southern Western Australia. General models of interactions that might affect distributions of invertebrates in rivers in the wheatbelt's disturbed landscapes (Kay et al. 2001) reflect the increased impacts of land use. Thus, salinity can be considered a link between geology and water chemistry but also with land use because land clearing has mobilised salt stored in the soil and that has later affected waterways. As Halse et al. (2003) put it 'salinisation exacerbates habitat simplification'. They predicted that around a third of aquatic invertebrate diversity in the wheatbelt area will disappear as a consequence of this secondary salinisation. Likewise, Pinder et al. (2005) suggested that up to 100 locally restricted species there could become extinct. By 2009, more than a million hectares were affected, and this is expected to increase by two-fourfold by 2050 (Carver et al. 2009). Colonisation of aquatic habitats in relation to their salinity showed that some salt-sensitive insects selected the less saline sites, implying the importance of such selective behaviour in the distribution of some taxa, here most clearly in the mosquitos *Culex australicus* and *Anopheles annulipes*.

Pinder et al. (2005) suggested that freshwater invertebrates of the wheatbelt area might be relatively salt-tolerant: 46% of the 752 freshwater species from a total 230 wetlands sampled were collected from salinities higher than 3 g/L$^{-1}$, and 17% above 10 g/L$^{-1}$. That tolerance may help buffer against mild additional salinisation, but greater changes may still increase risks of extinction. The insects sampled included 72 beetle species, mostly in mildly saline waters and with few species demonstrably halophilous; 145 flies, also with few recorded from concentrations of >10 g/L$^{-1}$, and relatively few other taxa. These included two mayfly species (*Tasmanocoenis tillyardi*, *Chloeon* sp.), both rare in lower saline waters and absent from those above 4–7 and 6 g/L$^{-1}$ respectively. Likewise, for Hemiptera (six freshwater species) and Odonata (23 species) with few salt-tolerant taxa, and Trichoptera appeared to be more sensitive to salt than the other orders, with 10 of 18 species restricted to fresh water and only two found in regimes above 10 g/L$^{-1}$.

In the Hunter River, New South Wales, Muschal (2006) noted numerous sources of salinity and focused on the importance of point sources (such as from mining and power stations) in causing toxicity and increased environmental salinity in the water. However, only two insect taxa were listed amongst the numerous susceptible biota likely to occur in that system. Earlier trials on *Asmicridea edwardsii* (Trichoptera) and adult elmid beetles revealed increased mortality at toxicity concentrations of 2400 and 3600 mg/L$^{-1}$, respectively.

Tolerance of hypersaline waters has also evolved independently in several lineages of beetles, as have movements to inhabit subterranean habitats such as the calcrete aquifers of Australia (p. 22), each of which supports endemic species as the outcome of a particular colonisation event (Bilton et al. 2019). Elsewhere, saline streams are viewed as a conservation priority for local waterbeetle faunas (Abellan et al. 2005, for the Iberian Peninsula).

Increasing salinity is thus of wide concern for its effects on aquatic organisms, and investigations of salinity tolerances of various species can indicate the limits of the environments in which they can thrive. As with much other 'pollution', the early stages of invertebrates are commonly the most sensitive to changes, and reliance on older stages alone for those estimates can be misleading—not least because experimental exposures are often of very limited duration (Kefford et al. 2003, 2004). Studies on a range of crustacean and insect species from Victoria (Kefford et al. 2004) confirmed that the tolerances of younger stages (eggs, hatchling larvae) can indeed be far less than those of older stages. In a saline lake in Nevada, United States, early instars of a damselfly (*Enallagma clausum*) survived for only short periods under increased salinity whilst later instars were more tolerant (Herbst et al.

2013). Experimental trials with *Enallagma* and larvae of two chironomid midges showed that salinity of 20–25 g/L either exceeded lethal limits or inhibited growth and survival to the extent that existing populations would be reduced considerably. For both fish and benthic invertebrates in that study, desirable salinity was in the range of 10–15 g/L of total dissolved salts.

In general, water salinity is a major factor constraining the habitat ranges of aquatic invertebrates, so that marked declines in taxon richness occur as salinity increases. In Australia, Coleoptera: Dytiscidae and some Diptera are the predominant taxa that can thrive in high salinity regimes—some, indeed, tolerate virtually the complete gradient from hyposaline to hypersaline conditions. Communities of saline waters typically include endemic species with high habitat specificity, restricted ranges and that often form isolated populations (Arribas et al. 2015), as has occurred for Mediterranean saline water bodies where surveys have disclosed 'pockets of endemism and diversity' that merit conservation. Progressive investigations in Australia may reveal additional diversity, as has occurred for subterranean aquifers (p. 22).

## 6.6 Exploitation

Increasing interests in using insects as food for people and feed for stock have focused mainly on terrestrial species. However, a number of aquatic insects are parts of traditional human diets in various parts of the world—they include giant waterbugs in south-east Asia, larvae of Megaloptera in Japan, and water beetles in parts of Africa, Asia and central America. In a broad survey that encompassed all major biogeographical regions, aquatic species comprised only about 13% of known edible insect taxa: 261 aquatic species, compared with 1921 edible terrestrial insect species (Jongema 2014). The latter include a variety of trophic levels, but a high proportion of the aquatic species are predators. Some large water beetles (such as *Cybister*, Dytiscidae) are sold for the aquarium trade (Jach and Balke 2008).

In a detailed overview Williams and Williams (2017) explored the potential for aquatic insects to contribute to human food supply and identified six orders as having potential for bulk harvesting and rearing operations toward that purpose. Table 6.8 summarises that potential, and the harvesting measures used. However, most of the methods are limited individual operations, rather than integrated or more organised market-scale exercises, and are undertaken with little regard for sustainability (Williams and Williams 2017). Species such as some large dytiscid beetles (they cited *Cybister tripunctatus*) and Giant water bugs (*Lethocerus*) have gained the greatest attention. *Lethocerus indicus* is attracted to light and is becoming rarer in nature in many places, leading to increased interests in rearing the bugs in tanks, with potential to increase the scale of such operations by aquaculture. At up to 12 cm long, this bug is one of the largest edible aquatic insects and is a 'desirable' food item. However, Williams and Williams also commented that some representatives of the other orders perhaps subject to increased exploitation in the future are 'facing local extinction if harvesting methods are not broadened to include population sustainment'. The major

**Table 6.8** Summary of existing and potential use of aquatic insects in entomophagy, with possible harvesting protocols (Williams and Williams 2017)

| Order/family | Existing | Potential | Harvesting protocol |
|---|---|---|---|
| Ephemeroptera | Low | Could be higher | Netting mass emergences of adults; possible breeding exercises |
| Odonata | Medium | Could be higher | Individual adult capture; possible tank rearing |
| Hemiptera | Medium/high | Could be higher | Netting and attraction to lights; laboratory rearing |
| Coleoptera | Medium/high | Could be higher | Wild collection of adults; some captive breeding |
| Diptera | | | |
| Tipulidae | None | Could be viable | Collective for all: netting of adults, wild collection of larvae; rearing in tanks |
| Culicidae | None | Could be viable | |
| Chaoboridae | Medium | Viable | |
| Chironomidae | Low | Very high | |
| Simuliidae | Very low | Could be viable | |
| Trichoptera | Low | Could be higher | Wild caught: collection at lights; possibility of rearing lentic species in tanks |

key to increasing entomophagy involving Ephemeroptera, Trichoptera and Chironomidae appears to be the need for improved methodology for harvesting and possible rearing. Other Diptera (Tipulidae, Simuliidae, Culicidae), scarcely used at present, could perhaps also be exploited more effectively. This may not always be feasible: although Odonata (both as larvae and adults) are eaten—in some cases commonly—in parts of the Oriental and Neotropical regions, potential for mass rearing is likely to be limited by the long periods needed for development and that their role as predators necessitates providing prey. Consequently, with rare exceptions it seems unlikely that most Odonata will be threatened by increased levels of human entomophagy.

Reviewing uses of aquatic Coleoptera for food, Ramos-Elorduy et al. (2009) found records of 78 species in 22 genera, with the highest number (45 species in nine genera) being Dytiscidae, followed by Hydrophilidae (19 species in six genera), and smaller numbers of Gyrinidae, Elmidae, Histeridae, Haliplidae and Noteridae. Perhaps reflecting their primary survey information, the highest number of beetle species eaten (36) was in Mexico, followed by China (26) and Japan (15). They also noted that individual large beetles (*Dytiscus*, *Cybister*) in Mexico can cost up to US$ 3.50 each, whilst smaller species were sold by the package, such as a sardine can full.

In China, edible diving beetles have become rare through habitat changes, and some species are now bred in captivity to satisfy the domestic food market. More direct concerns from overharvesting wild individuals have been advanced for giant water bugs (*Lethocerus*) in several south-east Asian countries.

## 6.7 Electrofishing

Samples of fish in streams and lakes are frequently obtained by the use of electrofishing, and collecting the stunned fish from the water surface. Any possible non-target effects of the technique are usually not acknowledged. Thus, Kruzic et al. (2005) noted that few studies had appraised the impacts of electrofishing on stream macroinvertebrates.

However, the importance of any effects on invertebrates, as prime constituents of fish diet and the supply possibly affecting fish numbers and distribution in areas sampled repeatedly, was discussed by Elliott and Bagenal (1972) for an English Lake District stream. Driftnet samples taken before and after electrofishing showed massive increase in invertebrate numbers during the fishing (Table 6.9), with more than 50 invertebrate species captured in the drift, or in benthos samples taken for comparison. Mean numbers of many taxa in benthos decreased after fishing, but the total drift accounted for only a small proportion (5%) of the benthos in the area sampled. The great increase in drift numbers was attributed to both the direct effect of the electrofishing and the substrate disturbance from the two operators of the electrofisher as they walked upstream. Only Ephemeroptera and Plecoptera were dislodged in large numbers by the shocker. Rapid re-settling after being dislodged may have reduced changes in local populations. Larger invertebrates, such as some stonefly larvae, can re-settle very rapidly whilst smaller, lighter taxa tend to drift further (Kruzic et al. 2005).

Similar findings of large increases in invertebrates in drift in New Zealand were made by using the same methods as those of Elliott and Bagenal (Fowles 1975). Mayflies predominated in the drift, with *Deleatidium* sp. by far the most abundant species as comprising 53% of total drift numbers. Blackfly larvae (Simuliidae) dislodged from the upper surfaces of stones were also numerous. Conversely, larvae of a caddisfly (*Pycnocentrodes aureola*) were by far the most abundant insects in benthic samples, but very few entered the drift, and elminthid beetles were also strongly under-represented. As in Britain, many invertebrates appeared not to drift far before re-settling.

Invertebrate drift was induced deliberately by electrofishing as a sampling method for benthic fauna in Colorado (Taylor et al. 2001). Advantages over more conventional substrate sampling methods included that the drift was free of detritus, and so considerably cheaper and more rapid to sort. The difference in invertebrate densities

**Table 6.9** Impacts of electrofishing on abundance of benthic invertebrates in drift. Numbers of individuals captured in drift net samplers entering (E) and leaving (L) a fished area of an English Lake District stream on three occasions (Elliott and Bagenal 1972)

| | April 1970 | | July 1970 | | May 1971 |
|---|---|---|---|---|---|
| | E | L | E | L | L |
| Before electrofishing | 17 | 35 | 9 | 26 | 15 |
| During electrofishing | 17 | 2457 | 9 | 2328 | 1552 |

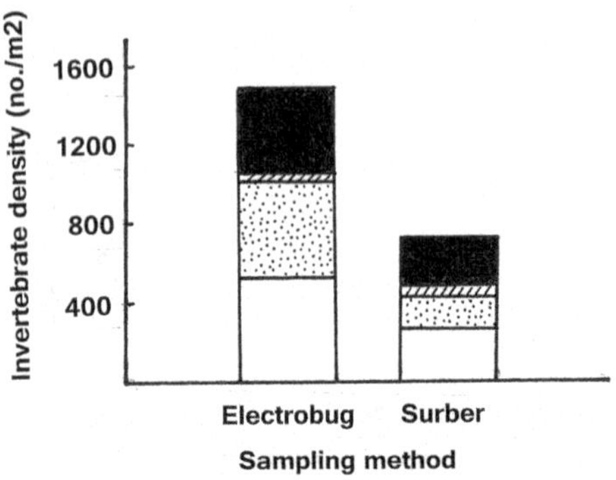


**Fig. 6.2** Electrofishing impacts: the density (no./m$^2$) of invertebrates in driftnet catches from area-restricted Electrobug and Surber sampling techniques from three streams in Canada in 1997 (dotted, Plecoptera; open, Ephemeroptera; diagonal hatch, Trichoptera; black, all other taxa) (Taylor et al. 2001)

between driftnet net catches for electrofishing and the more usual Surber sampling, for example, was considerable (Fig. 6.2). Those savings can justify the considerable additional costs of the electrofishing equipment. In addition, Taylor et al. claimed that the technique 'can provide accurate estimates of population size and diversity and minimise disturbance to the benthic habitats'.

## 6.8 Changes to Riparian Vegetation

Riparian zones are 'interfaces' between aquatic and terrestrial ecosystems, both providing biota that largely depend on the intermediate conditions present, and some of which rarely occur elsewhere. They have also been termed 'hotspots of ecological function in many landscapes' and 'hotspots of interactions between plants, soil, water, microbes, and people' (Groffman et al. 2003), and are subject to strong changes from human interventions. Whilst changes to riparian vegetation have received considerable attention, the functional changes in riparian zones from hydrological changes such as increased sedimentation may also have far-reaching ecological impacts. Those zones vary enormously in width and structure, but their importance in water body protection is recognised widely, together with the flooding regimes and vegetation that largely characterise any particular riparian area (Naiman and Decamps 1997). Changes to aquatic macroinvertebrate communities with loss or change of riparian vegetation have been reported especially among benthic fauna, but the zones also function as refuges or movement corridors—although those roles are better documented for birds and mammals than for most insects. Functionally, riparian vegetation may also reduce pollutant runoff into water and otherwise affect hydrological pathways (such as flood waters) that could otherwise more directly threaten aquatic environments.

Anthropogenic changes to waterside vegetation are a primary way through which changes to adjacent freshwater environments occur. Consequences such as increased erosion and sedimentation, increased exposure through loss of shade and associated with increased water temperatures, changed nutrient inputs and modifications to aquatic food webs, and decreased general 'buffering' of influences from shores, all constitute disturbances and may become possible threats in some form. In some heavily urbanised catchment areas with limited forest or other riparian vegetation, the remaining trees may provide the only sources of shading and organic inputs for stream ecosystems (Roy et al. 2006). Losses of vegetation near water bodies intergrade with wider changes, such as creation of agricultural landscapes in which natural vegetation is largely replaced by monocultures or alien plant species, with attendant protective agrichemical uses. Indeed, the small streams in European agricultural landscapes are recognised as among the most threatened habitats in the region. Many insect species found only or predominantly in those streams are listed in the European Community Habitats and Species Directive. One such species, the odonatan *Coenagrion ornatum*, in the Czech Republic was rediscovered there (in 2001) after more than 40 years (Harabis and Dolny 2014). Earlier ignorance of the habitat needs of this species led to management that caused extinction, rather than the anticipated enhancement of populations. Rediscovery of *C. ornatum* resulted, not from its recolonisation of managed habitats but, rather, because all the newly-found populations are in highly altered habitats that are usually overlooked in conservation management. These included semi-natural and artificial sites (p. 270), on which *C. ornatum* depended progressively as its natural habitats were lost or degraded. The largest populations were in artificial drainage ditches. Correlations with variety of stream and channel features led Harabis and Dolny to suggest that although habitat restoration for *C. ornatum* might be feasible, a number of less extensive management measures would also be worthwhile: (1) in this case, eliminating riparian vegetation and over-growing shrubs (largely to reduce shading); (2) encouraging development of species-rich vegetation; (3) establishing small pools and overflows along the stream channel as microhabitats for both larvae and adults; and (4) establishing buffers to reduce pollutant contamination of the channel.

Although riparian vegetation contributes to conservation of the freshwater ecosystems it flanks, even substantial buffers (with 30 m wide undisturbed buffers mandated in some areas—for example parts of Brazil (Cunha and Juen 2017)—may not prevent influences from changes landward from these. In that study, plantations of oil palm (*Elaeis guineensis*) still caused substantial changes to the adjacent stream environments, reflected in the complements of aquatic and semi-aquatic Hemiptera. Richness of both groups was lower in streams bordered by oil palm (with riparian buffers) than by natural forest (Fig. 6.3). Only one species (*Rhagovelia brunae*) was an indicator for oil palm sites, but 10 species so for native forest sites. The streams draining oil palm areas have less intensive cover for invertebrates, and less diverse substrates. The plantations are also far more uniform in structure than the natural forests, with far lower heterogeneity.

Streams running through forests, including areas of commercial forestry, may be susceptible to severe over-shading from either native or alien trees (Kinvig and Samways 2000). In such situations, dramatic declines of Odonata can occur, irrespective of the tree species involved. In that study in Kwa Zulu Natal, South Africa, the shade regime impacts led to suggestion that no plantation trees should shade the stream ecosystem and that plantings should be at least 30 m from the water.

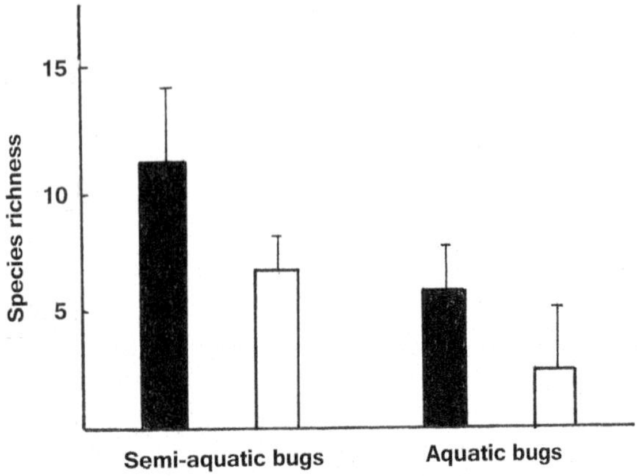


**Fig. 6.3** Species richness of semi-aquatic and aquatic Hemiptera in Brazil: native forest (black bars) and oil palm plantations (open bars) compared; s.d. shown (Cunha and Juen 2017)

Scanlon et al. (2007) and others have referred to riparian deforestation as one of the largest impacts on freshwater ecosystems, and Canning et al. (2019) remarked 'it is well known that riparian deforestation can alter macroinvertebrate assemblages'. In New Zealand, their comparative surveys of benthic invertebrates in paired forested/grassland margin sites on ten streams showed EPT (p. 44) to be dominant in forested sites, but Chironomidae and molluscs predominant in grass-bordered streams. The former were dominated by filter-feeding taxa exploiting particulate organic matter in the water column. The major insect species in grassland sites (Chironomidae, Tipulidae) fed more on diatoms and freshwater algae dominating in these unshaded streams. Assemblage differences reflected differences in predominant feeding modes—so, to different food web interactions and stability. The generalist mayflies *Deleatidium* spp. and *Coloburiscus humeralis* were abundant in both communities.

Likewise, mayflies were clearly sensitive to clearcut logging of pine plantations near perennial streams in New Zealand, where Quinn et al. (2004) found three common species to be less abundant at clearcut sites. In contrast, the micro-caddisfly *Oxyethira albiceps* (Hydroptilidae) responded positively to logging. As an algal-piercing species, this probably reflected high algal biomass after forest clearing. However, more generally, clearcut stream reaches had lower taxon richness, decreased relative abundance and richness of EPT, with half the clearcut reaches surveyed regarded as 'severely impacted' when compared with native forest reference sites. Retention of forest riparian buffers during *Pinus radiata* plantation logging provided strong environmental benefits for stream macroinvertebrates in this region. In contrast, loss of riparian forest can have strong and selective impacts. It can, for example, favour 'true dragonflies' over damselflies, reflecting their generally larger size and stronger flight activity. In the Amazon region of Brazil, damselflies (Zygoptera) were sensitive to the physical integrity of streams whilst dragonflies (Anisoptera) appeared to benefit from disturbance, as more species were found at disturbed streams (Miguel et al. 2017). The smaller-bodied Zygoptera may be more sensitive to higher temperatures so that greater exposure of water after vegetation removal may increase adverse impacts. They are also often categorised as poor dispersers and colonisers.

In the Western Cape of South Africa, abundance of adults of some Odonata was lower at sites with high or moderate shade cover from riparian vegetation than at sites with no shade (Remsburg et al. 2008), whilst the variety of perches provided at each site did not influence abundance. The two most abundant species (both species of *Trithemis*, Libellulidae) strongly avoided shade, but the key factors influencing this behaviour are not clear—possible influences include limits to prey availability, mate attraction, thermoregulation, or hunting efficiency. Nevertheless, aggressive alien riparian trees in South Africa have displaced indigenous vegetation through their superior competitive ability. Removal of that vegetation leads to greater exposure to sunlight and also to regrowth of marginal vegetation as succession proceeds from sedges and grasses, so increasing the variety of microhabitats suitable for Odonata (Samways and Sharratt 2010). Recovery of dragonfly assemblages was then rapid, and even partial clearing was beneficial—initially to widespread species but later also to more restricted endemic taxa as plant structural diversity increased. The structure of larval assemblages of Odonata can be influenced strongly by habitat selection by adults (p. 176), so that structure of waterside vegetation may be highly relevant to their conservation. Trials on the impacts of canopy cover in Canada involved using artificial cover (dark green shade cloth reducing light penetration by 50%) and simulated open canopy (using open-weave netting with little impact on light penetration) (French and McCauley 2018) to compare visits by adult dragonflies to these regimes, and also under natural canopy cover. Both numbers and richness of visitors declined with increased canopy cover and shade, with likely ensuing differences in larval populations as a direct outcome of this behaviour. In South Africa, some dragonflies were sensitive to the shade imposed by alien *Acacia* trees which formed dense riparian canopies. Remsburg et al. (2008) used 'shade plots' of nursery shade cloth imposing 30, 50 or 75% shade on reservoir shorelines, together with natural acacia cover, to demonstrate that the strong shade-avoidance behaviour of *Trithemis*. Notwithstanding lack of understanding the detailed mechanisms of this, that study led to recommendations for removal of alien riparian trees as a measure to enhance native dragonfly biodiversity.

Simply removing alien riparian trees can facilitate restoration of native vegetation structure and functions, with this trend enhanced if the larger fallen wood is removed from the riparian zone (Holmes et al. 2008). Any surviving native trees should be protected during alien removal, and the management pathways to restore ecological functions, vegetation structure and diversity projected for South Africa by Holmes et al. (2008) (Fig. 6.4) may be useful models for consideration in Australia.

Reversing the undesirable impacts of alien riparian vegetation can pose benefits that extend far beyond those to dragonflies alone. In the Limpopo Province of South Africa, changes to benthic macroinvertebrate assemblages with dense alien vegetation as well as to adult dragonfly richness, were substantial (Magoba and Samways 2010). Comparisons across sites assessed as 'natural' (15% alien vegetation cover, or less), 'alien' (dense cover of >75% aliens) and 'cleared of invasive alien trees' showed that some insect groups survived well (Gyrinidae), but some fly families (Dixidae,

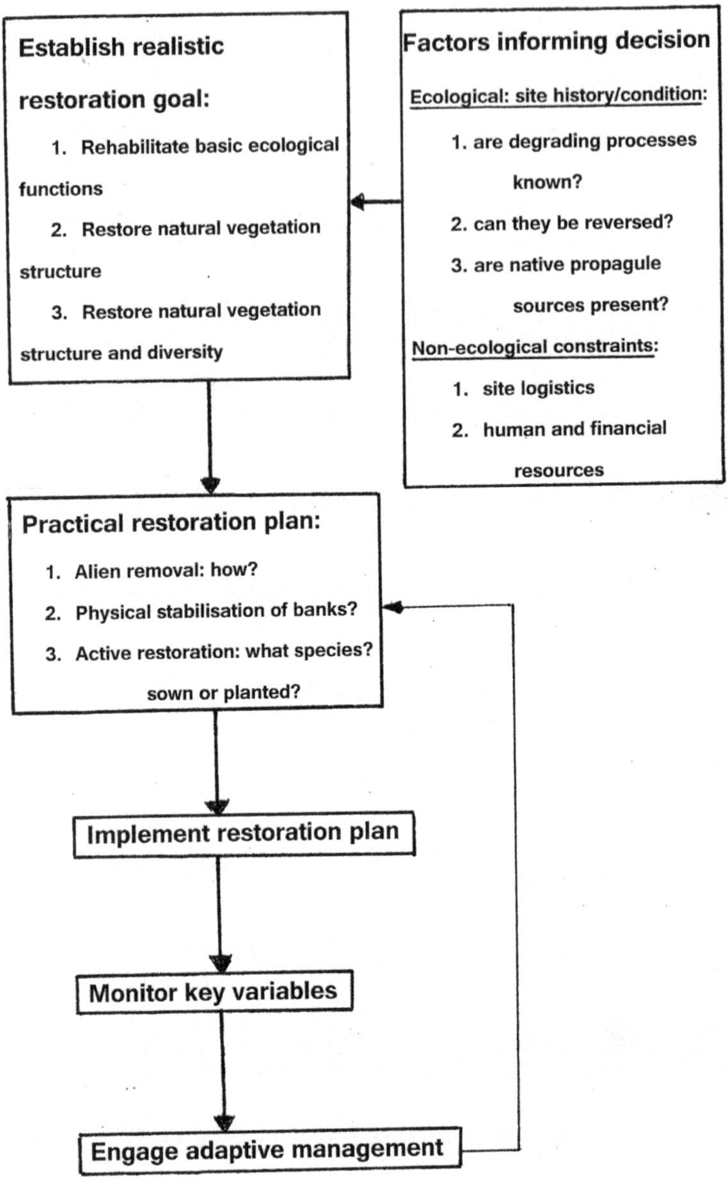


**Fig. 6.4** A conceptual framework for ecosystem repair in riparian zones invaded by alien species (after Holmes et al. 2008)

Simuliidae) benefited from the shady conditions provided by the alien trees. Assemblages at the cleared sites approached those at the natural sites, but numbers of the most sensitive taxa (larvae of Ephemeroptera, Plecoptera, Trichoptera and Odonata [EPTO]) had generally higher richness at natural sites, whilst a number of other sensitive taxa were lost from the alien sites. Focus on EPTO orders alone implied that these are a sensitive indicator in rivers with different riparian conditions.

Concerns over alien trees in Australian riparian zones are exemplified by the predominance of naturalised willows (*Salix* spp.) in the Murray-Darling Basin. *Salix babylonica*, for example, earlier dominated over about a third of the 830 km of the River Murray below its junction with the Darling (Schulze and Walker 1997). More than a hundred kinds—species, cultivars, hybrids and varieties—of willows are present in Australia, most regarded as 'weeds of national significance' and so subject to efforts to removal them and prevent further spread. As elsewhere, initial plantings were primarily to stabilise river banks or define boating channels, but the willows subsequently commonly intruded into the water and eventually led to increased erosion. The willows are thus a target for active removal and replacement by native vegetation (Zukowski and Gawne 2006). The dense extensive root mats of willows form important habitats and retreats for numerous invertebrates, and their emoval can have both direct and more indirect impacts—the latter by causing changes in sedimentation and flow rates and substrate characters. The seasonal inputs of deciduous willow foliage are also important factors in food supply, as demonstrated in New Zealand by Parkyn and Winterbourn (1997), and may even become a preferred food source for macroinvertebrates (Yeates and Barmuta 1999). The extent to which riparian willows and the native River red gum (*Eucalyptus camaldulensis*) may modify invertebrate habitats in the adjacent littoral zone has been compared by examining breakdown rates of foliage in mesh bags deployed in the water. After eight weeks, distinctive invertebrate assemblages had formed at sites with one or other treatment (Schulze and Walker 1997), with some of the less abundant taxa more site-specific. Although data were limited, the caddisfly *Hellyethira* sp. was found only at red gum sites and tabanid larvae only at willow sites, with abundance of some other taxa differing between the treatments. Willow leaves were broken down much more rapidly than eucalypt leaves, and it is possible that some specialised invertebrates were less able to utilise willow leaves, in turn possibly related to the bacterial decomposition that dominates in eucalypt foliage breakdown rather than the fungal decomposition more typical for willows. The major differences in invertebrates were perhaps related to differing utilisation of foliage of the two trees as food. Willow removal from riparian zones, followed by revegetation with native species, is a widespread restoration approach for Australian rivers (Becker and Robson 2009).

The headwaters of many Australian streams and rivers are in forest, with the implication that any forestry operations in those areas can probably affect all downstream areas—so giving such operations seemingly disproportionate importance in relation to the, nevertheless severe, primary impact of tree removal and associated local disturbances. The impacts stem from a range of activities, including gaining access to the timber, felling and removal of trees, disposal or discard of waste and, increasingly, establishing regrowth (Campbell and Doeg 1989). Major effects included deposition of sediment and debris, and changes of riparian vegetation affecting light regimes and increased runoff. Further, natural regrowth vegetation may not always be suitable to stream invertebrates.

The most widely recommended or mandated measure to counter such impacts is the retention of defined buffer strips of undisturbed vegetation along watersides. However, Campbell and Doeg remarked on the then 'complete absence' of studies of their effectiveness in maintaining the ecological integrity of streams subject to adjacent timber harvest.

Logging vegetation close to streams has many impacts leading to changed diversity and composition of invertebrate assemblages, from primary disturbances such as sedimentation, hydrology and affected nutrient regimes. Intensity of any such impact is related to proximity and timing of the operation, the kind of equipment used, and the pattern of logging (such as clearfelling or more selective take, or the practice known as 'clearfell, burn and sow' ['CBS']), as well as the proportion of the catchment—although it is acknowledged widely that a forestry impact anywhere in a catchment has some potential to affect stream/river integrity. Some effects of logging are changed significantly by width of buffer vegetation strips (Davies and Nelson 1994), with difference in riffle-inhabiting macroinvertebrates between reaches 'upstream' and 'downstream' from logging in Tasmania, and abundance decreased at lower buffer widths. Significant declines in abundance of Ephemeroptera (Leptophlebiidae) and Plecoptera occurred at sites with buffers of about 30 m wide. Small buffers (those 10 m or less wide) did not give any significant protection to those streams, whilst buffers 30–100 m wide did so. The regulatory requirement (under Tasmania's Forest Practices Guide) for buffers to be at least 30 m wide was not universally evident, with some of the streams assessed by Davies and Nelson having far less buffer. Integrity of riparian buffers is also important, with avoidance of penetration during forestry reducing chances for pesticide drift and sediment runoff.

In a later account, Davies et al. (2016) noted that prescriptions to manage and mitigate such impacts are increasing in Tasmania. Compliance with regulations of site practices, such as coupe location and size, have increased and in turn increased the protection afforded to first order streams, as well as moves to reduce pollutant sediment inputs.

In sites adjacent to upstream CBS, only, trends amongst EPT were significant. The proportion of EPT families to all taxa (families plus species) declined as the proportion of land subjected to CBS rose above 40% (Fig. 6.5, Davies et al. 2016). Changes in the macroinvertebrate fauna across the gradient were attributed to increase in benthic fine sediments caused by upstream CBS activities. The '40% level' parallels closely an estimate for impacts of forest harvesting on New Zealand stream macroinvertebrates (Reid et al. 2010).

Forest harvesting near streams links with a range of disturbances to those streams, with consequent changes to macroinvertebrate (predominantly, benthic) communities and ecosystem functions. Deforestation involving riparian or adjacent vegetation can lead to large changes in water pH, dissolved oxygen levels and temperature regimes, and natural forest canopy is a major source of food material for aquatic

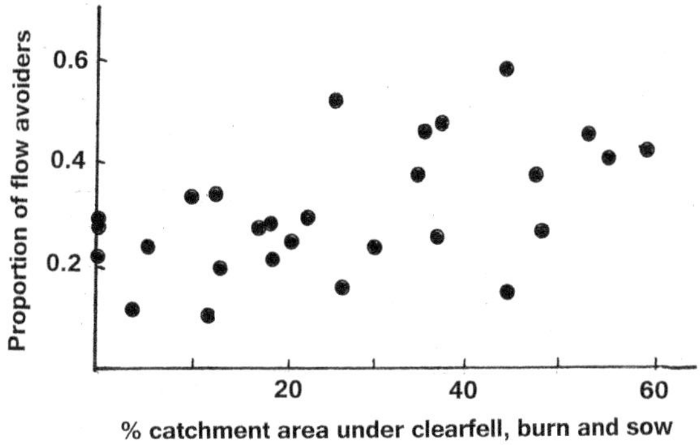


**Fig. 6.5** The relationship between the proportion of total macroinvertebrate abundance in the 'flow avoider' group of taxa and the proportion of catchment area under clearfell, burn and sow forestry practice (Davies et al. 2016)

insects, notably shredders (p. 166). Especially for clearcut harvesting, increased sedimentation from the exposed slopes is commonly associated with losses of sensitive insects (including most EPT taxa) in favour of more disturbance-tolerant taxa such as many Diptera (New Zealand: Reid et al. 2010). Intact riparian vegetation buffers can ameliorate these impacts, and alternative harvesting approaches such as 'progressive logging' of smaller coupes is likely to cause less intensive disturbances. Reid et al. also noted that communities in narrow streams, which are more dependent on riparian vegetation for shade and nutrient inputs, may derive greater benefits from buffering than communities in wider streams. Impacts on benthic invertebrates increased with the proportion of upstream catchment harvested, or of banks with riparian vegetation. In particular, Ephemeroptera densities declined and, as Reid et al.'s surveys extended over 17 years and across 15 streams, these findings come from one of the most extensive such surveys. In contrast to mayflies, densities of Diptera increased.

In Western Australia, a buffer zone of 100 m wide appeared effective in preventing stream disturbance from forest clearfelling (Growns and Davis 1991), and for clearfelled streams, impacts on macroinvertebrate communities persisted for up to eight years after logging disturbances. Impacts of commercial forestry operations in many parts of the world are associated with the condition of riparian buffers. In upland streams in Wales and Scotland, buffer strips of broadleaved trees or moorland/grassland vegetation had rather different effects on macroinvertebrates (Ormerod et al. 1993), but aspects of stream chemistry were perhaps more generally influential, as richness of all EPT groups declined considerably with increased acidity and aluminium concentrations. Management of conifer plantations near streams can strongly influence the macroinvertebrates through increased acidity, and maintaining a variety of streamside habitats may be important in countering such local effects (Britain: Weatherley et al. 1993).

Adult aquatic insects in the bankside vegetation of grazed or ungrazed reaches of chalk streams in England were strongly associated with species-rich herbaceous vegetation (Harrison and Harris 2002). Lower numbers on simple intensively grazed vegetation implied that this was far less suitable, or reduced insect survival. Early stages of some taxa may also be affected—many aquatic beetle larvae leave water to pupate in moist soils above the water line, and soils of grazed or trampled banks will often become waterlogged or, in dry weather, desiccated. Microclimates along more naturally vegetated banks are likely to provide more suitable conditions for taxa such as many elmid beetles to develop. Intensive cattle trampling of the banks should generally be avoided wherever possible.

The values of diverse waterside vegetation in agricultural areas extend to considerations of conservation biological control in nearby crops, as a major supplier of pollen/nectar and sources of food plants, prey or hosts that can sustain a range of terrestrial insect predators and parasitoids in proximity to crops into which they can move as pest attack proceeds. As well as direct removal of riparian vegetation, as above, other waterside disturbances such as cattle grazing (or, more precisely, overgrazing) have a range of impacts that are not always immediately evident (Fig. 6.6). Inputs of excrement can raise concentrations of inorganic nitrogen and phosphorus

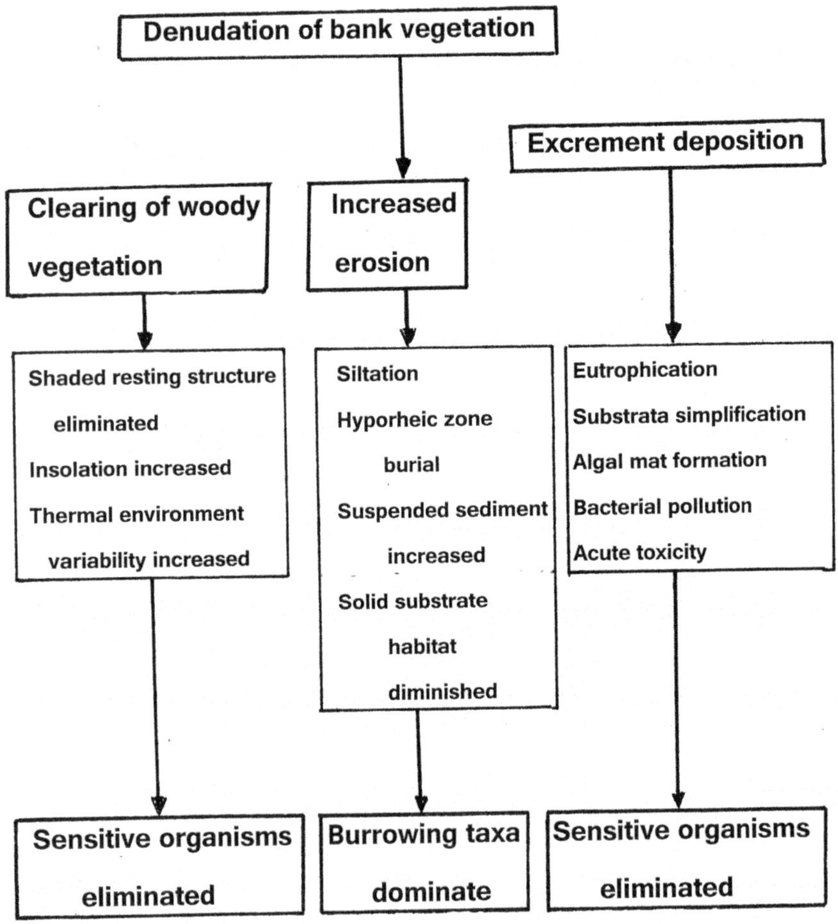


**Fig. 6.6** Some common effects of grazing by cattle on the aquatic environment (after Strand and Merritt 1999)

in the water, and in combination with the removal of palatable emergent vegetation this can lead to accumulations of filamentous green algae that replace the diverse array of photosynthetic plants present earlier. Although some insects can feed on the algae, many herbivorous species cannot do so, and decline (Strand and Merritt 1999). With excessive excrement reaching the water, ammonia concentrations may increase to levels that harm sensitive insects.

Because many riparian areas of major rivers are privately owned (as in the Murray-Darling Basin), it is often not feasible to exclude cattle grazing. However, improved riparian conditions can be pursued by measures such as lower stocking rates in upper reaches of catchments, resting waterside paddocks to enable recovery and reduce impacts to the water, and providing off-river watering points as alternatives to need for direct river access (Jansen and Robertson 2001). Periods of water scarcity exacerbate the widespread effects of stock (grazing, trampling, pollution) as stock progressively concentrate close to remaining water supply. Many different factors and roles have been nominated for riparian zones (Table 6.10). For each of these, particular components perform those functions, and visual indicators for their integrity can be designed to evaluate condition. These are the foundation of a rapid 'riparian condition index' that may have widespread applications in indicating where some focused restoration may be a conservation priority.

**Table 6.10** Functions of the riparian zone at different levels of organisation and the components contributing to those functions (from Jansen and Robertson 2001)

| Function | Components |
|---|---|
| *Physical* | |
| Reduction of bank erosion | Roots |
| Sediment trapping | Roots, aquatic coarse woody debris |
| Controlling stream microclimate/discharge/water temperatures | Riparian forest |
| Filtering of nutrients | Vegetation, soil, leaf litter |
| *Community* | |
| Provision of organic matter to aquatic food chains | Vegetation |
| Retention of plant propagules | Terrestrial coarse woody debris, leaf litter |
| Maintenance of plant diversity | Regeneration of dominant species, presence of important species, dominance of natives versus exotics |
| Provision of habitat for aquatic and terrestrial fauna | Aquatic and terrestrial coarse woody debris, leaf litter, standing dead trees, hollows, riparian forest, habitat complexity |
| *Landscape* | |
| Provision of biological connections in the landscape | Riparian forest (cover, width, connectness) |
| Provision of refuge in droughts | Riparian forest |

Condition of riparian vegetation can influence the input of terrestrial insects to streams and rivers and, as these may be significant contributions to water-based food webs and components of the diet of stream fishes, their decline with loss of riparian vegetation may have far-reaching effects. In a New Zealand river, the terrestrial invertebrates entering reaching water were measured by drift net catches. An upstream drift net was used to exclude drift into the test reach of 10 m, before entries were intercepted in a second drift net, and catches compared in reaches bordered by forest, tussock grassland grazing and more intensive pasture grass grazing (Edwards and Huryn 1996) to reflect different land use regimes. Of the three regimes compared, the pasture grassland supplied the lowest biomass of invertebrates. Because some drift-feeding fish select larger prey, the generally larger invertebrates from the tussock grass might comprise more suitable food than the smaller-bodied taxa from pasture grass—but interpretation is difficult, as is assessing any impacts on fish populations. In this study, the terrestrial invertebrate biomass entering streams differed little between forest and tussock grassland reaches, but both were considerably higher than the pasture grass.

### *6.8.1 Emergent Vegetation*

Emergent vegetation, especially when in dense monospecific stands, reduces open water surface area and might exclude insects such as water beetles, and modify their assemblages. Thus Cattails (*Typha angustifolia*, *T. latifolia*, Typhaceae) in Hungary were postulated to have effects of (1) shading the water and so reducing water temperature and (2) weakening the polarised light signals from the water surface and reducing colonisation by polarotactic insects (p. 138) (Molnar et al. 2011). Comparisons of beetles in intact plots and those in which *Typha* was mown to water level showed increased abundance in the latter, although species richness in the treatments was similar. Both the above predictions were satisfied, and mowing was considered a useful method in conservation management. However, mowing might also have other influences on water beetles: Molnar et al. (2011) listed (1) increased detrital food resources from remaining *Typha* after mowing and raking, with this perhaps insignificant; (2) higher light intensity affecting algal production and so oxygen content and productivity, again considered a minor influence; (3) possible increase in other plant species to replace the cattails, with submerged vegetation possibly beneficial to macroinvertebrates; and (4) effects on food resources, such as increased density of Chironomidae benefiting predaceous species.

In seasonally flooded marshes in California, amounts of emergent Salt grass (*Distichlis spicata*, Poaceae) affected macroinvertebrate colonisation (de Szalay and Resh 2000). Culicidae, brineflies (Ephydridae) and hoverflies (Syrphidae) were all positively correlated with the amount of plant cover whilst, in contrast, Chironomidae, hydrophilid beetles and water boatmen (Corixidae) showed negative correlations. Species assemblages also differed seasonally because different taxa colonised wetlands at different times of the year. Management of the structure of emergent vegetation can influence invertebrate communities by changes to colonisation and subsequent oviposition.

## 6.9 Alien Species

Accidental or deliberate introductions of non-native species to Australian waters, ranging from casual discards (such as of surplus aquarium fish or ornamental weeds) to planned releases are widespread, often undocumented and difficult to detect until the organisms have become well-established, equally difficult to eradicate, and accepted widely as threats to the receiving environments. Many introduced species can become naturalised and become highly invasive, especially in lowland streams from where upstream spread is not uncommon. In addition, terrestrial taxa can pose further pressures on water bodies.

Whilst alien species in aquatic environments and their neighbourhoods continue to exert great changes to their receiving environments, including the insects present, very few of those aliens are themselves insects. Strayer (2010) commented that insects 'are almost unrepresented in lists of alien species' whilst, in contrast, alien crustaceans, molluscs, plants and vertebrates (notably fish) are all strongly represented. Many waterfowl and aquatic mammals have established populations far beyond their native ranges, but their impacts on native aquatic insects are rarely considered.

The paucity of alien aquatic insects was treated as a 'paradox' by Fenoglio et al. (2016), who advanced several reasons for this trend, especially in relation to the greater representation of some other invertebrate groups. The various factors discussed included: (1) very limited practical interests in deliberately moving aquatic insects for human food supply or (other than a very few semi-aquatic beetles and moths) as biological control agents—or, indeed, any commercial use; (2) true obligate or specific associations between aquatic insects and host plants are rare and, again in contrast to terrestrial taxa, limitations of host plants are unlikely to play a significant role in movements of dependent insects; (3) in general, aquatic insects lack many of the attributes that would adapt them to passive dispersal (a major exception is some mosquitos, which have indeed been spread widely by factors such as desiccation-resistant eggs and rapid development in very small amounts of water); (4) almost complete absence of traits such as parthenogenesis (some Chironomidae are exceptions) and haplodiploidy, that can favour colonisation by some terrestrial insects; (5) complications from needing both aquatic and terrestrial environments for their different life stages, a marked increase in complexity from taxa needing only one of these contrasting regimes; and (6) complications from many species needing the highly heterogeneous environments of running waters, perhaps inhibiting their ability to spread. Fenoglio et al. also pointed out that very few aquatic insects have any clear potential to exploit human transportation systems, so that a complex of different factors may contribute to their low representation among invasive aquatic organisms. Nevertheless, many aquatic insects share some of the traits that mark successfully invading species—but these general attributes differ markedly in detail among large taxonomic groups. Despite the rather low general incidence of alien freshwater insects in relation to other taxa, they could reasonably be expected to participate in the environmental impacts on native biodiversity and the fate of specialised endemic taxa. Impacts, summarised by Strayer (2010), are predicated on the richness and variety of alien species and that some gain enormous abundance through competitive advantage and displace native taxa. Strayer's comment that 'at least hundreds of species have been moved outside their native ranges' so that 'many bodies of fresh water now contain dozens of alien species' is a realistic summary of the global scenario. Many impacts have been recorded as changes in abundance, distribution and assemblage composition of native species, but far wider background is available for terrestrial taxa and environments (New 2016).

As a relatively unusual example, the Australian dragonfly *Hemicordulia australiae* (Corduliidae) is a strongly dispersing species that colonised New Zealand from about 1930. It apparently partially displaced the native *Procordulia grayi* (Armstrong 1978) and has now become the most abundant anisopteran in the North Island (Rowe 1987). Isolation of many oceanic islands and the atypical nature of many of their ecosystems can lead to evolution of highly endemic fauna of aquatic insects and also thwart immigrations of alien aquatic insects simply by the distances involved and restricted receptor habitats available and that adults of many species in these groups are naturally rather weak fliers and unlikely to reach such areas unaided. The Hawaiian archipelago is some 4000 km from the nearest continent, and has no native species of Ephemeroptera, Plecoptera or Trichoptera (Howarth and Polhemus 1991), whilst endemic Odonata (*Megalagrion* spp., p. 117) have diversified considerably.

In an unusual case, several alien mayfly species were introduced deliberately to Hawaii to supplement the restricted benthic invertebrate prey naturally available for introduced trout in the Kokee State Park. Both *Nixe rosea* and *Epeorus lagunitas* were introduced from California in 1961, collectively by more than 200,000 eggs, but both the introductions failed completely (Englund and Polhemus 2001).

However, an introduced caddisfly is among the dominant aquatic insects in forested waterbodies on Oahu (Brasher 2003), outnumbered only by Diptera, notably the alien chironomid *Cricotopus bicinctus*. The North American *Cheumatopsyche pettiti* (Hydropsychidae) was introduced unintentionally to Hawaii in the 1960s, collected there first in 1965 on Oahu. It is widely distributed in North America and tolerates a wide range of water temperatures and stream types. It has become frequent in most perennial streams of all major islands of Hawaii (Kondriateff et al. 1997) and is a significant component of the diet of some native fish as one of the most abundant benthic insects in some systems. Two smaller alien caddisflies (*Oxyethira maya*, *Hydroptila arctica*, Hydroptilidae) also occur in Hawaiian streams. These several alien caddisflies were all apparently unplanned introductions and may have occurred through transfer of eggs on imported aquarium plants, but all had become abundant by the early 1970s.

Accidental introductions of Odonata to Hawaii have led to concerns for native *Megalagrion* spp. but despite distributions of native and alien species overlapping on several islands (Oahu, Lanai, Hawaii, Molokai), Englund (1999) found little or no evidence of any adverse impacts. However, in contrast, poeciliid fish (such as

Green swordtail [*Xiphosurus helleri*] and Shortfin molly [*Poecilia mexicana*]) had severe impacts on damselflies, with losses especially high in lowland and coastal areas (p. 117).

Despite globally wide trends for increasing biotic homogenisation in freshwater faunas as alien species are spread widely, two factors have retarded this somewhat in Australia—although the other two key drivers of this process noted by Rahel (2002), namely (1) habitat alterations that facilitate losses of native insect species and (2) the actual losses of those species, can only tentatively compensate for the relative lack of introductions of non-native insect species. Rahel noted that constructions of dams and reservoirs can contribute to homogenisaton because local riverine species may be replaced by widespread lentic species. In functional terms, the geographical isolation of Australia is likely to have thwarted 'natural' arrivals of alien aquatic insects, and increasingly sophisticated biosecurity measures also guard against many imported organisms. Those trends do not minimise the importance and ecological impacts of the many alien animal species already present, amongst which fish have received the greatest attention (p. 117).

### 6.9.1 Plants

Deliberate introductions of alien aquatic plants as livestock forage ('ponded pasture') are regarded as a threat to many freshwater ecosystems. Some species have potential to colonise water >1 m deep, and some have also proved to be more significant as invasive weeds than as useful pasture species.

Para grass (*Urochloa mutica*) had little impact on macroinvertebrates in northern Australia (Douglas and O'Connor 2003), whilst Olive hymenachne (*Hymenachne amplexicaulis*) affected many components of benthos in central Queensland (Houston and Duivenvoorden 2002). However, benthic invertebrates in hymenachne differed somewhat from those in Para grass and rice, in part related to the deeper water of sites in which *Hymenachne* occurred. *H. amplexicaulis* has been listed as one of worst environmental weeds in Australia and produces dense plant beds that reduce native plant richness in rivers. Comparisons of invaded sites with nearby stands of native vegetation in the Fitzroy River, Queensland, showed significantly lower abundance of Ephemeroptera, Hemiptera and Odonata, whilst Coleoptera were more abundant. Changes in fish assemblages were also attributed to changed plant community structure.

Red sweet-grass (*Calyceria maxima*) impacts in upland streams in Gippsland, Victoria, were surveyed by comparative sampling of invertebrates across paired (invaded and non-invaded) sites, demonstrating that invasion by *C. maxima* was associated with changes from more diverse to a more uniform fauna, through compositional changes and changing balance of functional feeding groups (Clarke et al. 2004), although no significant changes in lotic invertebrate abundance occurred.

Amongst the more uniform aquatic macrophytes cultivated for human use, rice fields often support considerable aquatic biodiversity that may be influenced strongly by the crop management regimes imposed—including measures to suppress damaging pests. Comparisons of aquatic macroinvertebrates of rice fields in Yanco, New South Wales, yielded 90 recognised morphospecies, 66 of them insects, across three commercial management regimes. These regimes were (1) conventional-aerial (agrochemicals applied, aerially sown with pregerminated seed); (2) conventional-drill (agrochemicals applied, directly drill-sown); and (3) organic-drill (agrochemical free; directly drill-sown). In general, macroinvertebrate communities were more diverse and with higher morphospecies richness in the last regime. Over the rice-growing season, the communities of the three regimes became more similar, with their initial differences in part taken to reflect the imposition of agrochemicals to protect the early seedlings.

Weedy alien aquatic macrophytes may, however, sometimes be a valuable resource for Odonata, as demonstrated for Water lettuce (*Pistia stratiotes*, Aracaceae) and Water hyacinth (*Eichhornia crassipes*, Pontoderiaceae) in the Kruger National Park, South Africa (Stewart and Samways 1998). They provided good perching and oviposition sites, and shade and protection from predators—so, acting as 'microrefuges' for some small-bodied species. Higher dragonfly richness was found at infested sites but all species benefitting were widespread opportunistic species and none was among those of high conservation importance.

### 6.9.2 Fish

Harm to specialised native aquatic insects in Australia and New Zealand from predation or habitat disturbance by introduced fish is invoked frequently as a threat. Tillyard (1926), for example, commented that releases of trout into New Zealand streams in the early twentieth century had caused severe losses of native mayflies to the point that some species 'are now extinct or nearly so'. He supposed that Oniscigastridae and other mayflies in New Zealand or Australia were being rapidly affected by introduced Brown and Rainbow trout. Historical background to fish introductions to Australia recognises arrivals from aquarium and aquaculture interests leading to unplanned escapes and casual releases to deliberate introductions for sport, bait, commercial or pest control purposes (Fletcher 1986) and, despite increased awareness of undesirable impacts, such releases continue. Three major phases of fish introductions to Australia can be recognised (Arthington and McKenzie 1997), commencing with activities of 'acclimatisation societies' in the second half of the nineteenth century, and whose legacy includes ten major established species (three trout, European perch, goldfish, carp, tench, roach, mostly recreational angling targets). They were followed by introductions of mosquito fish (*Gambusia* spp.) from the 1920s and also in the 1940s (below) for control of mosquito populations. Most recently, from the later twentieth century proliferation of aquaculture and the aquarium/ornamental species trade led to considerable potential for escapes and unauthorised releases of many species, some of which have 'enormous potential for range expansion' (Arthington and McKenzie 1997). International trade in aquarium fish, augmented by local breeding of many species, has led to more efficient aquaculture operations for some. Release concerns span several poeciliids, cichlids and the Oriental weatherloach (*Misgurnius anguillicaudatus*).

Concerns for native fish arise from chances of hybridisation, or from them being prey or inferior competitors of alien species and by far outweigh any 'official' concerns over parallel impacts on invertebrates, despite widespread suggestions of impacts from (in particular) predatory trout and mosquito fish, with several species of native Galaxiidae implicated as having declined or being locally lost. As elsewhere, predation on aquatic invertebrates by introduced fish is commonly inferred but rarely proven as a significant threat. However, macroinvertebrates and other faunal assemblages had for long been known to have only low resistance to introduced fish (whether of native or alien species, with the former a frequent occurrence in Australia, p. 264), especially when living in naturally fishless waters. Classic examples from elsewhere include studies of the impacts of introduced trout in more than 500 lakes in California (Knapp et al. 2001) and of native fish in lakes in Scandinavia (Nilsson 1972)—and Burrows (2004) commented that deleterious effects from similar exercises had been documented frequently. Direct predation is the most obvious and most frequently cited possible threat to native insects from fish and can also apply to contexts in which either native or alien fish may hamper insect colonisation of pools. Possible more subtle interactions, whereby some predatory insects may use chemical cues to assess habitat quality in relation to risks of predation by fish, have been suggested (Abjornsson et al. 2002), but are not yet a focus in conservation.

The North American Nevares Spring naucorid bug, *Ambrysus funebris*, is a flightless heteropteran restricted to a single spring system in Death Valley, California. It coexists with several other endemic aquatic invertebrates, endorsing the prevalence of spring-endemic taxa amongst those aquatic insects that have become cases for listing under the United States Endangered Species Act. It may be the most critically threatened of the three endemic naucorids in the Amargosa River system, and was suggested as a potential 'umbrella species' for protection of those desert springs (Whiteman and Sites 2008). In contrast to the Ash Meadows naucorid, *Ambrysus amargosus* (p. 265), *A. funebris* does not coexist with any vertebrate of equivalent or dominating conservation interest, and focus on its protection could help to conserve its parent community. Suitable environmental conditions for the bug, noted in its original description, were that it was distributed 'along a 100-yard reach of Cow Creek', especially where spring outflow removed silt from the substrate but flow was not strong enough to remove coarse gravel. It is the only naucorid in this spring system and occupies water with the relatively constant temperature range of 35–40 °C. That system is susceptible to introduced fish. Conservation measures (Whiteman and Sites 2008) could include removal of non-native fish, and surveillance for crayfish and other alien species, together with removal of non-native vegetation, augmentation of native vegetation and institution or maintenance of suitable flow regimes. Because the region's spring systems each harbour unique invertebrate assemblages, each may merit conservation, and comparative assessment is needed on a case-by-case basis, including incidence of non-native fish predators.

A variety of predatory fish have been introduced into South African rivers, to lower reaches and also, with the aid of helicopters, into headwater streams (de Moor 1992) where—despite then lack of objective information—severe impacts on native invertebrates have been implied. De Moor noted that some complexes of endemic insects in largely fish-free rivers in the Cape region had evolved behaviours that rendered them highly visible—such as walking exposed and displaying openly on open substrates—that render taxa such as larvae of *Chlorolestes* spp. (Odonata, Synlestidae) and *Athripsodes* spp. (Trichoptera, Leptoceridae) clearly visible to predators. Introductions of fish in this region have 'undoubtedly had a major impact on those conspicuously visible insects'. Circumstantial evidence for that claim arose from comparisons of earlier collections from a pool in the southern Cape without introduced fish (yielding 12 species of *Athripsodes*) and a pool in the western Cape with various alien fish that yielded only five species of these caddisflies. De Moor noted the richness (27 regional species) and ecological variety of this genus and suggested that the diversity rendered them vulnerable to fish predation, through both specialisation and conspicuousness.

Numerous attempts have been made to use biological control by fish predators to suppress mosquito larvae, but the great variety of experimental investigations and field trials renders strict comparisons of the impacts of different fish species, and those of the same species in different environments, difficult. Water temperature, for example, may be an important influence on predator impacts, and is often overlooked (Lawrence et al. 2016). The introduction of Mosquito fish, *Gambusia holbrooki*, into Australia in 1925 for mosquito control has led to it becoming a widespread predator, implicated directly in the declines of many small native fish, together with trout which are reared commercially for releases for recreational angling, and their impacts on macroinvertebrates are believed substantial—but are largely inferred through correlations of fish presence and diminished invertebrate richness or abundance, and the presence of some insects only in areas that remain free from introduced fish. *G. holbrooki* is a subtropical fish, originating from Georgia (United States).

However, and despite its reputation for controlling mosquitos, laboratory trials have implied that some native fish might 'outperform' *Gambusia* in this role (Lawrence et al. 2016). In warm waters, the Crimson-spotted rainbowfish (*Melanotaenia duboulayi*) has been shown to eat more mosquito larvae (Hurst et al. 2004), and other native species had similar predation rates. Declines of native fish in south-western Australia may contribute to increased abundance and spread of mosquitos, and those losses (exemplified by declines of Western minnow [*Galaxias occidentalis*], Western pygmy perch [*Nannoperca vittata*] and Nightfish [*Bostokia porosa*]) in the Swan Coastal Plain also reflect seasonal drying of formerly permanent water bodies from decreased rainfall, aided by features such as deforestation, increased groundwater extraction, and climate change. Combined with increased establishment of artificial ponds and lakes, mosquito populations can proliferate, with likely increased disease incidence and spread.

The great variety of studies on impacts of introduced *Gambusia* (Pyke 2008) focuses largely on impacts on mosquitos, with many surveys failing to reveal any clear attributable effect. However, Pyke also cited cases of backswimmers, water beetles and dragonfly larvae being reduced in *Gambusia*-infested waters compared with control sites. Consequently, and together with evidence that some native fish are effective predators of mosquito larvae, recommendations have increased that such water bodies should be stocked with indigenous fish rather than with *Gambusia* for mosquito control. Lawrence et al. (2016) noted the likely values of those sites as 'conservation repositories' for those fish species known to be increasingly vulnerable in the wider landscape, citing as an example the critically endangered Spotted minnow (*Galaxias trutescens*) which has a very narrow distribution in southern Western Australia. Hatchery production of several such fish species is now possible, and prospects for their use for mosquito control are perhaps increasingly likely.

The coastal mangrove saltmarshes in eastern Australia have also been a focus for mosquito control, in which the potential of native fish predators rather than *G. holbrooki* has been advanced. The Pacific blue-eye (*Pseudomugil signifer*) had similar consumption rates of mosquito larvae, and the additional advantage of longer daily activity periods. Griffin (2014) recommended that *Pseudomugil signifer* could be included in an integrated mosquito control programme that incorporates native mangrove-associated fish.

In a rather different context, discussed by Polhemus (1993), Hawaiian endemic *Megalagrion* damselflies, especially those in lowland streams, declined appreciably after introduction of mosquito fish. Elaborated further by Polhemus and Asquith (1996), *Megalagrion* is a notable flagship taxon in Hawaii, and many of the 23 taxa are very restricted in distribution. Many are montane, but others are more typically lowland species, in waters more susceptible to disturbance. Substantial declines of some species (such as *M. xanthomelas*, formerly abundant and widespread, but now occurring only in water bodies without introduced fish) in lowlands are clearly attributable to those fish. Fish introduced for mosquito control are deemed the cause of extinctions of *M. pacificum* on most major islands of the archipelago, as well as of *M. xanthomelas*. Those relatively early introductions, in the first half of the twentieth century, are reasonably well understood, but many more recent introductions—some clandestine and most from the aquarium trade—have invaded stream systems, leading Polhemus and Asquith to comment that some damselflies now occur only in drainages on the higher parts of stream systems, where alien fish have not yet become established. In essence 'alien fish now determine the distribution of native aquatic damselflies in Hawaii'. Their future also reflects the extensive development of resorts, urban and wider agricultural and residential development in the lowland coastal plains, so that many lentic habitats have been lost or changed severely—to the extent that losses were estimated by Polhemus and Asquith as 'probably approaching 80–90%' of freshwater habitats.

In contrast to these impacts from poeciliid fish, the commonly presumed high impacts of Rainbow trout (*Oncorhynchus mykiss*) on *Megalagrion* and other Hawaiian aquatic biota were far harder to evaluate. Examination of trout stomach contents revealed only 7.5% of prey numbers to be native insects (Englund and Polhemus 2001), amongst a very wide range of organisms consumed. Numerically, the most important dietary items were introduced Trichoptera (p. 198).

Harmful direct impacts on native insects from alien predatory fish are strongly inferred for Australia. The five lakes of the Snowy Mountains area collectively harbour a unique invertebrate assemblage (p. 219), amongst which some changes over time have been documented. Two of the lakes (Cootapatamba and Albina) did not contain fish, and their assemblages differed from those in lakes with fish. Thus, a chironomid (*Chironomus ?oppositus*) was common only in two lakes with fish, but some conspicuous zooplankton (*Daphnia nivalis*, *Boeckella montana*) were abundant in fishless lakes, but absent from Blue Lake. Timms et al. (2015) recorded changes in Lakes Cootapatamba and Albina following the advent of fish (the native Mountain galaxias, *Galaxias olidus*) to the former following fire in 2003, by comparisons of the two lakes in 2012. Changes in isopods and amphipods, both documented foods of the galaxiid, suggested impacts on these groups, and abundance of *C. ?oppositus* had 'definitely increased' in Lake Cootapatamba since fish arrived.

Removal of alien predatory fish from waterways is very difficult, especially when also attempting to avoid further threats to native species. Attempts to eradicate them from farm ponds in Japan by draining the ponds (Maezono and Miyashita 2004) led to losses of macrophytes and changes to invertebrate assemblages—in their example of dragonfly richness reflecting the decreased use of emergent vegetation. However, the dragonfly *Pseudothemis zonata* increased, but this species does not use macrophytes for oviposition. That fish and aquatic plants influence invertebrate assemblages is well-known (Hanson et al. 2015), but the precise mechanism(s) involved are often difficult to clarify because many influences act simultaneously on the environment and interactions. In shallow lakes in Minnesota, association between fish and macrophytes was much stronger than between either fish and macroinvertebrates or fish and zooplankton. However, increasing the variety and/or abundance of benthic-feeding fish has some potential to limit benthic invertebrate assemblages.

### 6.9.3 Mammals

Grazing cattle and other farm stock can create substantial disturbance to water bodies through trampling, fouling and erosion, and pressures of native mammals, notably larger macropods, can increase substantially during dryer periods. Impacts of feral mammals are less frequently appraised.

Large feral herbivores are a major threat to many remote water bodies in central Australia, and protection from such intruders is a key need in their protection. The region of the Northern Territory studied by Brim-Box et al. (2016) is highly significant to local Aboriginal culture, and water bodies there have ceremonial, social and economic importance—with their integrity assured through long-term traditional management. However, the region now harbours some of the highest densities of feral camels (*Camelus dromedarius*) in the country and is a focus for camel management. Camels have severe impacts on water bodies: Brim-Box et al. listed 'trampling, fouling, muddying, destabilising, drinking, grazing and browsing', and can also reach the more remote water bodies not previously accessible to domestic stock or feral horses. 'Drinking' reflects the large water intake of camels compared with native fauna: McBurnie et al. (2015) cited a camel drinking about 200 L/day, in comparison with a red kangaroo needing only about 100 ml/day. Comparisons of aquatic macroinvertebrates in water bodies (rock holes, springs) accessible or inaccessible to camels yielded a total of 83 taxa, 62 of them insects. Biodiversity was significantly less at the accessible sites, in part linked with lower water quality and thus fewer taxa that were sensitive to habitat degradation. A mayfly (*Cloeon* sp.) was an indicator for sites without camel impacts, as part of a wider trend for taxa with gills to be better represented under those conditions. Both tolerant and non-tolerant species declined in springs where camels were present, implying direct extirpation from their impacts.

Impacts of camel dung were investigated experimentally using artificial 'pools' (100 L plastic containers, 1 × 0.5 m) of size resembling that of many small natural arid zone pools. Nine of the 18 pools were untreated, whilst the others each had 2 kg of camel dung added; macroinvertebrates were sampled weekly for eight weeks (McBurnie et al. 2015). Although pollution-tolerant taxa such as some mosquito larvae occurred in the dung-treated pools, more sensitive taxa (larval Ephemeroptera and Odonata) were more common in untreated controls. The taxa were typical of those occurring in temporary pools in the region, but this study indicated that camel dung indeed had adverse effects and that preventative measures may be needed to conserve their invertebrate faunas.

## 6.10 River Regulation

Deliberate manipulations of water resources for industrial, agricultural and domestic demands for water often involve direct alterations of rivers, such as by constructing reservoirs and dams—broadly 'regulation' of flow. Macroinvertebrates have proved to be useful tools in assessing wider effects of dams (Wang et al. 2019), through studies in many parts of the world, but with needs for studies to broaden from species-levels to more embracing community perspectives in order to provide a more holistic scenario. River regulation encompasses several distinct measures that affect riparian and aquatic habitats, as (1) removal from reaches; (2) diversion of water flow into reaches; (3) impoundment; and (4) upstream regulation (Bates et al. 2009). Reservoirs store water during high flow periods, for release when needed for irrigation or other purpose. Effects on downstream water flow and conditions can be substantial and influential, with many riverine biota long-adapted to natural flow regimes becoming more vulnerable as their environments change. Whether high or low flows, those changes can be dramatic and, whilst most emphasis on their impacts in Australia has focused on fish, recognition of 'Alteration to the natural flow regimes of rivers and streams' and 'Prevention of passage of aquatic biota by instream structures' as formally designated Potentially Threatening Processes in Victoria, with Action Statements prepared for their amelioration (Doeg and Heron 2003a, b, O'Brien and Blackburn 2003) attests to their wider conservation relevance. The two themes are linked, because dams and weirs are among the structures that can impede passage by fish and other animals; they can be functional barriers to migration and affect local community structure and dynamics. O'Brien and Blackburn (2003) noted that an inventory of potential barriers to fish movement had identified 2438 such structures in Victoria alone, with many culverts and road crossings not included in that total. In New South Wales, more than 5000 weirs impound water and regulate water levels on rivers (Thoms and Sheldon 2000), with 'weir pool' environments encouraging some forms of algal/bacterial blooms. More widely, Arthington et al. (2010) cited a global estimate of about 45,000 dams more than 15 m high, that collectively hold back >6500 $Km^3$ of water—about 15% of total global annual river runoff. River regulation may invoke any of three very general responses—species may show no perceivable impact, or their abundance may increase or decrease (Growns and Growns 2001). Further, the changed novel conditions may enable other species, not previously present, to establish.

Impacts of dams and weirs are claimed widely to disrupt continuity along rivers, fragment previously more continuous ecosystems, and isolate insect populations by disrupting key dispersal processes such as drift. Weirs can create wide and deep pools with slow-moving water that may differ from conditions in physically similar more natural pools. Studies of drift rates in a tributary of the Snowy River (Mowamba River), New South Wales, compared incidence of several key insect species in riffle and weir pool regions (Brooks et al. 2018). Those taxa—*Offadens* spp. (Ephemeroptera, Baetidae), *Simsonia* spp. (Coleoptera, Elmidae), *Cheumatopsyche* spp. (Trichoptera, Hydropsychidae) and *Austrosimulium* spp. (Diptera, Simuliidae)—are commonly found in drift and are more prevalent in flowing riffle lengths than in pools. The weir pool reduced numbers of *Offadens*, *Simsonia* and *Austrosimulium* in drift, despite their differing biology and body sizes, suggesting that the weir might be a significant barrier to drift, and disrupt downstream insect movement during moderate flow conditions. Such factors might impede recovery of such insects in rivers that are essentially fragmented by dams and weirs. Macroinvertebrate samples along the impounded Cotter River (Australian Capital Territory, and with three dams) showed relatively more chironomid larvae, but fewer EPT and Coleoptera at sites immediately below the dams (Nichols et al. 2006), but recovery of the assemblages took place within 4 km downstream of one dam.

Most attention to faunal changes has been to those downstream from dams, or to the effects of the dams themselves in interrupting connectivity through thwarting insect dispersal. Upstream impacts, by contrast, have been studied relatively little (Jonsson et al. 2013), but the reservoirs formed above dams essentially transform rivers or streams into lake-like waterbodies that permanently submerge previously terrestrial or intermittently flooded areas. The most obvious changes include reduced water velocity and increased sedimentation, both likely to lead to lower numbers of aquatic invertebrates and generally lower production of benthic taxa—essentially moving from lotic to lentic ecosystems. Some changes may be reasonably predictable—loss of Simuliidae that depend on flowing water, for example—but others are far less so (Jonsson et al. 2013). However, comparisons of the aquatic insects emerging into adjacent forests from regulated and free-flowing rivers in Sweden and Finland implied lower numbers from regulated rivers, so possibly lessening food supply for terrestrial forest taxa. Those changes were implied to persist for several decades after the initial regulation. In order to assess the full impoundment impacts on aquatic insects, changes in composition, distribution and abundance above, at, and below an impoundment need investigation over an extended period.

River restoration aims, in part, to reinstate those natural flow regimes to which local biota have adapted, acknowledging that those regimes are a major driver of river ecosystem functions and integrity. Flow regimes are indeed one of the most universally influential processes in rivers—Walker et al. (1995) referred to flow regimes as the 'maestro' that orchestrates river pattern and process. River regulation brings to the fore a modification to the River Continuum Concept (p. 15), as the 'Serial Discontinuity Concept' (SDC: Ward and Stanford 1983, 1995). This idea proposes that the river continuum can be re-set by discontinuities such as impoundments that create a departure from the reference condition, with a longitudinal (usually downstream) change associated with a perhaps considerable 'recovery distance' to regain this state. Changes in many variables can occur.

Benthic macroinvertebrates, in particular, are influenced strongly by flow regulation in rivers (Ellis and Jones 2013), and changes in abundance, richness and balance of functional feeding groups can extend for many Km below a dam or other flow-changing structure. However, unregulated tributaries that enter regulated rivers downstream may have strong ameliorating effects (Milner et al. 2019). They can be sources of colonists for the main flow channel and contribute to transitional habitats at the junction. The additional habitat variety they provide can be associated with community composition downstream from the junction, but also provide areas for larval upstream movements from the river and adult flight dispersal. They may provide a refuge from high flow influences and in some cases perhaps favourable breeding sites replacing those lost from flow changes. Despite need for fuller evaluation, Milner et al. considered that unregulated tributaries should be 'safeguarded through conservation management and be promoted as valuable links in the landscape for enhancing biodiversity conservation'.

Development of the 'Lotic-invertebrate Index for Flow Evaluation' (LIFE) from surveys of benthic invertebrates in rivers in England (Extence et al. 1999) was a forerunner for similar applications elsewhere. It is based on the occurrence of definable associations of different macroinvertebrate species and families at different flow regimes, of which six such groupings were defined (Table 6.11). Each species was allocated to a single flow group, with allocations sometimes reflecting current velocity more than habitat type. For Group VI, velocity is clearly zero. Family level data (although with consequent lack of precision, p. xx) can be used if species information is not available. Some very widespread taxa, such as Chironomidae, are excluded because they provide only minimal local information.

The index is calculated as 'LIFE = $\sum$ fs/n', where '$\sum$fs' is the sum of individual flow scores for all the taxa in the sample, and 'n' is the number of those taxa. Performance of the index was tested on lowland streams in England and Denmark, for both of which habitat features were easily assessed through comprehensive local surveys (Dunbar et al. 2010). In Denmark, the LIFE score responded additively to three main predictors: high flow in the previous four months (positive), substrate composition, and whether the stream channel meandered or was straight. The England fauna also

**Table 6.11** The freshwater benthic macroinvertebrate flow groups, ecological associations and defined current velocities proposed by Extence et al. (1999)

| Group | Ecological flow association | Mean current velocity |
|---|---|---|
| I | Taxa primarily associated with rapid flows | Typically > 100 $cm/s^{-1}$ |
| II | Taxa primarily associated with moderate to fast flows | Typically 20–100 $cm/s^{-1}$ |
| III | Taxa primarily associated with slow or sluggish flows | Typically < 20 $cm/s^{-1}$ |
| IV | Taxa primarily associated with flowing (usually slow) and standing waters | |
| V | Taxa primarily associated with standing waters | |
| VI | Taxa frequently associated with drying or drought impacted sites | |

**Table 6.12** Variable habitat components used in scoring for surveys of river habitat modification in Denmark and Britain (Dunbar et al. 2010)

| Habitat modification | Attribute scored |
|---|---|
| Culverts | Number and extent |
| Bank and bed reinforcement | Extent |
| Bank and bed resectioning | Extent |
| Berms and embankments | Extent |
| Weirs, dams and sluices | Number and extent |
| Bridges | Number and extent |
| Fords | Number and extent |
| Outfalls and deflectors | Number and extent |

responded to three predictors: high and low flows in the preceding six months (positive), and extent of the re-setting of the channel (negative). The two sets of habitat features differed somewhat, but the variety summarised in Table 6.12 illustrates the range of variables that may influence the invertebrates present, in conjunction with flow. Major inferences of the data were (1) taxa in the lower LIFE flow groups (I, II) become more predominant in less modified habitats, and (2) those in higher LIFE flow groups (III—VI) are favoured by more modified reaches. The more specialised taxa thereby become more abundant in the more natural channels, with both habitat and flow modifications influential.

Especially for 'bottom-release reservoirs' water temperatures downstream can be reduced considerably by those releases. Decreases of around 10 °C are not uncommon for south-east Australia in summer (Pardo et al. 1998) and may affect developmental rates and abundance of insects. Larval abundance of several mayflies (*Coloburiscoides* spp., *Baetis* spp.) declined during summer water releases, and recruitment of the former genus was delayed. However, the mayfly fauna of a regulated river in Victoria (Mitta Mitta, with 11 species) differed somewhat from that of its unregulated tributary (Snowy Creek, with 17 species). Differences largely reflected absences of several of the rarer Snowy Creek species from the Mitta Mitta River, and relative abundance of the predominant species also differed. Effects of river regulation on particular taxa may vary greatly, with studies reporting either increase or decrease close to dams, and gradients of downstream recovery to 'normal' conditions of abundance and biomass also very variable. Cases noted by Ellis and Jones (2013) spanned those responses. Thus, Plecoptera are especially susceptible to some regulatory measures that affect temperature, substrate character and food resources, and can be largely absent from rivers below impoundments. In contrast, some filter-feeding taxa of Diptera and Trichoptera are enhanced below dams, and increased downstream, and mayflies have very varied responses. Enhanced growth of algae and mosses can increase abundance of some taxa, but also exclude those other species that need clean surfaces for attachment. Many mayflies also have very restricted thermal requirements. As emphasised later (Ellis and Jones 2016), more details on recovery gradients below dams are needed in seeking general understanding. However, in the Canadian rivers surveyed the richness of benthic invertebrates increased with distance downstream, with clear changes within the first 5 km and strong recovery around 10 km from the dam. That study also confirmed variations between taxa.

Allied to considering effects of dams and other regulating changes, their removal as a component of restoration has also prompted interest in management. Thus, in Wisconsin, United States, re-evaluations of many dams exposed factors such as lessened economic returns, structural weaknesses and environmental effects leading to preference for their removal over continuing repair and re-licensing (Stanley et al. 2002), in trends that have become more widespread for smaller dams. Accompanying one such project (involving removal of three dams over a seven Km reach on the Baraboo River, the dams differing from 2.5 to 5 m high), surveys of macroinvertebrates showed considerable differences in impounded lentic and free lotic river sections, implying that even such relatively small dams can induce changes in upstream benthic communities. After dam removal, rapid changes occurred in channel structure (above dam sites) and substrate sediment (below dam sites), and lentic invertebrate assemblages at two upstream sites being displaced by more typical lotic assemblages only a year after removal. The previously imposed changes from sedimentation were thus considered only moderate, and short-lived (Stanley et al. 2002).

Outcomes of impounding lakes were illustrated by the highly controversial flooding of Lake Pedder, Tasmania from late 1972 for hydroelectric power generation, an action that was opposed strenuously by many environmentalists. During a prolonged campaign attempting to prevent inundation, biologists urged consideration for the lake's biota (Bayley et al. 1972) within the wider case for conservation (Johnson 1972). In turn, these claims for harmful impacts of flooding on lake fauna were opposed strongly as 'exaggeration'. Lake (2001) subsequently monitored the benthos of the flooded lake over 12 sites annually from 1975 to 1989, finding 73 species—but also that four lake-endemic taxa had disappeared together with five locally endemic taxa. Insects (60 species) dominated the littoral macrofauna. One species, a case-building leptocerid caddisfly (*Notalina parkeri*) was by far the most abundant species over this period but after peak abundance in 1977 declined rapidly to almost disappear toward the end of this extended sampling period (Fig. 6.7). Similarly, the bug *Diaprepocoris pedderensis* (Corixidae) also increased in range and

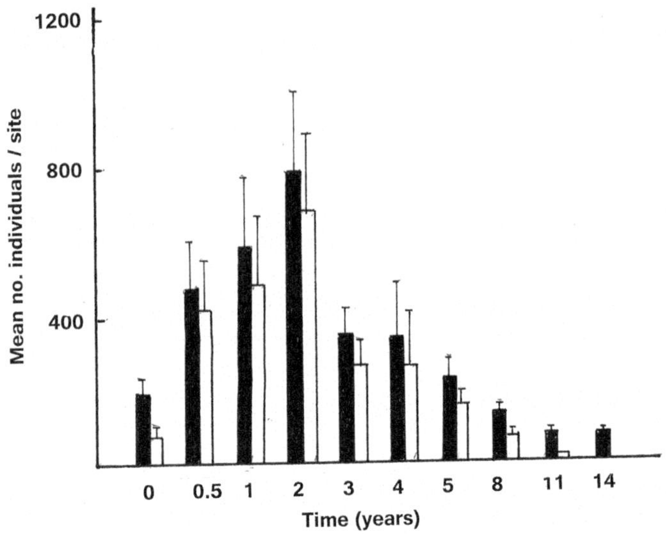


**Fig. 6.7** Lake Pedder, Tasmania: fauna 1975–1989. Mean number of individuals/site of all species (black bars) and the caddisfly *Notalina parkeri* (open bars) in samples over that period (Lake 2001)

abundance over 1975–1979 but had disappeared from all collection sites by 1983. At that later time the most abundant species was the amphipod *Austrohittonia australis*, and the slow change from insect-dominated to non-insect domination of lake fauna has been reported elsewhere (Lake 2001). Rapid increase of insects after impoundment has been attributed to increased production from nutrient inputs from drowned soil and vegetation, and abundant detritus from the latter. At that time, also, the invertebrate 'boom' in the Lake Pedder impoundment led to the area's reputation as a haven for large trout—and removal of introduced trout is a clear need for restoring the lake biota.

## 6.11 Fire

Wildfire and deliberate burning of waterside areas cause disturbances that can change aquatic environments and the communities they support. Immediate direct effects on aquatic insects—other than perhaps cremating adults resting in waterside vegetation—are largely negligible (Malison and Baxter 2010), but some other consequences may be far more serious. Short-term intensive heating of water, exposure to smoke or toxic chemicals, and deposits of ash and charcoal may lead to considerable insect mortality. In the longer term (at around 5–10 years post-fire: Minshall 2003), changed runoff, scouring and sedimentation regimes can lead to assemblage changes, with those impacts linked to reduced re-colonisations from the modified terrestrial environment. The latter contribute to changes in community structure and food web structures. Some are clearly long-term: salvage logging of burned trees may affect the supply of 'snags' in waterways because their sources are harvested and removed, with additional disturbances arising from machinery roadage and access. Minshall noted that inputs of snags provide habitat but also increase nutrients and retard runoff and channel changes during the critical stages of stream and riparian vegetation recovery.

Surveys of benthic macroinvertebrates in the western United States implied a relatively constant pattern of response that Minshall (2003) suggested could help to guide management.

Whilst fire is a common natural disturbance, more frequent and intensive anthropogenic fires may have more substantial and more 'unnatural' impacts and, in more natural waterside riparian vegetation or commercial forests can have profound—and mostly not fully understood—effects. Many studies on stream benthic macroinvertebrates have implied changed drift patterns, changed abundances and richness, and corresponding changes among terrestrial stages and emergences following fires, and the underlying mechanisms are not fully known (Mellon et al. 2008). In Washington, United States, comparisons of macroinvertebrates in burned and unburned (control) streams showed the former to have greater drifting and lower diversity. For more than two years after burning, headwaters of burned streams contained four times as many invertebrates in drift, and more than twice as many in benthos compared with unburned sites. Biomass did not differ across the treatments. As in other such surveys, Chironomidae predominated strongly. In Idaho, streams in the more severely burned catchments did not increase invertebrate similarity to reference streams over three years (Arkle et al. 2010).

In addition to responses to 'time-since-fire', as a frequently studied theme, the intensity/severity of the fire also mediated impacts, with greater severity having greater and more enduring effects and consequences for aquatic ecosystems varying greatly (Malison and Baxter 2010). Assessing those disturbances to invertebrates necessitates studying both the benthic early stages and the emergences of adult insects to the land. Malison and Baxter used Surber samplers and floating emergence traps for these purposes, to demonstrate that high severity fires might drive increased productivity through shifting the assemblages to include more taxa that develop rapidly (as 'r-strategists', such as some Ephemeroptera [*Baetis*], Chironomidae and Simuliidae) and multivoltine taxa, at the expense of more specialised species. In addition, Viera et al. (2004) had shown that unexpected post-fire flash floods helped to determine community responses in a burned stream. Burned organic matter in streams may not be suitable food for many invertebrates, limiting opportunities for changes in food webs to exploit this (Mihuc and Minshall 1995). Generalist feeders, as above, are important components of post-fire communities, and can potentially exploit, and switch between, different food materials. Mihuc and Minshall suggested that this prevalence might caution against uncritical use of functional feeding groups to infer trophic dynamics in stream invertebrate communities.

Heterogeneity across the macroinvertebrate assemblages of Australian alpine streams, where each stream supports a largely distinct assemblage, masked clear differences between burned and unburned streams (Crowther and Papas 2005) and so added to the difficulties of detecting impacts of fire. Impacts are also confounded because processes such as grazing, recreational skiing and drought were also present. Because fire impacts may have long-term consequences, continuing monitoring programmes were recommended to assist detailed interpretation.

## 6.12 Urbanisation

Urban developments have affected vast numbers of streams and creeks and led to direct loss of numerous water bodies in the increasing areas affected. Urbanisation is associated with degraded water quality and an associated array of chemical, physical and ecological changes, with the outcome of reduced biodiversity and the loss of numerous species sensitive to such changes. One wide outcome is that urban waterways are occupied by impoverished invertebrate communities dominated by rapidly colonising species tolerant of such changes, and the loss of many of the more specialised or intolerant taxa which are commonly those of greater conservation concern. Associated riparian vegetation in urban areas is commonly also degraded. However, whilst macroinvertebrate communities may suffer extensively from urbanisation, surveys in the Georges River catchment, Sydney, implied that odonate larvae do not do so, and endorsed other suggestions that habitat structure, especially riparian conditions, may be more influential than water quality in sustaining dragonfly diversity (Tippler et al. 2018), and that those features should receive priority attention in moves to conserve urban Odonata—but also not to neglect more direct responses of dragonflies to intensifying urbanisation (Caitling 2005). In Rhode Island, United States, although overall dragonfly richness and diversity did not change along an urbanisation gradient (Lubertazzi and Ginsberg 2010), several species showed more restricted distributions, with relatively rare species mostly found toward the rural end of the gradient. Urban wetlands, even when artificial constructions in newly developed housing areas, are important dragonfly habitats at all grades of urbanisation, and can support species of notable conservation concern, such as Hine's emerald dragonfly (p. 215).

More widely, in urban green spaces, dragonfly species richness is associated with heterogeneity of ponds and their surrounds. The human cultural appreciation of urban ponds and other waterbodies fostered through perceived values for recreation, aesthetic features, education and 'connection with nature' can be an important focus of conservation awareness (Ngiam et al. 2017).

Most information on impacts of urbanisation on macroinvertebrates is from lotic systems and Gal et al. (2019a, b) found that, in general, insufficient information was available from lentic ecosystems. They, as others have done (Chap. 4), urged the need for appraisals of specific taxonomic groups rather than the entire macroinvertebrate community, and to undertake this across a variety of habitats and in understudied parts of the world, with one purpose being to develop sound use of indicator groups. Urban expansion is a major threat to lotic macroinvertebrates, principally from direct habitat destruction and water pollution, and continued and accelerating urban growth with its attendant demands for solid waste disposal, storm water and sewage management generating much information on changes in water quality. Several rather different symptoms follow from the widespread degradation of urban waterways. Collectively, they constitute the 'Urban Stream Syndrome' (Meyer et al. 2005) and include (1) loss of sensitive macroinvertebrates; (2) decline of water quality; (3) invasions by alien/pest species; (4) modifications of flow regimes; and (5) reduction of habitat values from a wide range of disturbances. Characteristics of urban streams implied in the Syndrome include (1) high stormwater flows; (2) degraded and/or disconnected riparian zones; (3) homogeneous habitats replacing greater natural variety (including stormwater homogenising substrate sediments); and (4) elevated nitrogen content (Sudduth et al. 2011). Simply reflecting their proximity to many and varied sources of extraneous materials, urban streams also tend to accumulate a large range of pollutants (p. 92).

The term was developed at a symposium in Melbourne (Cottingham et al. 2004), to reflect the responses to increasing catchment urbanisation by four interrelated sets of disturbance/degradation that directly or indirectly affect stream ecology. The four groups distinguished were (1) disturbance of hydrological and hydraulic patterns; (2) disturbances to stream geomorphology; (3) degradation of water quality; and (4) habitat degradation or simplification. However, assessing details of these impacts, necessitating comparisons of urban waterways with natural reference waterways in the same region, can become difficult, as illustrated for the Ku-ring-gai area of northern Sydney. There, intensification of human activities over almost 200 years had essentially eliminated pristine 'reference' sites (Wright et al. 2007). They noted that it is 'practically impossible' to conduct valid before/after control/impact (BACI) surveys, and that comparisons of treatment with multiple reference sites may partially compensate. In that comparison, of benthic EPT taxa, the low numbers in urban waterways in comparison with local forested streams were taken to reflect their poor ecosystem health. Figure 6.8 exemplifies the trends found for urban and reference sites, together with the SIGNAL scores (Chessman 1995, p. 52). In such contexts, a 'reference site' is essentially a 'near natural, minimally impacted or best available site' (Metzeling et al. 2004).

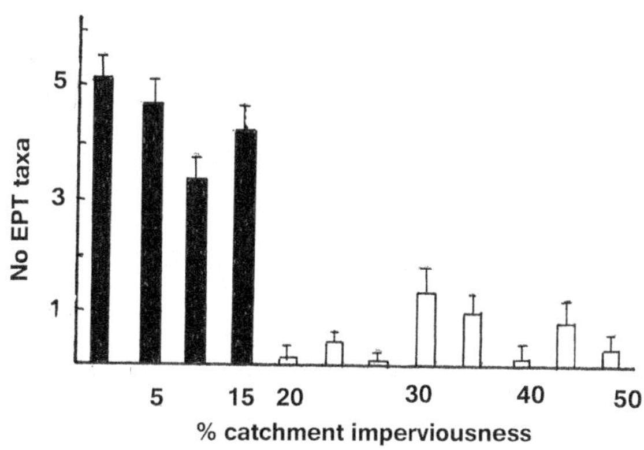


**Fig. 6.8** Macroinvertebrates and water quality along an urban gradient: the richness of EPT orders at reference sites (black bars) and urban sites (open bars) with increasing catchment imperviousness (%) in Ku-ring-gai, New South Wales (after Wright et al. 2007)

Macroinvertebrate assemblages are an important tool in assessing condition of urban streams (Nichols et al. 2016), but selection of particular traits and parameters, and the complex interactions these undergo with anthropogenic changes can become very dificult. Urbanisation has major impacts on stream environments and invertebrates, often reducing biodiversity (Paul and Meyer 2001) and interrupting connectivity between streams both through hydrological changes and increased isolation. These trends emerge from many comparative studies. One common outcome is simply increased homogenisation of the previously varied environments and communities. For example, in a New York stream system, comparisons of upstream and urban reaches over two years clearly showed loss of insect richness and abundance with urbanisation (Lundquist and Zhu 2019). However, Lundquist and Zhu found that heterogeneity could persist, reflecting 'habitat filtering' (such as through chemical or physical variations) and differential impacts on insect dispersal. That persistence suggested more complex interactions than the usually implied direct relationships between increased urbanisation and community decline, and also that impacts might be ameliorated by riparian vegetation, local parks, stretches of undeveloped stream sides and similar features within the urban environment. Attributing losses of urban aquatic insects to any particular cause is difficult. Urban et al. (2006) explained that because urbanisation impacts can occur at multiple scales, local extinctions may be attributed to both local habitat degradation and regional changes in land use. The most common conservation presumption has been that within-stream effects are responsible, leading to focus on restoration of the stream to near-undisturbed conditions. However, if effects from wider spatial scales prevail, together with landscape fragmentation preventing or limiting insect recolonisation, wider conservation measures within the catchments may be vital. If construction of new water bodies (p. 270) is contemplated, the density and functional connectivity of these may be manipulable to favour conservation aims. For example, the density of ponds within urban landscapes may markedly influence the richness of aquatic macroinvertebrates (Gledhill et al. 2008). In agricultural landscapes, also, proximity of drainage ditches to each other and to other waterbodies may strongly affect community composition (Herzon and Helenius 2008).

Urbanisation impacts on streams are very diverse, but also many are are not immediately obvious and are difficult to appraise, not least because of their interactions and synergistic or confounding effects. Thus, sublethal effects on biota are rarely considered other than in direct pollution monitoring. Current best practice management is almost inevitably pursued with very incomplete understanding (Wenger et al. 2009)—indeed, some of the key questions in urban stream ecology might, in their words 'never be assessed fully from a scientific perspective, and streams nevertheless must be managed now'. Wenger et al. (2009) suggested three steps toward this, as (1) identifying the desired stream ecosystem state (with streams targeted for management assessed as 'minimally altered, 'moderately altered' or 'severely altered'); (2) identifying the major streams or stream sources and select appropriate management actions, with emphasis on treating causes rather than just symptoms, and seeking to prevent further impacts from arising; and (3) identifying appropriate indicators for monitoring, and manage adaptively as progress toward goals is progressively assessed.

The Urban Stream Syndrome embraces the many factors associated with degradation of streams that drain urban lands (Walsh et al. 2005), so representing a very wide range of stressors. The three major roles of urban streams, namely (1) habitats for potentially diverse and productive biota; (2) carriers of water and processors of materials in that water; and (3) important social and aesthetic foci for people in those catchments, are each influenced strongly by impacts of urban developments. Invertebrates may be affected by aspects of water flow, water quality, structure of the streams, and organic materials present, all these both directly and through their effects on food and other resources. Urban development is often associated closely with losses of riparian vegetation (p. 100), as another strong contributor to changes. Three major categories of studies on urban aquatic invertebrates were distinguished by Paul and Meyer (2001), as (1) studies along a gradient of increasing urbanisation in a single catchment; (2) comparisons of an urbanised and a reference catchment; and (3) larger studies of gradients and responses across several catchments. Surveys in the first category, perhaps that most frequently undertaken, reveal declines in invertebrate diversity as urbanisation increases, and similar trends are widely evident in comparative surveys across catchments. More widely, correlates of this trend are with the extent of impervious surface, total discharge of effluents, human population density, and housing density. For most studies, presence/absence data or relative abundance information have been reported, with correlations with one or more of the above features or with subcategories such as sedimentation or turbidity. Many studies have demonstrated the absence or near-absence of sensitive invertebrate species from urban streams, leading to global description of those streams being 'characterised by species-poor assemblages, consisting mostly of disturbance-tolerant taxa' (Walsh et al. 2005), amongst which Chironomidae are usually the most diverse insects represented. The consistency of this pattern implied that benthic macroinvertebrates comprise 'arguably the most useful' group through which to compare variations in response to land use, with potential also to evaluate recovery measures as degraded streams are remediated. In parallel, surveys in Virginia (United States) demonstrated increasing relative abundance of Diptera (notably Chironomidae) and declines of most other insect orders with increasing levels of urbanisation (Jones and Clark 1987).

Degradation of urban stream systems from stormwater runoff reflects two major drivers of urbanisation: (1) the creation and increasing extent of impervious surfaces, and (2) the connection of those surfaces to receiving watercourses through pathways such as pipes or constructed ditches, often with the transfers aided by channelisation activities (Fletcher et al. 2014). Both drivers have severe hydrological and ecological impacts, and the core of management for biodiversity conservation involves measures to restore more natural magnitude, timing and variability of water flow—although these levels may be difficult to both define and achieve. Many urban drainage systems were designed with the primary purpose of collecting and removing water runoff from the area as rapidly and effectively as possible, in part to avoid local urban flooding.

Water is thus treated simply as 'a nuisance' (Zhou 2014), and biodiversity values of the drainage systems largely overlooked.

Urban drainage systems can support macroinvertebrate communities similar to those in rural waterbodies in the same area (the Netherlands: Vermonden et al. 2009). Four categories of water bodies were distinguished in that study, differing in their invertebrate assemblages and values of putative indicator taxa, but the key factors for invertebrate conservation were nitrate concentration, sediment composition, water transparency, and nymphaeid and submerged vegetation. Two red-listed invertebrates (one an insect: the caddisfly *Leptocerus tineiformis*, the other a planarian) were found. In this example, management to promote invertebrate biodiversity could include lowering nutrient levels, stimulating aquatic vegetation growth, and increasing water transparency such as by developing natural banks.

However, freshwater systems in urban areas have been relatively poorly studied when compared with terrestrial environments, and details that clarify the widespread acceptance of decreased richness and abundance of their many invertebrate inhabitants are fragmentary. Even for dramatically expanding areas such as western Sydney, the accompanying effects on aquatic biodiversity are only very incompletely understood (Chessman and Williams 1999). Drained largely through the Hawkesbury-Nepean River, lotic macroinvertebrates of this area comprised 443 recognisable species and morphospecies and, whilst some were deemed 'vulnerable' the status of most could not be assessed reliably because information was lacking. Synopsis of the taxa reported by Chessman and Williams (Table 6.13) is perhaps not atypical of the relative abundance of major taxa in such systems: Coleoptera (given as a conservative figure because some larvae could not be identified fully) and Diptera (more than a third of them Chironomidae) predominated, followed by Trichoptera (with representatives of 13 families) and Odonata larvae of 11 families.

Melbourne's streams, in common with those of many other urban areas are 'not well adapted to the abrupt frequent delivery of polluted stormwater' (Danger and Walsh 2008). The condition of these stream ecosystems is correlated with the proportion of each catchment covered by impervious surfaces that are connected directly to

**Table 6.13** Numbers of lotic species of insect orders recorded from rivers and streams in urban western Sydney, New South Wales (Chessman and Williams 1999)

| Order | No species/morphospecies |
|---|---|
| Ephemeroptera | 24 |
| Odonata | 35 |
| Plecoptera | 3 |
| Hemiptera | 23 |
| Coleoptera | 69 |
| Neuroptera | 2 |
| Megaloptera | 2 |
| Diptera | 115 |
| Lepidoptera | 11 |
| Trichoptera | 50 |

**Table 6.14** The categories of urban tolerance for animal species associated with streams of the Melbourne water region, Victoria (Danger and Walsh 2008) (DCI = proportion of the catchment covered with impervious surfaces, such as roofs and paved surfaces, connected directly to the stream)

| Category | Definition |
|---|---|
| A. Urban tolerant species | Abundance or occurrence is either not correlated or positively correlated with DCI |
| B. Transient species | Recorded occurrences in urban streams are likely to be transient, either during migration or for non-resident use of urban stream habitat |
| C. Urban sensitive species | Abundance or occurrence is negatively correlated with DCI (i.e. while they might occur in urban streams, they are less abundant or common than in rural streams) |
| D. Urban intolerant species | Species with near zero probability of being observed at a site with >1% DCI |

the waterway by sealed drains or pipes: that proportion has sometimes been referred to as 'effective imperviousness' but is less ambiguous as 'Directly connected Catchment Imperviousness' (DCI). Danger and Walsh found this to link strongly with ecological condition of Melbourne's streams. DCI as low as 1–2% may lessen stream biodiversity, and above 5–10% DCI, streams are severely degraded. The working threshold used by Danger and Walsh (2008) was >1% DCI, above which ecological impacts were observed in all streams studied: those streams were designated 'urban'. Four categories of urban tolerance were recognised among the 66 macroinvertebrate species (mostly non-insects), based on the presence of those species (Table 6.14)—so that *Leptoperla kallistae* (p. 222) was ranked as 'D' (as absent from streams with DCI > 1%), and the two families of Megaloptera (p. 195) were both also urban intolerant and restricted to streams with very low DCI and absent from all those with higher levels (Sialidae DCI > 0% and entirely absent from the metropolitan area, Corydalidae DCI > 2%).

An important finding from this Melbourne survey was that streams with much higher levels of imperviousness—up to 12% Total Imperviousness (TI) —were still in good ecological condition as long as those surfaces were not connected to streams by pipes. Danger and Walsh (2008) gave the example of Sassafras Creek (10% TI) being in good condition because most roads were unsealed or drain to an earthern drain, most homes drain to gardens or rainwater tanks, and for the roads that do channel into pipes those pipes drain to hillsides well above the stream. Sassafras Creek has near-zero DCI. Better drainage systems in urban areas, notably by reducing drainage connections to streams by stormwater pipes, are a key management solution to protecting stream biota in urban areas (Walsh 2004).

Aquatic insect richness in urban areas can be high. In 30 cities of central Europe, all but six of the 81 species of the regional dragonfly fauna were represented (Goertzen and Suhling 2015), but not all cities supported high richness. Most of the city species present occurred also in the city hinterlands and the 'missing' species were either very rare in Europe, or restricted—such as high elevation species unlikely to be found in the city areas surveyed. Those present included 14 species of conservation concern, ten of them establishing breeding populations and six abundant in, at least, single cities. Most species of conservation concern had specific habitat needs, and some, indeed, were common in city environments—so that cities offer valuable conservation possibilities for these. Half the species of conservation concern were found in built up areas.

Assemblage changes of damselflies (Zygoptera) along a 50 km length of the River Tiber as it flowed through Rome and suffered increasing impacts from urbanisation, notably losses of water quality, showed progressive domination of generalist lentic taxa rather than more typical riverine species, with the latter most poorly represented at the most polluted sites (Solimini et al. 1997). That domination (mostly by *Ischnura elegans* and *Cercion lindeni*, together comprising 1308 of the 1422 individuals captured) reflected tolerance of low oxygen conditions, a relatively long reproductive period, and absence of diapause. These two species were present in all larval samples, with four other species found only in low numbers.

Effects of urbanisation on both aquatic and terrestrial stages of water-breeding insects must be addressed, often initially as separate issues, in conservation. Historically stronger focus on the aquatic systems themselves has led to some under-representation of needs of the adult stage, in particular of the needs and changed ability for adults to disperse (p. 231) across highly altered urban landscapes (Smith et al. 2009) that must be considered in restoration. Very broadly, local changes to urban streams and their adjacent areas can prevent completion of life cycles, limit adult dispersal, and so lower chances of the populations persisting (Smith et al. 2009). Presence of ecological traps (p. 138) may exacerbate those losses. Street lights, for example, can have profound impacts on adult aquatic insects (p. 239). Movements in 'drift' (p. 234) can be decreased by artificial light in urban areas, but the trend is more pronounced in larger than in smaller streams (Henn et al. 2004), validating the epithet 'light pollution' as major driver of changed insect behaviour in urban areas. Riparian buffer zones may decrease the amount of artificial light reaching stream environments and inducing such responses.

The extent of local urbanisation near a pond is more influential on Odonata than the simple position of a pond along an urban–rural gradient (Jeanmougin et al. 2014, for Paris). There, urban and peri-urban ponds supported rather similar assemblages, and this was attributed in part to good dispersal capability of adults so that any pond with adequate resources could be colonised easily. The variety and naturalness of vegetation is a major influence on dragonfly richness at urban ponds (Goertzen and Suhling 2013), and creation and protection of these has important conservation roles (p. 270). Features of both aquatic vegetation (used by larvae for foraging, shelter and emergence sites) and terrestrial/emergent vegetation (as adult resources for oviposition, perching and territorial activity, and other behaviours) are both significant. A key design need, therefore, is to minimise the proportion of artificial shorelines, such as mown grass or paved borders, and reduce removal of pondside vegetation for 'sanitation'—and adding vegetation where necessary. Relatively small changes to existing management may be sufficient: planting native macrophytes may significantly increase conservation value, for example (Hill et al. 2017).

In the area of Dortmund, Germany, both vegetation and a collective high diversity of pond types were endorsed as promoting high urban dragonfly richness. Presence of waterfowl and fish also led to declines in species richness, but Goertzen and Suhling (2013) noted the considerable opportunities to be gained for dragonfly conservation by guiding gardening, recreational activities and 'nature experience' even in inner city areas. In surveys of Odonata across 22 central European cities, with an average representation of 33 of the pool of 64 species obtained, richness was related also to city area (Willigalla and Fartmann 2012), with a general pattern of richness increasing from the city centre toward more rural regimes. Both garden ponds and stormwater ponds (p. 277) can help to augment city Odonata richness—and the latter, indeed, can be important habitats for some rare pioneer species such as *Ischnura pumilio*. Land-use patterns, involving comparisons of Odonata in landscapes dominated by urban, agricultural or more natural uses (Goertzen and Suhling 2018) imply that urban landscapes maintain species diversity better than agricultural landscapes, but lead to modifications of assemblage composition at this landscape level. Variations among odonatan abundance, richness and community composition, together with plant community structure, occurred across series of stormwater and natural ponds around Ontario, Canada (Perron and Pick 2019) with significant correlations between dragonfly abundance and diversity of obligate natural wetland plants that could be enhanced readily in management of stormwater ponds.

Summary of the actual and suspected drivers of urban dragonfly biodiversity (Fig. 6.9) from increasing urbanisation impacts reveals the multitude of effects that may occur. Some have not been studied in detail, but Villalobos-Jimenez et al. (2016) also noted that 'carry-over' effects across life stages (such as impacts on larvae affecting the ensuing adults) are likely, and also difficult to evaluate. Generalist species of Odonata predominated in disturbed sites in urban areas of Manaus, Brazil (Monteiro-Junior et al. 2014), in a study that again demonstrated the values of riparian vegetation as a 'protective filter' from many disturbances. The more specialised species, most of them Zygoptera, were likely to become extinct locally due to deforestation leading to increased water exposure and temperature, and increased pollution. 'Habitat integrity' was especially important for many Zygoptera, whilst some larger Anisoptera apparently benefited from the changes. A similar trend in South Africa (Samways and Steytler 1996) was reflected in many Anisoptera occurring in exposed sunny areas with a high proportion of exposed macrophytes, as an important and influential variable for Odonata. That study showed clear differences in odonate assemblages in different biotopes in relation to vegetation and temperature regimes.

The changes to aquatic biota from urban development can be severe, and rehabilitation of individual streams or ponds is a widespread community focus in urban conservation (Oertli and Parris 2019), especially in the wealthier parts of the world. In many developing countries, conflations from rural migration, urban expansions and needs for sanitation and potable water clearly take priority over environmental considerations. Elsewhere, amenity and environmental values can become confused, but the interface between conservation and the differing demands of burgeoning

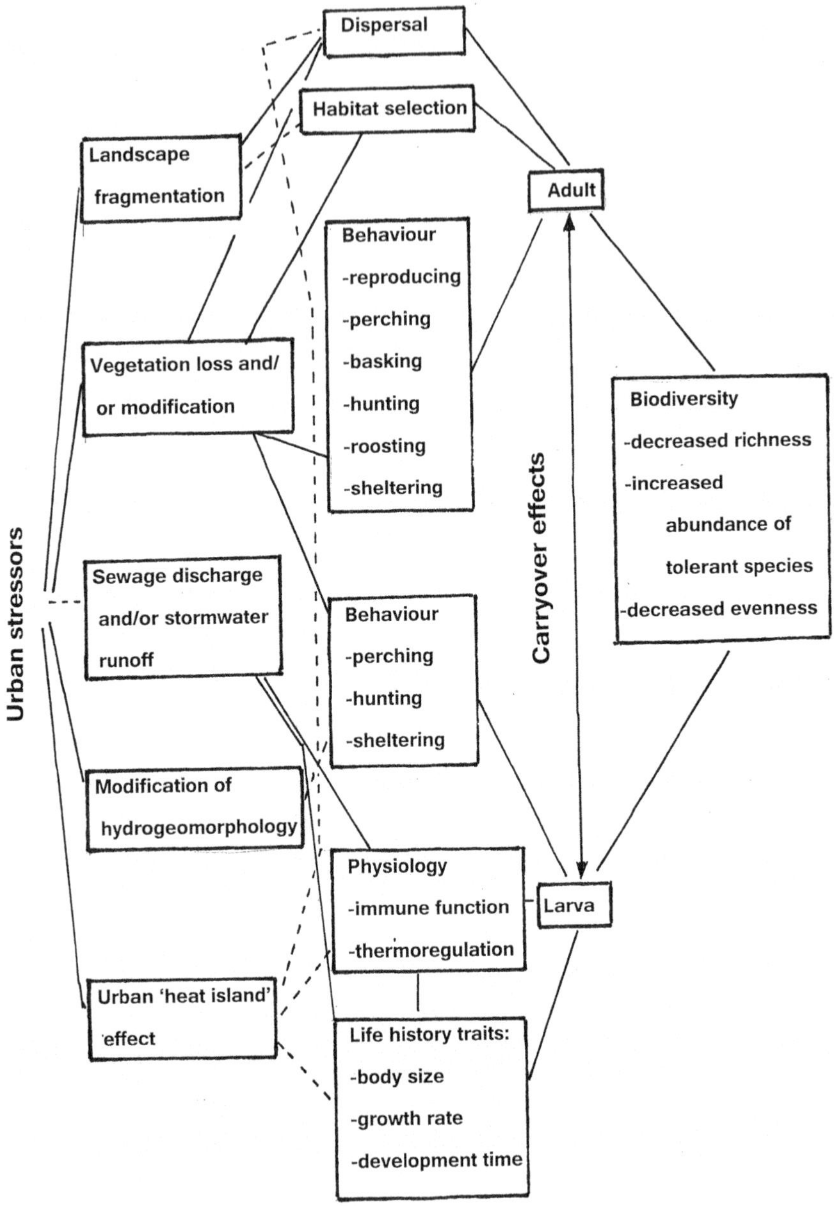


**Fig. 6.9** A summary of the drivers of odonate biodiversity in cities due to heavy management, as reviewed by Villalobos et al. (2016). Dashed lines represent hypothetical effects, as no studies were found that investigated associations between the specified stressor and corresponding odonate trait(s)

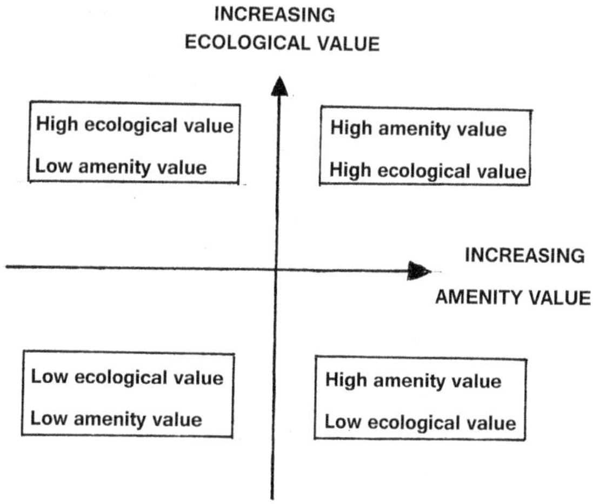


**Fig. 6.10** Urban waterways: hypothetical distribution of quality in the Melbourne region plotted by ecological and amenity values (Cooper et al. 2017, used with permission from Taylor and Francis Ltd, on behalf of Environmental Institute of Australia and New Zealand; www.tandfonline.com)

human urban populations continue to challenge management to satisfy all those demands. Using Melbourne's urban waterways as a model framework, Cooper et al. (2017) recognised the four categories noted in Fig. 6.10 as subjects for management to improve ecological and amenity values. The latter reflected a range of cultural values and characteristics that contribute to appreciation of pleasure, aesthetic worth, and encouragement to tourism and recreation. The values are largely context-specific but are viewed as gains for people. In Melbourne's 'Healthy Waterways Strategy' (2013), 'amenity' is defined in part as 'the pleasantness of a waterway to visitors', which commonly includes an appearance of being natural (Cooper et al. 2017). Alternative restoration strategies are usually available and may be driven by relative costs and local priority—but a planner may need to know how to assess the relative impacts and changes to both amenity and ecological values.

Major activities include replantings of riparian vegetation (sometimes following widespread removal of undesirable weeds and alien vegetation), removal of debris from the water, and reestablishment of the stream bed substrates and of pool-riffle sequences in streams. However, urban stream communities have commonly failed to recover even under such improved conditions (Blakely et al. 2006). Restoration of recruitment and colonisation, linked with assured conditions for oviposition by repopulating insects, may be difficult—the key substrate and vegetation states must be available, and the insects able to reach these and establish. Both 'longitudinal' (along streams) and 'lateral' (overland, between streams) movements may be interrupted—but in urban areas the latter may effectively be eliminated because of the highly modified intervening landscapes, and along stream dispersal becomes the predominant, even only, pathway available.

That dispersal, however, may be disrupted by roads, culverts (p. 231), bridges and other constructs that can constitute barriers—together with instream barriers such as constructed dams, weirs, and piped stream lengths, and patches of unusually dense riparian vegetation. Studies in Christchurch, New Zealand, explored some of these impacts through surveys of Trichoptera upstream and downstream from bridges and culverts on tributaries of the Avon and Heathcote rivers (Blakely et al. 2006). Numbers of adult caddisflies declined upstream; numbers in Malaise traps were about 2.5 times higher in traps just below than just above five culverts. Bridges had no significant effects on catch sizes upstream or downstream, but culverts were partial barriers. Series of culverts can have a cumulative effect, as traps near three successive culverts in this survey each yielded fewer adults upstream than downstream (Fig. 6.11). Roads and their underlying culverts clearly impeded upstream flight—but were not absolute barriers to movements. As another example, road crossings (bridges and culverts) had negative effects on richness and abundance of native macroinvertebrates in Hungary, where Gal et al. (2019a, b) compared assemblages

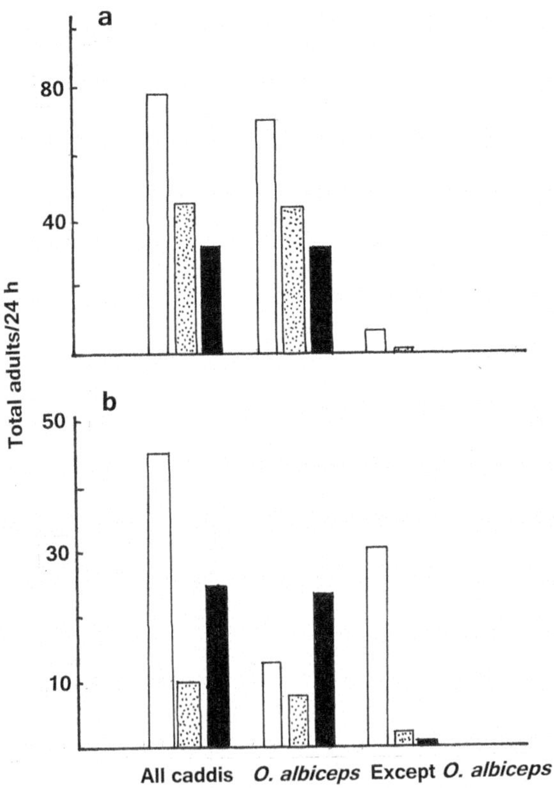


**Fig. 6.11** Numbers of adult Trichoptera (all caddis, *Oxyethira albiceps*, all except *O. albiceps*) in Malaise traps over 24 h. Traps directly upstream of three successive culverts (open, dotted, black bars) on two streams in New Zealand: **a** Okeover; **b** Waimari (Blakely et al. 2006)

**Table 6.15** Conceptual framework for considering impacts of highway construction and subsequent changes on stream habitats (Wheeler et al. 2005)

| | Development stage | | |
|---|---|---|---|
| Impact characteristics | Highway construction | Highway presence | Urbanisation |
| Temporal extent | Temporary | Chronic | Chronic |
| Spatial extent | Local | Regional | Regional |
| Primary nature | Physical | Physical/Chemical | Physical/Chemical |
| Degree of investigation | Low | Moderate | High |

upstream, at and downstream from crossings within the Danube river system. They also found that alien species were more abundant at the road crossings which, they suggested, might contribute to spread of those species. The major patterns were attributed to habitat modifications caused by the road crossings.

The impacts of paved road and highway constructions may be especially severe on stream ecosystems and, as with many other disturbances, impacts from a local disruption can spread far downstream—a markedly different effect from many equivalent primary disturbances confined to more circumscribed terrestrial areas. Reviewed by Wheeler et al. (2005), the impacts of highways (defined as roads with four or more traffic lanes) on streams occur in three successive stages. Angermeier et al. (2004) termed these: (1) initial highway construction, with all the short-term construction impacts that are primarily local, but have potential to spread downstream; (2) highway presence, with the variety of secondary impacts such as chemical pollution from traffic and runoff, and channel alterations; and (3) landscape urbanisation, with the variety of impacts that become progressively widespread and chronic. These may be viewed as a temporal gradient, with features summarised in Table 6.15, and which have received very uneven attention; the first has been studied considerably more than the later phases. However, Wheeler et al. demonstrated that the last (landscape urbanisation) may generally be the most severe threat.

Disturbances can become severe in urban navigational waterways, from hydraulic disturbances such as boat washes and return currents and consequent bank destructions and change. Rehabilitation measures have been investigated, for example in Berlin, where artificial bank structures created shallow littoral habitats protected from hydraulic disturbance (Weber et al. 2017). These led to substantial improvements in aquatic and riparian vegetation and generated some invertebrate habitats (such as organic mud) that are often rare in urban waterways.

## 6.13 Recreation

Intensive recreational boating activities can accompany increased human populations and may flow from urbanisation, as can increased recreational angling leading to bank and shoreline disturbances. The former can include marina developments and, very broadly, impacts can embrace physical effects (such as changes in wave intensity and direction related to speed and shape of a boat and its distance from shore, with effects on shoreline structure and vegetation), chemical (fuel pollution and other spillage or discards), and biotic (such as facilitating increased movements of invasive species).

Studies of the impacts of recreational boating on Odonata in the Georgian Bay Region of Lake Huron, Canada, involved surveys of exuviae and adults from 17 islands in this large archipelago (Hall et al. 2015) and implied that boating did indeed have some harmful impacts, in part through actively dislodging larvae and emerging adults and affecting vegetation. Reduction of those impacts might be approached, at least in principle, through measures such as (where possible) re-locating boat channels away from important dragonfly habitats and/or reducing boat speeds (Hall et al. 2015).

## 6.14 Ecological Traps

Urbanisation is associated with a range of 'ecological traps' for aquatic insects, most notably the diversion of oviposition from natural waters by polarised light impacts (p. 239). The reflectance patterns of water surfaces are resembled by that from many motor vehicles, black plastic sheeting, some road surfaces, and others, from which decoy effects induce female insects to attempt to lay on these surfaces. The trap effect occurs essentially because the environmental cues used by insects, and with which they have evolved to inform decisions over (in this case) oviposition site selection, have been rendered unreliable and substituted negative for positive outcomes by responding to them. In this context, water-seeking aquatic insects are preferentially induced to lay eggs by supernormal attraction to polarised light surfaces, on which oviposition is futile. For rarer species, such effective population mortality may exacerbate other threats in accelerating declines. Several examples (Villalobos-Jimenez et al. 2016) suggest that such losses of reproductive effort may be widespread. Ecological traps for aquatic insects are likely to be more diverse than generally believed, as illustrated by some recent examples, mainly for dragonflies. Each fulfills the basic definition of an ecological trap as being a low quality habitat that is actively selected and preferred over available habitats of higher quality, and that cannot sustain a population (Battin 2004). The wider need is to recognise these and to find ways of mitigating their impacts—for birds, Battin suggested measures such as decreasing trap attractiveness and increasing habitat quality to decrease deleterious effects, measures widely applicable for other taxa and based in the reality that ecological traps will persist and diversify and cannot be effectively eliminated on any realistic scale. Urban stormwater ponds have been suggested to be ecological traps, with attraction of reproductive insects in areas with few alternative habitats leading to reduced survival of offspring because of low quality water conditions associated with the ponds being 'sinks' for urban contaminants (Perron and Pick 2019). A contrary view is perhaps more prevalent—that the values of stormwater ponds as additional habitat in urban areas outweigh the risks of losses through increased exposure to toxins. From either viewpoint, stormwater ponds can become a target for management to increase their conservation values (p. 277).

Behavioural studies on mayflies and dragonflies, in particular, have repeatedly demonstrated their polarotactic attraction to suitable reflective surfaces—with vehicles, road surfaces and black plastic sheeting, pools of waste oil and perspex sheets cited amongst the surfaces that can provide supernormal stimuli for the insects and cause them to attempt oviposition there. More unusually, polished black gravestones in a Hungarian cemetery were attractive to five species of *Sympetrum* (Libellulidae) of both sexes (Horvath et al. 2007). Individuals perched on or near gravestones, and defended those perches. Some repeatedly touched the stones with the ventral side of their abdomen, and tandem pairs circled over the stones—all behaviours typical for true aquatic habitats. Oviposition attempts were not reported in this study, despite strong inference of it occurring, but has been reported in some other parallel cases of dragonflies attracted to polarising surfaces. However, Horvath et al. postulated that, as breeding sites of *Sympetrum* can often be temporary pools, oviposition in a large number of sites might be beneficial, and those potential sites might include scattered ecological traps, even when far from real wetlands.

The European caddisfly *Hydropsyche pellucidula* swarms before sunset and riverside buildings with differing tints of glass cladding and windows included darker glass panels that acted as swarm markers by reflecting horizontally polarised light. Females laid eggs on the glass, but those swarming caddisflies are also food for many insectivorous birds. In Budapest, swarms on the shore of the River Danube were exploited also by some more unusual birds such as Great spotted woodpeckers (*Dendrocopos major*) and Hooded crows (*Corvus cornix*) (Pereszlenyi et al. 2017).

Recent attention to photovoltaic solar panels, which are increasing rapidly in both urban and more rural areas, has also revealed their attractiveness to polarotactic insects. However, Horvath et al. (2010) demonstrated that increasing fragmentation of the polarising surfaces by interposing a white grid reduces their attractiveness and eliminates the ecological trap effect. Without such modification, solar panels attracted mayflies, caddisflies and dolichopodid and tabanid flies, all exhibiting behaviour associated with oviposition. The design of solar panels and their collectors, as well as their proximity to aquatic habitats, may each potentially affect aquatic insect populations. Designs to combine polarising and non-polarising white surfaces can 'disarm' the trap effect (Black and Robertson 2020), and can be pursued with relatively small losses of the solar-active areas.

Each of the above examples of putative or proven effects of ecological traps concerns anthropogenic artefacts that reflect polarised light. A rather different form of ecological trap for dragonflies was described for the European *Sympetrum depressiusculum* by Sigutova et al. (2015) who identified intensive fishponds as a possible trap. *S. depressiusculum* is a habitat specialist, and fish ponds—although apparently suitable for oviposition—expose larvae to factors such as fish predation, eutrophication, and inappropriate water regimes, collectively with considerably lessened suitability for development, a situation paralleled there for the widespread *Sympetrum sanguineum*. Shoreline transect counts of adult dragonflies and of exuviae from random quadrats along shorelines of the ponds over the entire emergence period of the two species, showed that *S. depressiusculum* developed only in one pond (where the greatest number of adults was observed) and implying that the other five ponds investigated were ecological traps (Fig. 6.12). In parts of northwest Germany, within the northerly range of the species, *S. depressiusculum* is confined to shallow artificial ponds in which water temperatures can rise above the levels typical for the area. Schmidt (2008) suggested that artificial carp-breeding ponds were attractive for the dragonfly. Carp need water of at least 20 °C for egg and larval development, and the shallow spawning ponds (fundamentally, flooded grazing meadows surrounded by ditches) are kept dry from autumn to late spring, when they are flooded with water warmed in other ponds. At that time, the carp larvae are too small to eat the dragonfly larvae, and a range of Odonata species are common there. With abandonment of some ponds from rearing carp because of high levels of predation on the fish, accumulated mud was associated with loss of *S. depressiusculum*, but later resumption of carp rearing enabled the dragonfly to reestablish rapidly.

Although not conventionally treated as an ecological trap, direct road mortality of adult dragonflies from vehicle impacts can become high (Soluk et al. 2011), particularly for some species that fly at low levels. Roads do not appear to constitute a significant barrier to dispersal by dragonflies, and it is possible that some polarotactic species may be attracted more directly to road surfaces. Long adult life may predispose some dragonflies to traffic mortality, and Soluk et al. showed this to be

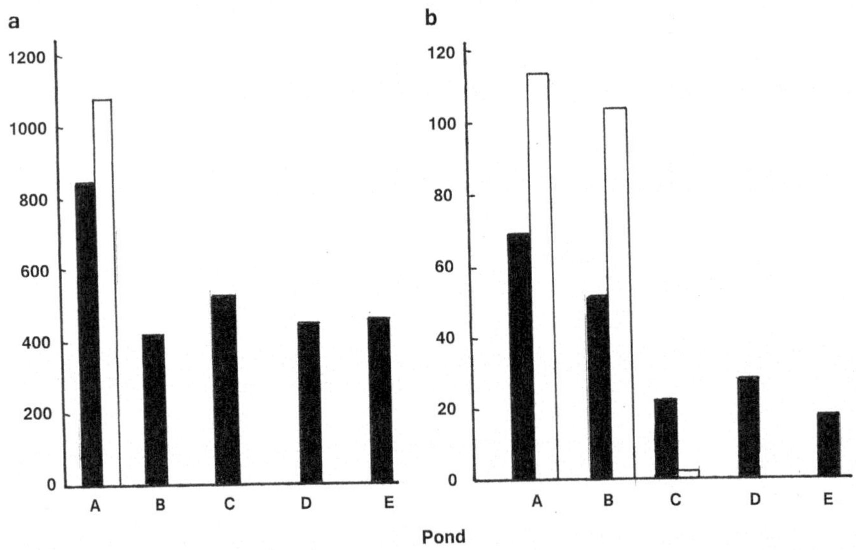


**Fig. 6.12** Numbers of individuals of **a** *Sympetrum depressiusculum* and **b** *S. sanguineum* ovipositing (black bars) in individual ponds in 2014 compared with the numbers of individuals that emerged that year (open bars). A, B, C are fish-breeding ponds and D, E are intensive fish ponds; only pond A allowed development of *S. depressiusculum*, suggesting that others were ecological traps; the generalist *S. sanguineum* could develop in all fish-breeding ponds, but the intensive ponds did not allow development of either species (Sigutova et al. 2015)

substantial in Illinois, and that few studies had attempted to quantify mortality rates. These might also become biased by the two sexes behaving differently. Roads can be used by dragonflies as dispersal corridors, for example by Hine's emerald dragonfly (p. 215). Some other *Somatochlora* spp. had far more casualties among females than males (Riffell 1999). However, some other genera displayed male-biased mortality and, although numbers of some were low, 20 of 25 species collected had male-biased road mortality, with this statistically significant for six species. Riffell suggested the need for further detailed surveys to clarify the extent of mortality, but also that careful consideration may be needed before constructing further roads in or near key dragonfly breeding sites and that alternatives should be considered when habitats of known threatened species may be affected.

Wider considerations of how ecological traps can be affected, and generated or mitigated, by management procedures emphasise the variety and complexity of situations that can occur in urban freshwater environments (Hale et al. 2015).

## 6.15 Climate Change

Widespread agreement exists that future climate changes with uncertain rates and extents of modifications to current regimes represent major challenges for practical conservation in all major global ecosystems. Climate change fundamentally involves changes in temperature and patterns of precipitation, and these are likely to affect all aspects of inland waters and their hospitality for insects and other biota and their relationships with enveloping environments. Both variables are among the current major drivers of biotic changes, and their interactions with other drivers and threats will continue to diversify. Despite numerous models and practical investigations of environmental needs and tolerances, predicting the severity and extent of changes on the species and wider ecosystems (as well as the broader enveloping landscapes) remains highly uncertain and speculative. Freshwater biodiversity is likely to be particularly sensitive to some changes (Woodward et al. 2010), and there is urgent need to devise effective conservation strategies to counter species losses and changed ecological integrity (Bush et al. 2014). In a wide survey of the traits of aquatic insects in relation to susceptibility to climate changes, species preferring very cold temperatures showed the highest responses (Bhowmik and Schafer 2015). These, together with caddisflies preferring moderate temperatures, had highest potential for distributional changes. Several traits had earlier been associated with climate change—so that range contractions were widespread among cold water taxa with low dispersal powers, as for many ecological specialists that had small distributions and short flight periods. For Germany, Bhowmik and Schafer anticipated considerable longitudinal changes in distributions of species along streams as conditions changed.

Many freshwater species are regarded widely as highly vulnerable to climate changes, from increased water temperatures, changed hydrological regimes, newly fragmented environments impeding dispersal, and the altered impacts of other threatening processes. Unspecified 'climate change' is noted commonly as a threat to aquatic insect species of conservation concern, but details can only rarely be suggested. Impacts of climate change have been predicted to be especially strong at high latitudes and high elevations, regions in which the duration and timing of snow cover and snow melt can strongly affect water temperature and flow. In Australia, many of the cool-adapted Gondwanan invertebrates may be especially susceptible. Many endemic Ephemeroptera and Plecoptera, in particular, are largely restricted to localised cool upland waters in the south-east (Chessman 2012), and range contraction is perhaps their most likely response to changing conditions. Opportunities for compensatory upward movements are clearly limited with, essentially, 'nowhere to go' and southward movements restricted by Bass Strait forming a largely impenetrable barrier to reaching Tasmania from the mainland. Many of these taxa already occupy the coolest habitats in the region. Chessman also implied that rheophilous taxa, representing specialisations to living in fast currents through traits such as streamlining and attachments to substrate surfaces, may be especially susceptible to range contractions. Collectively, Chessman's information accumulated over many invertebrate families and 16 years from New South Wales and the Australian Capital Territory inferred range changes through (1) cool-edge expansion with warm-edge contraction (71 families); (2) wet-edge expansion plus dry-edge contraction (71); (3) contractions from both cool and warm extremes (36); and (4) contractions from both dry and wet extremes (28). A preference for water flow was associated with the second of these categories—but a considerable variety of responses occurs. In contrast to many other insect groups less tractable to study, studies on adult Odonata have clearly demonstrated range changes to incorporate higher elevations and latitudes but also that local community diversity and function may be affected (Flenner and Sahlen 2008), and that habitat adoption may become more selective near the edge of a distribution than near the centre of the range. Thus, in Sweden, species near the northern (coolest) limit of a range may be sensitive to choosing the warmest habitats available but should climate change increase the range of options available at higher latitudes, choice may become wider.

Distributional knowledge of many Odonata facilitates predicting future changes with climate alteration, and Collins and McIntyre (2017) emphasised the importance of identifying 'climate refugia' as more species become affected. In addition to rising temperatures, they noted that increased rainfall intensity and flooding may harm riverine species in North America. Refugia may become especially important for those taxa that cannot track climate change by active dispersal (p. 279). In common with findings from some European surveys, many lotic Odonata appear to be less capable than many lentic species of range changes.

Incremental changes to current 'average conditions' are the most frequently considered components of climate change, but increased frequency or intensity of 'extreme events' may have even greater consequences as sudden or catastrophic occurrences (Woodward et al. 2016). More frequent droughts, floods, fires and others expose freshwater ecosystems to conditions without historical precedent, with uncertainties also in predicting changes for the systems themselves and for how human needs may be influenced. For the Murray-Darling Basin, for example, such impacts may lead to changes in catchment properties. Woodward et al. suggested greater use of drains from increased storm events and greater irrigation needs during droughts. In short, the 'stress' on aquatic ecosystems is likely to increase and also include novel components as innovations to counter these are sought.

Some long-term changes in freshwater macroinvertebrate assemblages have been attributed to long-term climate changes, together with phenological effects (Dingemanse and Kalkman 2008, on Odonata). Earlier appearance dates of 37 species of Dutch Odonata over 1995–2004 reflected increase in spring temperature affecting time of adult emergence. This seasonal response contrasted with other seasons, for which no such response was evident, so casting doubt over earlier implication (Hassall et al. 2007) that more generalised increasing environmental temperatures would lead to advancement of phenology—in England, the earliest flight date for many species had advanced over the previous 40 years, but Dingemanse and Kalkman's data supported the more complex theme that both qualitative and quantitative differences in effect occur between seasons.

Both community structure and diversity may change (Burgmer et al. 2007). Abundance and life history phenology changes with climate have most commonly been examined for individual study species, and far fewer studies have elucidated the wider shifts in diversity and composition of macroinvertebrate communities. Burgmer et al. (2007) showed that two rather different associations between biodiversity and climate change can occur, reflecting the scales of influence. These are (1) local interactions, with essentially local communities changed by climate influences; and (2) much more broadly, with variables such as temperature and evapotranspiration becoming strong predictors of both aquatic and terrestrial biodiversity and leading to changes in distribution and shifts in elevation and/or latitude varying across species. Those shifts lead to largely unpredictable changes in species interactions. Despite considerable difficulties in detecting any simple or direct relationships, Burgmer et al. concluded that benthic aquatic invertebrates 'are likely to show strong responses to climate warming'.

Individual species differ greatly in their actual or predicted responses to changing climate, and many inferences are based on observations over rather short intervals, perhaps too short for real enduring change to occur. Somewhat exceptionally a 25-year period was assessed by Durance and Ormerod (2007). In New South Wales over the period 1994–2007, Chessman (2009) recorded changes among 124 families/family groups of stream macroinvertebrates, to reveal significant trends of increase (33 taxa), decrease (37 taxa) or no significant trend (54 taxa, many of them collected only rarely and any real trends uncertain). Over that period, air and water temperatures increased, and rainfall and river flows declined so that both water temperature and water flow changed, and detailed interpretation of the causes of invertebrate changes remained uncertain, especially as climate change most commonly interacts with other anthropogenic influences such as water withdrawal and loss of riparian vegetation.

Both phenological and distributional changes may be induced. Changed water temperatures (p. 82) are perhaps among the most easily documented influences, but with the consequences still complex. In the relatively few cases in which adequate long-term data sets are available, some shifts in species' features and community structure consistent with climate change effects have been reported—but inadequate sampling can limit detecting wider trends (Bush et al. 2013). One need is the capability to detect and track changes in assemblages, but in most contexts, historical data are inadequate to form a sound foundation against which more recent changes can be assessed.

Examining the distributions of European caddisflies (including 1134 species or subspecies, across 23 ecoregions) and their potential susceptibility to climate change, Hering et al. (2009) found five parameters that largely described a species' sensitivity. These were (1) endemism; (2) preference for springs; (3) preference for cold water temperatures; (4) short emergence periods; and (5) having restricted feeding habits and ecological breadth. Interpreting changes in diversity, even in this largely well-documented fauna, was limited by still incomplete taxonomic knowledge of the rich Trichoptera fauna of southern Europe—from where numerous species continue to be described, and larval recognition to species level is sometimes uncertain. In contrast, the relatively sparse northern European caddisflies are far better known. Of the parameters listed, species preferring or restricted to springs cannot move further upstream if waters warm (so their situation parallels that of many alpine biota driven upward in that they have 'nowhere to go', p. 232), and preference for cold water can also limit future options. Both feeding specialisation and short emergence periods are dimensions of ecological limitation that can influence the extent of more general adaptations to changing environments.

Examination of the likely consequences of climate change for European Plecoptera (de Figueroa et al. 2010) showed that at least 324 (almost 63%) of the taxa fell into one or more of the relevant susceptible categories, namely being restricted to particular stream zones, preferring a (defined) restricted elevational range, preferring a particular current regime, having a restricted or preferred temperature range, and scale of endemism, as well as being 'rare'. The outcomes (Fig. 6.13) included a high proportion of the taxa (252 of 383 for which information was available) present in the spring/spring-brook (crenal) zones, 39 of them restricted to this and considered vulnerable to change in longitudinal zonation. Likewise, 21 species occurred mainly at high elevations and many species (201 of the 266 for which data was available) were found mainly or solely in cold waters. Within these categories the small-scale endemic species (109) and 'rare' taxa (111) were considered the most endangered. Soberingly, de Figueroa et al. noted also that the total number at risk is even higher than their survey indicated, because some species (including several known to have declined) were not included and are also susceptible to combinations of climate and habitat changes. In contrast, perhaps only a few generalist and low elevation species are likely to increase—collectively a very small proportion of the European stonefly fauna.

Dragonflies have been dubbed 'climate canaries for river management' (Bush et al. 2013) because the distributions of many species seemed highly sensitive to climate factors. In eastern coastal New South Wales, surveys over 791 sites yielded 97 species of Odonata across 46 genera and 10 families. Distributions varied. For example, (1) some species were warm-adapted and declined substantially with increasing elevation or latitude, and (2) some others appeared cool-adapted and were more common

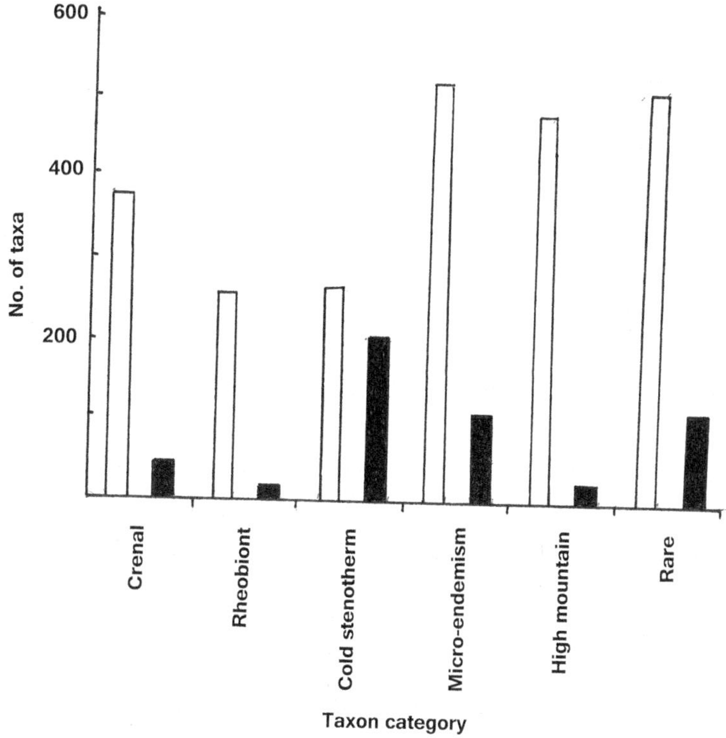


**Fig. 6.13** The numbers of vulnerable taxa (black bars) amongst the numbers of evaluated taxa (open bars) for European stonefly taxa, based on selected ecological data for 516 species/subspecies evaluated (de Figueroa et al. 2010)

at higher elevations. In this case, as others, identification to species level is a key to assessing and realising their potential as indicators of ecological responses to climate conditions.

The long history of recording dragonflies in Britain forms the most complete and reliable country-wide template for assessing distributional changes likely to be related to climate, but with trends also linked to habitat changes. Thus, Parr (2010) noted that three species had become extinct there over the previous 60 years, due to degradation of their habitats. In contrast, other species expanded their range northward in Britain following the more general and widely documented trend of moving to encompass higher latitudes as conditions warm, but also linked with habitat availability. The northward expansion of *Orthetrum caledonicum*, for example, was attributed in part to the spread of flooded gravel workings as excellent habitat for this previously rare species earlier confined to southern England. In eastern Australia, the limited mire habitats of petaltail dragonflies (*Petalura* spp., p. 187) may be susceptible to resultant changes in water table level and increased fire hazards, both viewed as threats by Baird and Burgin (2016).

In Sweden dragonflies responded rapidly to climate change, with both abundance and assemblage composition changing over a decade. Boreal forest lake assemblages showed trends of (1) formerly rare species of Odonata becoming more abundant and more widely distributed and (2) some formerly widespread species becoming more restricted in their habitat usage (Flenner and Sahlen 2008). With such changes in pattern, reflecting changes in individual species' ecology and in the habitats themselves, some doubts arose over continuing to use dragonflies as diversity indicators. Influences of climate change may extend far more widely to patterns of diversity and association on which the indicator premise is founded. Dragonflies are regarded widely as potentially useful indicators, as responders to climate changes, but their responses in eastern Australia differed from that of most other macroinvertebrates compared by Bush et al. (2013), not least because species recognition was relatively easy and interspecific confusions were largely avoided. Reliable species-level information was not available for most other invertebrate groups surveyed. In this study, climatic factors 'explained' three times as much variation in assemblages among odonatan species than either dragonflies or other macroinvertebrate assemblages assessed at family levels. Distribution of dragonfly assemblages can thus be most strongly associated with climate factors at the species level, giving this fine level of identification considerably more value than the family-level analyses more typical of such exercises. Any such survey with species-level interpretations has potential to provide a sound reference point for comparison with future studies. Analyses can draw also on the increasing number of reports of odonate range changes attributed to climate changes—notably the northerly range extensions of many species in the northern hemisphere and, on a different scale, the comparable range shifts of downstream running-water-adapted species moving upstream to nearer the higher headwaters (Domisch et al. 2011). In some environments (such as some montane streams), equivalent movements to upstream refuges may not be possible, so that proactive translocations to other, more similar, locations might be viable conservation measures (Heller and Zavaleta 2009).

Creating freshwater refugia for Odonata will be needed within a broader strategy to promote connectivity and facilitate adaptive range shifts as climate changes (Bush et al. 2014). In Australia, the generally low relief limits opportunities for elevational range shifts, and most species needing to track climate changes are more likely to undergo latitudinal range changes. The most vulnerable Odonata identified by Bush et al. are all endemic. Predicted range shifts showed likely abandonment of current northerly range areas and concentrations toward Victoria and Tasmania. In a broader review of environmental warming impacts on Odonata, many species were predicted to benefit from poleward range shifts, but differences in trophic, developmental and dispersal traits between species will also change assemblage compositions and lead to novel interactions (Hassall and Thompson 2008).

An important possible consequence of upstream range changes linked with climate changes is that the lower reaches may also become more accessible to invasive or non-native species. Domisch et al. (2011) noted that some of these may already be adapted to higher temperatures and/or low oxygen levels, and might exhibit characteristics of keystone/ecological engineer species able to induce major changes in community composition in those lower reaches. Their modelling across a wide variety of macroinvertebrates predicted that some headwater species would lose large amounts of suitable habitat, whilst those cold-adapted species would progressively be replaced by species adapted to the now warmer stream regimes. More precise predictions of consequences were much more difficult to assess. Different climate change scenarios (Domisch et al. compared outcomes from scenarios 'A2a' ['business as usual'] and 'B2a' ['moderate'] as detailed by IPCC 2001) gave rather different outcomes. These models implied likely changes in species' elevation of 122 m and 83 m, respectively. One insect, the chironomid fly *Rheocricotopus fuscipes*, was predicted to go extinct under A2a but able to survive under B2a.

Numbers of alpine insect species are regarded as likely to increase as climates warm. Aquatic insects, as those elsewhere, are likely to include both 'winners' (successful colonists) and 'losers' (those becoming extinct), and predicting the relative incidence of these trends, and of the taxa that will remain more-or-less unaffected, is difficult and somewhat speculative.

For small ponds in Switzerland, Rosset and Oertli (2011) included aquatic Coleoptera (both larvae and adults) and Odonata (adults) among the taxa investigated in attempting to assess winners and losers in two ecological groups—cold thermal specialists and cool thermal specialists for losers, and warm thermal generalists and warm generalists for winners. Their basic assumption, as in some other studies, was that species with restricted thermal ranges, particularly in cold or cool temperatures, are potential losers, and species with ranges restricted to warm temperatures or with large thermal ranges are potential winners. Elevational and latitudinal distributions of each species were used as a surrogate for temperature and, with inferred thermal preference, for indicating likely susceptibility to warming.

Rosset and Oertli also evaluated the 'resilience' of each species (Coleoptera 122 species, Odonata 54 species associated with ponds), using five criteria: (1) dispersal ability; (2) extent of habitat specialisation; (3) geographic extent in the study area; (4) future trend of that extent; and (5) future trend in habitat availability. For Coleoptera, 13 species were identified as cold thermal specialists, but all others could not be allocated confidently to category. The better-documented Odonata gave fuller inferences (Fig. 6.14), and only three species could not be categorised confidently. That study allowed strong suggestions to be made on which species were likely losers or winners under warming climates, with inferences on resilience from the above criteria augmenting these.

As is evident from the number of notable 'flagship' species of aquatic insects associated with Australia's highest elevation waterbodies (Chap. 8), many of the species in alpine streams or ponds are endemic or near-endemic. Further, these species participate in communities considerably less rich than those of equivalent waterbodies at lower elevations. Many of those species should in theory be vulnerable to climate changes, with warming driving them toward local extinctions. Even at lower elevations, headwater streams are indeed sensitive to climate change, but factors such as acidification and eutrophication may override climatic effects (Durance and Ormerod 2007), so that attributing faunal changes specifically to directional climate influences

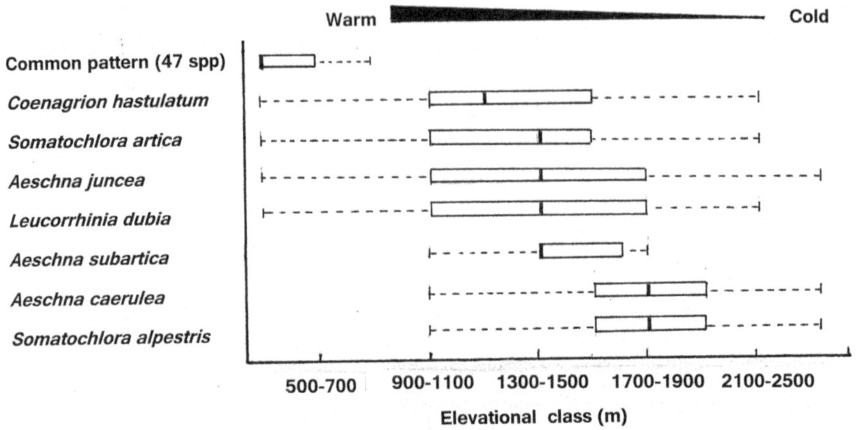


**Fig. 6.14** Patterns of elevational distribution of pond Odonata in Switzerland, using elevation (horizontal axis, mostly at 200 m intervals) as a surrogate for temperature (Rosset and Oertli 2011)

is, in their term, 'challenging'. That override effect from acidification might operate through reducing richness and thereby simplifying assemblages. Their comparisons of invertebrate assemblages in three major stream types in Wales over the 1981–2005 period revealed each stream category (in circumnatural moorland, acid moorland, acid forest) to have different most abundant taxa. Whilst some species were characteristic respectively of cooler or warmer years, others were 'core species' present over a wide temperature range. Increasing stream temperatures had direct effects—of moderate changes in assemblage composition and seasonal (April) declines of macroinvertebrate abundance. Those variations mostly affected the less common species that occurred under cooler or warmer extremes. Headwater stream ecosystems can become especially susceptible to change, but Durance and Ormerod also made two strong recommendations to increase practical understanding, as (1) need for greater efforts and commitment to assess climate change impacts on headwater organisms and the processes that affect them, and (2) increased research into adaptation and mitigation measures within stream ecosystems.

Reduced river macroinvertebrate richness at increasing temperatures has been detected in several contexts, and the fate of high elevation cool water specialists may be of especial conservation concern.

## References

Abellan P, Sanchez-Fernadez D, Velasco J, Millan A (2005) Assessing conservation priorities for insects: status of water beetles in southeast Spain. Biol Conserv 121:79–90

Abjornsson K, Bronmark C, Hansson L-A (2002) The relative importance of lethal and non-lethal effects of fish on insect colonisation of ponds. Freshw Biol 47:1489–1495

Allan JD, Abell R, Hogan Z, Revenga C, Taylor BW, Welcomme R, Winemiller K (2005) Overfishing of inland waters. BioScience 55:1041–1051
Allan JD, Castillo MM (2007) Stream ecology. Structure and function of running waters, 2nd edn. Springer, Dordrecht
Angermeier PL, Wheeler AP, Rosenberger AE (2004) A conceptual framework for assessing impacts of roads on aquatic biota. J Fisheries 29:19–29
Arkle RS, Pilliod DS, Strickler K (2010) Fire, flow and dynamic equilibrium in stream macroinvertebrate communities. Freshw Biol 55:299–314
Armstrong JS (1978) Colonisation of New Zealand by *Hemicordulia australiae*, with notes on its displacement of the indigenous *Procordulia grayi* (Odonata: Corduliidae). N Z Entomol 6:381–384
Arribas P, Abellan P, Velasco J, Millan A (2015) Evolutionary ecology, biogeography and conservation of water beetles in Mediterranean saline ecosystems. Limnetica 34:481–494
Arthington AH (2012) Environmental flows: saving rivers in the third millennium. University of California Press
Arthington AH, McKenzie F (1997) Review of impacts of displaced/introduced fauna associated with inland waters. State of the Environment Technical Paper Series, Department of the Environment, Canberra, Australia
Arthington AH, Naiman RJ, McClain ME, Nilsson C (2010) Preserving the biodiversity and ecological services of rivers: new challenges and research opportunities. Freshw Biol 55:1–16
Baird IRC, Burgin S (2016) Conservation of a groundwater-dependent mire-dwelling dragonfly: implications of multiple threatening processes. J Insect Conserv 20:165–178
Bates AJ, Sadler JP, Henshall SE, Hannah DM (2009) Ecology and conservation of arthropods of exposed riverine sediments (ERS). Terr Arthr Revs 2:77–98
Battin J (2004) When good animals love bad habitats: ecological traps and the conservation of animal populations. Conserv Biol 6:1482–1491
Bayley IAE, Williams WD (1981) Inland waters and their ecology, 2nd edn. Longmans Cheshire, Melbourne
Bayley IAE, Lake PS, Swain R, Tyler PA (1972) Lake Pedder: its importance to biological science. In: Pedder papers: anatomy of a decision, Australian Conservation Foundation, Melbourne, pp 41–49
Becker A, Robson BJ (2009) Riverine macroinvertebrate assemblages up to 8 years after riparian restoration in a semi-rural catchment in Victoria, Australia. Mar Freshw Res 60:1309–1316
Bhowmik AK, Schafer RB (2015) Large scale relationships between aquatic insect traits and climate. PLoS ONE 10(6):e0130025. https://doi.org/10.1371/journal.pone.0130025
Bilton DT, Ribera I, Short AEZ (2019) Water beetles as models in ecology and evolution. Ann Rev Entomol 64:359–377
Black TV, Robertson BA (2020) How to disguise evolutionary traps created by solar panels. J Insect Conserv 24:241–247
Blakely TJ, Harding JS, Mcintosh A, Winterbourn MJ (2006) Barriers to the recovery of aquatic insect communities in urban streams. Freshw Biol 51:1634–1645
Blyth JD, Doeg TJ, St Clair RM (1984) Response of the macroinvertebrate fauna of the Mitta Mitta River, Victoria, to the construction and operation of the Dartmouth Dam. 1. Construction and initial filling period. Occ Pap Mus Vic 1:83–100
Brasher AMD (2003) Impacts of human disturbances on biotic communities in Hawaiian streams. Bioscience 53:1052–1060
Brim-Box J, McBurnie G, Strehlow K, Guest T, Campbell M (and 10 other authors) (2016) The impact of feral camels (*Camelus dromedarius*) on remote waterholes in central Australia. Rangel J 38:191–200
Brittain JE, Campbell IC (1991) The effect of temperature on egg development in the Australian mayfly genus *Coloburiscoides* (Ephemeroptera: Coloburiscidae) and its relationship to distribution and life history. J Biogeogr 18:231–235

Brittain JE, Saltveit SJ (1989) A review of the effect of river regulation on mayflies (Ephemeroptera). Reg Rivers Res Manage 3:191–204
Brooks AJ, Wolfenden B, Downes BJ, Lancaster J (2018) Barriers to dispersal; the effect of a weir on stream insect drift. River Res Appl 34:1244–1253
Burgmer T, Hillebrand H, Pfenniger M (2007) Effects of climate-driven temperature changes on the diversity of freshwater macroinvertebrates. Oecologia 151:93–103
Burrows DW (2004) Translocated fishes in streams of the Wet Tropics Region, North Queensland: distribution and potential impact. Cooperative Research Centre for Tropical Rainforest Ecology and Management, Rainforest CRC, Cairns
Bush A, Theischinger G, Nipperess D, Turak E, Hughes L (2013) Dragonflies: climate canaries for river management. Divers Distr 19:86–97
Bush A, Hermoso V, Linke S, Nipperess D, Turak E, Hughes L (2014) Freshwater conservation planning under climate change: demonstrating proactive approaches for Australian Odonata. J Appl Ecol 51:1273–1281
Caitling PC (2005) A potential for the use of dragonfly (Odonata) diversity as bioindicator of the efficiency of sewage lagoons. Canad Field Nat 119. https://doi.org/10.22621/cfn.v119i2.111
Campbell IC, Doeg TJ (1989) Impact of timber harvesting and production on streams: a review. Aust J Mar Freshw Res 40:519–539
Canning AD, Death RG, Gardner EM (2019) Forest canopy affects stream macroinvertebrate assemblage structure but not trophic stability. Freshw Sci 38:40–52
Carver S, Storey A, Spafford H, Lynas J, Chandler l, Weinstein P, (2009) Salinity as a driver of aquatic invertebrate colonisation behaviour and distribution in the wheatbelt of Western Australia. Hydrobiologia 617:75–90
Chessman BC (1995) Rapid assessment of rivers using macroinvertebrates: a procedure based on habitat-specific sampling, family-level identification and a biotic index. Aust J Ecol 20:122–129
Chessman BC (2009) Climatic changes and 13-year trends in stream macroinvertebrate assemblages in New South Wales, Australia. Glob Change Biol 15:2791–2802
Chessman BC (2012) Biological traits predict shifts in geographical ranges of freshwater invertebrates during climate warming and drying. J Biogeogr 39:957–969
Chessman BC, Williams SA (1999) Biodiversity and conservation of river macroinvertebrates on an expanding urban fringe: western Sydney, New South Wales, Australia. Pacif Conserv Biol 5:36–55
Chessman BC, Robinson DP, Hortleck G (1987) Changes in the riffle macroinvertebrate fauna of the Tanjil River, southeastern Australia, during the construction of Blue Rock Dam. River Res Appl 1:317–329
Clarke A, Lake PS, O'Dowd DJ (2004) Ecological impacts on aquatic macroinvertebrates following upland stream invasion by a ponded pasture grass (*Glyceria maxima*) in southern Australia. Mar Freshw Res 55:709–713
Collins SD, McIntyre NE (2017) Extreme loss of diversity of riverine dragonflies in the northeastern U.S. is predicted in the face of climate change. Bull Amer Odonat 12:7–19
Conti L, Schmidt-Kloiber A, Grenouillet G, Graf W (2014) A trait-based approach to assess the vulnerability of European aquatic insects to climate change. Hydrobiologia 721:297–315
Cooper B, Crase L, Maybery D (2017) Incorporating amenity and ecological values of urban water into planning frameworks: evidence from Melbourne, Australia. Aust J Environ Manage 24:64–80
Cottingham P, Walsh C, Rooney G, Fletcher T (eds) (2004) Urbanization impacts on stream ecology—from syndrome to cure? (Outcomes of workshops held at the symposium on Urbanization and Stream Ecology, Melbourne). CRC for Freshwater Ecology, Canberra
Craig LS, Olden JD, Arthington AH, Entrekin S, Hawkins CP (and nine other authors) (2017) Meeting the challenge of interacting threats in freshwater ecosystems: a call to scientists and managers. Elem Sci Anth 5:72. https://doi.org/10.1525/elementa.256
Crowther D, Papas P (2005) Determining the impact of fire on invertebrate communities in Alpine streams in Victoria. Report to Department of Sustainability and Environment, Victoria, Melbourne

Cunha EJ, Juen L (2017) Impacts of oil palm plantations on changes in environmental heterogeneity and Heteroptera (Gerromorpha and Nepomorpha) diversity. J Insect Conserv 21:111–119

Danger A, Walsh CJ (2008) Management options for conserving and restoring fauna and other ecological values of urban streams in the Melbourne Water region. Report to Melbourne Water, Victoria.

Davies PE, Nelson M (1994) Relationships between riparian buffer widths and the effects of logging on stream habitat, invertebrate community composition and fish abundance. Aust J Mar Freshw Res 45:1289–1305

Davies PE, Cook LSJ, Mallick SA, Munks SA (2016) Relating upstream forest management to stream ecosystem condition in middle catchment reaches in Tasmania. For Ecol Manage 362:142–155

Davies-Colley RJ, Hickey CW, Quinn JM, Ryan PA (1992) Effects of clay discharges on streams. 1. Optical properties and epilithon. Hydrobiologia 248:215–234

de Moor FC (1992) Factors influencing the establishment of aquatic insect invaders. Tran R Soc S Africa 48:141–158

de Szalay F, Resh VH (2000) Factors influencing macroinvertebrate colonization of seasonal wetlands: responses to emergent plant cover. Freshw Biol 45:295–308

de Figueroa JMT, Lopez-Rodriguez MJ, Lorenz A, Graf W, Schmidt-Kloiber A, Hering D (2010) Vulnerable taxa of European Plecoptera (Insecta) in the context of climate change. Biodiv Conserv 19:1269–1277

Dingemanse NJ, Kalkman VJ (2008) Changing temperature regimes have advanced the phenology of Odonata in the Netherlands. Ecol Entomol 33:394–402

Doeg TJ (1984) Responses of the macroinvertebrate fauna of the Mitta Mitta River, Victoria, to the construction and operation of Dartmouth Dam. II Irrigation Release. Occ Pap Mus Vic 1:101–117

Doeg T, Heron S (2003a) Alteration to the natural flow regimes of rivers and streams. Action Statement No 177. Department of Sustainability and Environment, Melbourne

Doeg T, Heron S (2003b) Alteration to the natural temperature regimes of rivers and streams. Action Statement No 178. Department of Sustainability and Environment, Melbourne

Doeg TJ, Davey GW, Blyth JD (1987) Response of the aquatic macroinvertebrate communities to dam construction on the Thomson river, southeastern Australia. River Reg Appl 1:195–209

Domisch S, Jahnig SC, Haase P (2011) Climate-change winners and losers: stream macroinvertebrates of a submontane region in Central Europe. Freshw Biol 56:2009–2020

Douglas MM, O'Connor RA (2003) Effects of the exotic macrophyte, para grass (*Urochloa mutica*), on benthic and epiphytic macroinvertebrates of a tropical floodplain. Freshw Biol 48:962–971

Downes BJ, Lake PS, Glaister A, Bond NR (2006) Effects of sand sedimentation on the macroinvertebrate fauna of lowland streams: are the effects consistent? Freshw Biol 51:144–160

Dunbar MJ, Pedersen ML, Cadman D, Extence C, Waddingham J, Chadd R, Larsen SE (2010) River discharge and local-scale physical habitat influence macroinvertebrate LIFE scores. Freshw Biol 55:226–242

Durance I, Ormerod SJ (2007) Climate change effects on upland stream macroinvertebrates over a 25-year period. Glob Change Biol 13:942–957

Edwards ED, Huryn AD (1996) Effects of riparian land use on contributions of terrestrial invertebrates to streams. Hydrobiologia 337:151–159

Elliott JM, Bagenal TB (1972) The effects of electrofishing on the invertebrates of a Lake District stream. Oecologia 9:1–11

Ellis LE, Jones NE (2013) Longitudinal trends in regulated rivers: a review and synthesis within the context of the serial discontinuity concept. Environ Rev 21:136–148

Ellis LE, Jones NE (2016) A test of the serial discontinuity concept; longitudinal trends of benthic invertebrates in regulated and natural rivers of northern Canada. Riv Res Appl 32:462–472

Englund RA (1999) The impacts of introduced poecilid fish and Odonata on the endemic *Megalagrion* (Odonata) damselflies of Oahu Island, Hawaii. J Insect Conserv 3:225–243

Englund RA, Polhemus DA (2001) Evaluating the effects of introduced rainbow trout (*Oncorhynchus mykiss*) on native stream insects on Kauai Island, Hawaii. J Insect Conserv 5:265–281
Extence CA, Balbi DM, Chadd RP (1999) River flow indexing using British benthic macroinvertebrates: a framework for setting hydroecological objectives. Reg Riv Res Manage 15:543–574
Extence CA, Chadd RP, England J, Dunbar MJ, Wood PJ, Taylor ED (2013) The assessment of fine sediment accumulation in rivers using macroinvertebrate community response. Riv Res Appl 29:17–55
Fenoglio S, Bonada N, Guareschi S, Lopez-Rodriguez MJ, Millan A, de Figueroa JMT (2016) Freshwater ecosystems and aquatic insects: a paradox in biological invasions. Biol Lett 12:20151075
Flenner I, Sahlen G (2008) Dragonfly community re-organisation in boreal forest lakes: rapid species turnover driven by climate change? Insect Conserv Divers 1:169–179
Fletcher AR (1986) Effects of introduced fish in Australia. In: De Dekker P, Williams WD (eds) Limnology in Australia. CSIRO, Melbourne/Dr W Junk, Dordrecht, pp 231–238
Fletcher TD, Vietz G, Walsh CJ (2014) Protection of stream ecosystems from urban stormwater runoff: the multiple benefits of an ecohydrological approach. Prog Physic Geogr 38:543–555
Fowles CR (1975) Effects of electric fishing on the invertebrate fauna of a New Zealand stream. N Z J Mar Freshw Res 9:35–43
French SK, McCauley SJ (2018) Canopy cover affects habitat selection by adult dragonflies. Hydrobiologia 818:129–143
Gal B, Weiperth A, Farkas J, Schmera D (2019a) The effects of road crossings on stream macroinvertebrate diversity. Biodiv Conserv 29:729–745
Gal B, Szivak I, Heino J, Schmera D (2019b) The effect of urbanization on freshwater macroinvertebrates—knowledge gaps and future research directions. Ecol Indic 104:357–364
Glazaczow A, Orwin D, Bogdziewicz M (2016) Increased temperature delays the late-season phenology of multivoltine insect. Scient Reports 6:38022. https://doi.org/10.1038/srep38022
Gledhill DG, James P, Davies DH (2008) Pond density as a determinant of aquatic species richness in an urban landscape. Landsc Ecol 23:1219–1230
Goertzen D, Suhling F (2013) Promoting dragonfly diversity in cities: major determinants and implications for urban pond design. J Insect Conserv 17:399–409
Goertzen D, Suhling F (2015) Central European cities maintain substantial dragonfly species richness—a chance for biodiversity conservation? Insect Conserv Divers 8:238–246
Goertzen D, Suhling F (2018) Urbanization versus other land uses: diverging effects on dragonfly communities in Germany. Divers Distr 25:38–47
Griffin L (2014) Laboratory evaluation of predation on mosquito larvae by Australian mangrove fish. J Vector Ecol 39:197–203
Groffman PM, Bain DJ, Band LE, Belt KT, Brush GS (and four other authors) (2003) Down by the riverside: urban riparian ecology. Front Ecol Environ 1:315–321
Grove J, Turnbull J, Lawrence LP, Rutherfurd I, Silvester E, Colombi F (2019) Macklin M (2019) Mining to mud: a multidisciplinary approach to understanding Victoria's riverine landscape as a product of historical gold mining. Preview 200:44–56
Growns IO, Davis JA (1991) Comparison of the macroinvertebrate communities in streams in logged and undisturbed catchments 8 years after harvesting. Aust J Mar Freshw Res 42:689–706
Growns IO, Growns JE (2001) Ecological effects of flow regulation on macroinvertebrate and periphytic diatom assemblages in the Hawkesbury-Nepean River, Australia. Reg Riv Res Manage 17:275–293
Hale R, Coleman R, Pettigrove V, Swearer SE (2015) Review: identifying, preventing and mitigating ecological traps to improve the management of urban aquatic ecosystems. J App Ecol 52:928–939
Hall DN (1988) Effects of eductor dredging of gold tailings on aquatic environments in Victoria. Proc R Soc Vict 100:53–59

Hall AM, McCauley SJ, Fortin MJ (2015) Recreational boating, landscape configuration, and local habitat structure as drivers of odonate community composition in an island setting. Insect Conserv Divers 8:31–42

Halse SA, Ruprecht JK, Pinder AM (2003) Salinisation and prospects for biodiversity in rivers and wetlands of south-west Western Australia. Aust J Bot 51:673–688

Hanson MA, Buelt CB, Zimmer KD, Herwig BR, Bowe S, Maurer K (2015) Co-correspondence among aquatic invertebrates, fish and submerged aquatic plants in shallow lakes. Freshw Sci 34:953–964

Harabis F, Dolny A (2014) Necessity for the conservation of drainage systems as last refugia for threatened damselfly species, *Coenagrion ornatum*. Insect Conserv Divers 8:143–151

Harrison SSC, Harris IT (2002) The effects of bankside management on chalk stream invertebrate communities. Freshw Biol 47:2233–2245

Harrison ET, Norris RH, Wilkinson SN (2007) The impact of fine sediment accumulation on benthic macroinvertebrates: implications for river management. In: Wilson AL, Dehaan RL, Watts RJ, Page KJ, Bowmer KH, Curtis A (eds) Australian rivers: making a difference. Proceedings of 5th Australian stream management conference, Thurgoona, New South Wales

Hassall C, Thompson DJ (2008) The effects of environmental warming on Odonata: a review. Int J Odonatol 11:131–153

Hassall C, Thompson DJ, French GC, Harvey IF (2007) Historical changes in the phenology of British Odonata are related to climate. Glob Change Biol 13:933–941

Hawking JH, Theischinger G (2004) Critical species of Odonata in Australia. Int J Odonat 7:113–132

Healthy Waterways Strategy (Draft) (2013) Melbourne Water, Melbourne

Heller NE, Zavaleta ES (2009) Biodiversity management in the face of climate change: a review of 22 years of recommendations. Biol Conserv 142:14–32

Henn M, Nichols H, Zhang Y, Bonner TH (2014) Effect of artificial light on the drift of aquatic insects in urban central Texas streams. J Freshw Ecol 29:307–318

Herbst DB, Roberts SW, Medhurst RB (2013) Defining salinity limits on the survival and growth of benthic insects for the conservation management of saline Walker Lake, Nevada, USA. J Insect Conserv 17:877–883

Hering D, Schmidt-Kloiber A, Murphy J, Lucke S, Zamora-Munoz C (and three other authors) (2009) Potential impact of climate change on aquatic insects: a sensitivity analysis for European caddisflies (Trichoptera) based on distribution patterns and ecological preferences. Aquat Sci 71:3–14

Heron S, Doeg T (2003a) Degradation of native riparian vegetation along Victorian rivers and streams. Action Statement no 183. Department of Sustainability and Environment, Victoria, Melbourne

Heron S, Doeg T (2003b) Removal of woody debris from Victorian rivers and streams. Action Statement no 129, Flora and Fauna Guarantee Act 1988, Department of Sustainability and Environment, Victoria, Melbourne

Herzon I, Helenius J (2008) Agricultural drainage ditches, their biological importance and functioning. Biol Conserv 141:1171–1183

Hill MJ, Biggs J, Thornhill I, Briers RA, Gledhill DG (and three other authors) (2017) Urban ponds as an aquatic biodiversity resource in modified landscapes. Glob Change Biol 23:986–999

Hilzenhoff WL (1977) Use of arthropods to evaluate water quality of streams. Tech Bull Wisconsin Dept Nat Resources 100

Hilzenhoff WL (1987) An improved biotic index of organic stream pollution. Great Lakes Entomol 20:31–39

Hogg ID, Norris RH (1991) Effects of runoff from land clearing and urban development on the distribution and abundance of macroinvertebrates in pool areas of a river. Aust J Mar Freshw Res 42:507–518

Hogg ID, Williams DD (1996) Response of stream invertebrates to a global warming thermal regime; an ecosystem-level manipulation. Ecology 77:395–407

Holmes PM, Esler KJ, Richardson DM, Witkowski ETF (2008) Guidelines for improving management of riparian zones invaded by alien plants in South Africa. S Afr J Bot 74:538–552
Horvarth G, Malik P, Kriska G, Wildermuth H (2007) Ecological traps for dragonflies in a cemetery: the attraction of *Sympetrum* species (Odonata: Libellulidae) by horizontally polarizing black gravestones. Freshw Biol 52:1700–1709
Horvarth G, Blaho M, Egri A, Kriska G, Seres I, Robertson B (2010) Reducing the maladaptive attractiveness of solar panels to polarotactic insects. Conserv Biol 24:1644–1653
Houston WA, Duivenvoorden LJ (2002) Replacement of littoral native vegetation with the ponded pasture grass *Hymenachne amplexicaulis*: effects on plant, macroinvertebrate and fish biodiversity of backwaters in the Fitzroy River, Central Queensland, Australia. Mar Freshw Res 53:1235–1244
Howarth FG, Polhemus DA (1991) A review of the Hawaiian stream insect fauna. In: Proceedings of 1990 Symposium on freshwater stream biology and management, State of Hawaii, Honolulu, pp 40–51
Hurst TP, Brown MD, Kay BH (2004) Laboratory evaluation of the predation efficacy of native Australian fish on Culex annulicornis (Diptera: Culicidae). J Amer Mosq Control Assoc 20:286–291
Jach MA, Balke M (2008) Global diversity of water beetles (Coleoptera) in freshwater. Hydrobiologia 595:419–442
Jackson JBC, Kirby MX, Berger WH, Bjorndal KA, Botsford LW (and 14 other authors) (2001) Historical overfishing and the recent collapse of coastal ecosystems. Science 293:629–637
Jansen A, Robertson AI (2001) Relationships between livestock management and the ecological condition of riparian habitats along an Australian floodplain river. J Appl Ecol 38:63–75
Jeanmougin M, Leprieur F, Lors G, Clergeau P (2014) Fine-scale urbanization affects odonatan species diversity in ponds of a megacity (Paris, France). Acta Oecol 59:26–34
Johnson D (1972) Lake Pedder: why a national park must be saved, Camberwell, Victoria
Jones RC, Clark CC (1987) Impact of watershed urbanization on stream insect communities. Water Res Bull 23:1047–1055
Jongema Y (2014) List of edible insects of the world. University of Wageningen, The Netherlands
Jonsson M, Deleu P, Malmqvist B (2013) Persisting effects of river regulation on emergent aquatic insects and terrestrial invertebrates in upland forests. Riv Res Appl 29:537–547
Kay WR, Halse SA, Scanlon MD, Smith MJ (2001) Distribution and environmental tolerances of aquatic macroinvertebrate families in the agricultural zone of southwestern Australia. J N Am Benthol Soc 20:182–199
Kefford BJ (2018) Why are mayflies (Ephemerooptera) lost following small increases in salinity? Three conceptual osmophysiological hypotheses. Phil Trans R Soc B 374:20180021. https://doi.org/10.1098/rstb.2018.0021
Kefford BJ, Papas PJ, Nugegoda D (2003) Relative salinity tolerance of macroinvertebrates from the Barwon River, Victoria, Australia. Mar Freshw Res 54:755–765
Kefford BJ, Dalton A, Palmer CG, Nugegoda D (2004) The salinity tolerance of eggs and hatchlings of selected aquatic macroinvertebrates in south-east Australia and South Africa. Hydrobiologia 517:179–192
Kefford BJ, Zalizniak L, Dunlop JE, Nugegoda D, Choy SC (2010) How are macroinvertebrates of slow flowing lotic systems directly affected by suspended and deposited sediments? Environ Poll 158:543–550
Kinvig RG, Samways MJ (2000) Conserving dragonflies (Odonata) along streams running through commercial forestry. Odonatologica 29:195–208
Knapp RA, Corn PS, Schindler DE (2001) The introduction of nonnative fish into wilderness lakes: good intentions, conflicting mandates, and unintended consequences. Ecosystems 4:275–278
Knispel S, Sartori M, Brittain JE (2006) Egg development in the mayflies of a Swiss glacial floodplain. J N Amer Benthol Soc 25:430–443
Kondriateff BC, Bishop RJ, Brasher AM (1997) The life cycle of an introduced caddisfly, *Cheumatopsyche pettiti* (Banks) (Trichoptera: Hydropsychidae) in Waikolu Stream, Hawaii. Hydrobiologia 350:81–85

Kruzic LM, Scarnecchia DL, Roper BB (2005) Effects of electroshocking on macroinvertebrate drift in three cold water streams. Hydrobiologia 539:57–67

Lake PS (2001) The fauna of Lake Pedder—changes after the flooding and thoughts on restoration. In: Sharples C (ed) Lake Pedder: values and restoration. Occ Pap no 27, Centre for Environmental Studies, University of Tasmania, pp 87–98

Lawrence C, Rutherford N, Hamilton R, Meredith D (2016) Experimental evidence indicates that native freshwater fish outperform introduced *Gambusia* in mosquito suppression when water temperature is below 25 °C. Hydrobiologia 766:357–364

Lind PR, Robson BJ, Mitchell BD, Matthews TG (2009) Can sand slugs in rivers deliver conservation benefits? The biodiversity value of tributary junction plug wetlands in the Glenelg River, Australia. Mar Freshw Res 60:426–434

Lubertazzi MAA, Ginsberg HS (2010) Emerging dragonfly diversity at small Rhode Island (U.S.A.) wetlands along an urbanization gradient. Urb Ecosyst 13:517–533

Lundquist MJ, Zhu W (2019) Aquatic insect diversity in streams across a rural-urban land-use discontinuum. Hydrobiologia 837:15–30

Maezono Y, Miyashita T (2004) Impact of exotic fish removal on native communities in farm ponds. Ecol Res 19:263–267

Magoba RN, Samways MJ (2010) Recovery of benthic macroinvertebrate and adult dragonfly assemblages in response to large scale removal of riparian invasive alien trees. J Insect Conserv 14:627–636

Malison RL, Baxter CV (2010) Effects of wildfire of varying severity on benthic stream insect assemblages and emergence. J N Am Benthol Soc 29:1324–1338

McBurnie G, Davis J, Thompson RM, Nano C, Brim-Box J (2015) The impacts of an invasive herbivore (*Camelus dromedarius*) on arid zone freshwater pools: an experimental investigation of the effects of dung on macroinvertebrate colonisation. J Arid Environ 113:69–76

McKee D, Atkinson D (2000) The influence of climate change scenarios on populations of the mayfly *Cloeon dipterum*. Hydrobiologia 441:55–62

Mellon CD, Wipfli MS, Li JL (2008) Effects of forest fire on headwater stream macroinvertebrate communities in eastern Washington, U.S.A. Freshw Biol 53:2331–1343

Metzeling L, Wells F, Newall P, Tiller D, Reed J (2004) Biological objectives for rivers and streams—ecosystem protection. Information Bulletin, Environment Protection Authority, Victoria, Melbourne

Meyer JL, Paul MJ, Taulbee WK (2005) Stream ecosystem function in urbanizing landscapes. J N Am Benthol Soc 24:602–612

Miguel TB, Oliveira-Junior JMB, Ligeiro R, Juen L (2017) Odonata (Insecta) as a tool for the biomonitoring of environmental quality. Ecol Indic 81:555–566

Mihuc TB, Minshall GW (1995) Trophic generalists vs. trophic specialists: implications for food web dynamics in post-fire streams. Ecology 76:2361–2372

Milner VS, Yarnell SM, Peek RA (2019) The ecological importance of unregulated tributaries to macroinvertebrate diversity and community composition in a regulated river. Hydrobiologia 829:291–305

Minshall GW (2003) Responses of stream benthic macroinvertebrates to fire. For Ecol Manage 178:155–161

Molnar A, Hegedus R, Kriska G, Horvath G (2011) Effect of cattail (*Typha* spp.) mowing on water beetle assemblages: changes of environmental factors and the aerial colonization of aquatic habitats. J Insect Conserv 15:389–399

Monteiro-Junior CS, Juen L, Hamada N (2014) Effects of urbanization on stream habitats and associated adult dragonfly and damselfly communities in central Brazilian Amazonas. Landsc Urb Plann 127:28–40

Muschal M (2006) Assessment of risk to aquatic biota from elevated salinity—a case study from the Hunter River, Australia. J Environ Manage 79:266–278

Naiman RJ, Decamps H (1997) The ecology of interfaces: riparian zones. Ann Rev Ecol Syst 28:621–658

New TR (2016) Alien species and insect conservation. Springer, Cham
Ngiam RWJ, Lim WL, Collins CM (2017) A balancing act in urban social-ecology: human appreciation, ponds and dragonflies. Urban Ecosyst 20:743–758
Nichols S, Norris R, Maher W, Thoms M (2006) Ecological effects of serial impoundment on the Cotter River, Australia. Hydrobiologia 572:255–273
Nichols J, Hubbart JA, Poulton BC (2016) Using macroinvertebrate assemblages and multiple stressors to infer urban stream system condition: a case study in the central US. Urban Ecosyst 19:679–704
Nilsson NA (1972) Effects of introductions of salmonids into barren lakes. J Fish Res Board Canada 29:693–697
O'Brien T, Blackburn W (2003) Prevention of passage by aquatic biota by instream structures. Action Statement no. 129, Flora and Fauna Guarantee Act 1988, Department of Sustainability and Environment, Victoria, Melbourne
Oertli B, Parris KM (2019) Review: toward management of urban ponds for biodiversity. Ecosphere 10(7):e02810. https://doi.org/10.1002/ecs2.2810
Ormerod SJ, Rundle SD, Lloyd EC, Douglas AA (1993) The influence of riparian management on the habitat structure and macroinvertebrate communities of upland streams draining plantation forests. J Appl Ecol 30:13–24
Pardo I, Campbell IC, Brittain JE (1998) Influence of dam operation on mayfly assemblage structure and life histories in two south-eastern Australian streams. Reg Riv Res Manage 14:285–295
Parkyn SM, Winterbourn MJ (1997) Leaf breakdown and colonisation by invertebrates in a headwater stream: comparisons of native and introduced tree species. N Z J Mar Freshw Res 31:301–312
Parr AJ (2010) Monitoring of Odonata in Britain and possible insights into climate change. BioRisk 5:127–139
Paul MJ, Meyer JL (2001) Streams in the urban landscape. Ann Rev Ecol Syst 32:333–365
Pereszenyi A, Horvath G, Kriska G (2017) Atypical feeding of woodpeckers, crows and redstarts on mass-swarming *Hydropsyche pellucidula* caddisflies attracted to glass panes. Urban Ecosyst 20:1203–1207
Perron MAC, Pick FR (2019) Stormwater ponds as habitat for Odonata in urban areas: the importance of obligate wetland plant species. Biodiv Conserv 29:913–931
Pinder AM, Halse SA, McRae JM, Shiel RJ (2005) Occurrence of aquatic invertebrates of the wheatbelt region of Western Australia in relation to salinity. Hydrobiologia 543:1–24
Polhemus DA (1993) Conservation of aquatic insects: worldwide crisis or localized threats? Amer Zool 33:588–598
Polhemus D, Asquith A (1996) Hawaiian damselflies: a field identification guide. Bishop Museum Press, Honolulu
Pyke GH (2008) Plague minnow or mosquito fish? A review of the biology and impacts of introduced *Gambusia* species. Annu Rev Ecol Evol Syst 39:171–191
Quinn JM, Davies-Colley RJ, Hickey CW, Vickers ML, Ryan PA (1992) Effects of clay discharges on streams 2. Benthic invertebrates. Hydrobiologia 248:235–247
Quinn JM, Boothroyd IKG, Smith BJ (2004) Riparian buffers mitigate effects of pine plantation logging on New Zealand streams 2. Invertebrate Communities. For Ecol Manage 191:129–146
Rahel FJ (2002) Homogenization of freshwater faunas. Ann Rev Ecol Syst 33:291–315
Ramos-Elorduy J, Moreno JMP, Camacho VHM (2009) Edible aquatic Coleoptera of the world, with an emphasis on Mexico. J Ethnob Ethnomed 5:11. https://doi.org/10.1186/1746-4269-5-11
Reid DJ, Quinn JM, Wright-Stow AE (2010) Responses of stream macroinvertebrate communities to progressive forest harvesting: influences of harvest intensity, stream size and riparian buffers. For Ecol Manage 260:1804–1816
Relyea CD, Minshall GW, Danehy RJ (2012) Development and validation of an aquatic fine sediment biotic index. Environ Manage 49:242–252
Remsburg AJ, Samways OAC, MJ, (2008) Shade alone reduces dragonfly (Odonata: Libellulidae) abundance. J Insect Behav 221:460–468

Riffell SK (1999) Road mortality of dragonflies (Odonata) in a Great Lakes coastal wetland. Great Lakes Entomol 32:63–73

Roback SS (1974) Insects. In: Hart CW, Fuller SL (eds) Pollution ecology of freshwater invertebrates. Academic Press, New York, pp 314–376

Rosset V, Oertli B (2011) Freshwater biodiversity under climate warming pressure: identifying the winners and losers in temperate standing waterbodies. Biol Conserv 144:2311–2319

Rowe RJ (1987) The dragonflies of New Zealand. Auckland University Press, Auckland

Roy AH, Freeman MC, Freeman BJ, Wenger SJ, Meyer JL, Ensign WE (2006) Importance of riparian forests in urban catchments contingent on sediment and hydrologic regimes. Environ Manage 37:523–539

Ryan T, Webb A, Lennie R, Lyon J (2001) Status of cold water releases from Victorian dams. Report for Department of Natural Resources and Environment, Melbourne

Samways MJ (2003) Threats to the tropical island dragonfly fauna (Odonata) of Mayotte, Comoro archipelago. Biodiv Conserv 12:1785–1782

Samways MJ, Sharratt NJ (2010) Recovery of endemic dragonflies after removal of invasive alien trees. Conserv Biol 24:267–277

Samways MJ, Steytler NS (1996) Dragonfly (Odonata) distribution patterns in urban and forest landscapes, and recommendations for riparian management. Biol Conserv 78:279–288

Scanlon BR, Jolly I, Sophocleus M, Zhang L (2007) Global impacts of conversions from natural to agricultural ecosystems on water resources; quantity versus quality. Water Resour Res 43:W03437

Schmidt EG (2008) *Sympetrum depressiusculum* (Selys), a southern continental dragonfly depending on artificial habitats in Atlantic northwestern Germany (State of Northrhine-Westphalia) (Anisoptera: Libellulidae). Notul Odonatol 7:1–12

Schulze DJ, Walker KF (1997) Riparian eucalypts and willows and their significance for aquatic invertebrates in the River Murray, South Australia. Reg Riv Res Manage 13:557–577

Sheldon F, Boulton AJ, Puckridge JT (2002) Conservation value of variable connectivity: aquatic invertebrate assemblages of channel and floodplain habitats of a central Australian arid-zone river, Cooper Creek. Biol Conserv 103:13–31

Sigutova H, Sigut M, Dolny A (2015) Intensive fish ponds as ecological traps for dragonflies: an imminent threat to the endangered species *Sympetrum depressiusculum* (Odonata: Libellulidae). J Insect Conserv 19:961–974

Smith RF, Alexander LC, Lamp WO (2009) Dispersal by terrestrial stages of stream insects in urban watersheds: a synthesis of current knowledge. J N Am Benthol Soc 28:1022–1037

Solimini AG, Tarallo GA, Carchini G (1997) Life history and species composition of the damselfly assemblage along the urban tract of a river in central Italy. Hydrobiologia 356:21–32

Soluk DA, Zercher DS, Worthington AM (2011) Influences of roadways on patterns of mortality and flight behavior of adult dragonflies near wetland areas. Biol Conserv 144:1638–1643

Sondergaard M, Jeppersen E, Lauridsen TL (2007) Lake restoration: successes, failures and long-term effects. J App Ecol 44:1095–1105

Spencer CN, Gabel KO, Hauer FR (2003) Wildfire effects on stream food webs and nutrient dynamics in Glacier National park, USA. For Ecol Manage 178:141–153

Stanley EH, Luebke MA, Doyle MW, Marshall DW (2002) Short-term changes in channel form and macroinvertebrate communities following low-head dam removal. J N Am Benthol Soc 21:172–187

Stewart DAB, Samways MJ (1998) Conserving dragonfly (Odonata) assemblages relative to river dynamics in an African savanna game reserve. Conserv Biol 12:683–692

Stewart BA, Close PG, Cook PA Davies PM (2013) Upper thermal tolerances of key taxonomic groups of stream invertebrates. Hydrobiologia 718:131–140

Strayer DL (2010) Alien species in fresh waters: ecological effects, interactions with other stressors, and prospects for the future. Freshw Biol 5(Supplement 1):152–174

Strand M, Merritt RW (1999) Impacts of livestock grazing activities on stream insect communities and the riverine environment. Amer Entomol 45:13–29

Sudduth EB, Hassett BA, Cada P, Bernhardt ES (2011) Testing the field of dreams hypothesis: functional responses to urbanization and restoration in stream ecosystems. Ecol Appl 21:1972–1988

Suter PJ, Bishop JE (1990) Stoneflies (Plecoptera) of South Australia. In: Campbell IC (ed) Mayflies and stoneflies. Kluwer, Dordrecht, pp 189–207

Taylor BW, McIntosh AR, Peckarsky BL (2001) Sampling stream invertebrates using electroshocking techniques: implications for basic and applied research. Can J Fish Aquat Sci 58:437–445

Thoms MC, Sheldon F (2000) Lowland rivers: an Australian introduction. Reg Riv Res Manage 16:375–383

Tillyard RJ (1926) The insects of Australia and New Zealand. Angus and Robertson, Sydney

Timms BV, Morton J, Green K (2015) Temporal changes in the macroinvertebrate fauna of two glacial lakes, Cootapatamba and Albina, Snowy Mountains, New South Wales. Proc Linn Soc NSW 135:45–54

Tippler C, Wright IA, Davies PJ, Evans CR (2018) Are Odonata nymphs adversely affected by impaired water quality in urban streams? Austral Ecol 43:890–902

Urban MC, Skelly DK, Burchsted D, Price W, Lowry S (2006) Stream communities across a rural-urban landscape gradient. Divers Dist 12:337–350

Van Dyck H, Bonte D, Puls R, Gotthard K, Maes D (2015) The lost generation hypothesis: could climate change drive ectotherms into a developmental trap? Oikos 14:54–61

van Tol J, Verdonk MJ (1988) The protection of dragonflies (Odonata) and their biotopes. Nature and Environment Series, Council of Europe, Strasbourg

Vadher AN, Leigh C, Millett J, Stubbington R, Wood PJ (2017) Vertical movements through subsurface stream sediments by benthic macroinvertebrates during experimental drying are influenced by sediment characteristics and species traits. Freshw Biol 62:1730–1740

Velasco J, Suarez ML, Vidal-Abarca MR (1998) Factores que determinan la colonización de insectos acuaticos en pequeños estanques. Oecol Aquat 11:87–99

Vermonden K, Leuven RSEW, van der Velde G, van Katwijk MM, Roelofs JGM, Hendriks AJ (2009) Urban drainage systems: an undervalued habitat for aquatic macroinvertebrates. Biol Conserv 142:1105–1115

Viera NKM, Clements WH, Guevara LS, Jacobs BF (2004) Resistance and resilience of stream insect communities to repeated hydrological disturbances after wildfire. Freshw Biol 49:1243–1259

Villalobos-Jimenez G, Dunn AM, Hassall C (2016) Dragonflies and damselflies (Odonata) in urban ecosystems: a review. Eur J Entomol 113:217–232

Vorosmarty CJ, McIntyre PB, Gessner MO, Dudgeon D, Prusevich A (and six other authors) (2010) Global threats to human water security and river biodiversity. Nature 467:555–561

Walker KF, Sheldon F, Puckridge JT (1995) A perspective on dryland river ecosystems. Reg Riv Res Manag 11:85–104

Walsh CJ (2004) Protection of in-stream biota from urban impacts: minimise catchment imperviousness or improve drainage design? Mar Freshw Res 55:317–326

Walsh CJ, Roy AH, Feminella JW, Cottingham PD, Groffman PM, Morgan RP (2005) The urban stream syndrome: current knowledge and the search for a cure. J N Am Benthol Soc 24:706–723

Wang J, Ding l, Tao J, Ding C, He D (2019) The effects of dams on macroinvertebrates: global trends and insights. River Res Appl 35:702–713

Ward JV (1998) Riverine landscapes: biodiversity patterns, disturbance regimes, and aquatic conservation. Biol Conserv 83:269–278

Ward JV, Stanford JA (1982) Thermal responses in the evolutionary ecology of aquatic insects. Annu Rev Entomol 27:97–117

Ward JV, Stanford JA (1983) The serial discontinuity concept of lotic ecosystems. In: Fontaine TD, Bartell SM (eds) Dynamics of lotic ecosystems. Ann Arbor Scientific Publishers, Ann Arbor, MI, pp 29–42

Ward W, Stanford JA (1995) The serial discontinuity concept: extending the model to floodplain rivers. Reg Riv Res Manag 10:159–168

Watson JAL, Arthington AH, Conrick DL (1982) Effect of sewage effluent on dragonflies (Odonata) of Bulimba Creek, Brisbane. Aust J Mar Freshw Res 33:517–528

Weatherley NS, Lloyd EC, Rundle SD, Ormerod SJ (1993) Management of conifer plantations for the conservation of stream macroinvertebrates. Biol Conserv 63:171–176

Weber A, Garcia X-F, Wolter C (2017) Habitat rehabilitation in urban waterways: the ecological potential of bank protection structures for benthic invertebrates. Urban Ecosyst 20:759–773

Wenger SJ, Roy AH, Jackson CR, Bernhardt ES, Carter TL (and 14 other authors) (2009) Twenty-six key research questions in urban stream ecology: an assessment of the state of the science. J N Am Benthol Soc 28:1080–1098

Wheeler SAP, Angermeier PL, Rosenberger AE (2005) Impacts of new highways and subsequent landscape urbanization on stream habitat and biota. Rev Fish Sci 13:141–164

Whiteman NK, Sites RW (2008) Aquatic insects as umbrella species for ecosystem protection in Death Valley National Park. J Insect Conserv 12:499–509

Williams WD (1987) Salinization of rivers and streams: an important environmental hazard. Ambio 16:180–185

Williams DD, Hynes HBN (1976) The ecology of temporary streams. I. The faunas of two Canadian streams. Int Rev Ges Hydrobiol 61:761–787

Williams DD, Williams SS (2017) Aquatic insects and their potential to contribute to the diet of the globally expanding human population. Insects 8:72. https://doi.org/10.3390/insects8030072

Willigalla C, Fartmann T (2012) Patterns in the diversity of dragonflies (Odonata) in cities across Central Europe. Eur J Entomol 109:235–245

Wood PJ, Armitage PD (1997) Biological effects of fine sediment in the lotic environment. Environ Manag 21:203–217

Wood PJ, Greenwood MT, Armitage PD (2005) The response of four lotic macroinvertebrate taxa to burial by sediments. Arch F Hydrobiologie 163:145–162

Woodward G, Perkins DM, Brown LE (2010) Climate changes and freshwater ecosystems: impacts across multiple levels of organisation. Phil Trans R Soc B 365:2093–2106

Woodward G, Bonada N, Brown LE, Death RG, Durance I (and seven other authors) (2016) The effects of climate fluctuations and extreme events on running water ecosystems. Phil Trans R Soc B 371:20150274. https://doi.org/10.1098/rstb.2015.0274

Wright IA, Davies P, Wilks D, Findlay S, Taylor MP (2007) Aquatic macroinvertebrates in urban waterways: comparing ecosystem health in natural reference and urban streams. In: Wilson AL, Dehaan RL, Watts RJ, Page KJ, Bowmer KH, Curtis A (eds) Proceedings of 5th Australian stream management conference. Australian Rivers: making a difference. Charles Sturt University, Thurgoona, pp 467–472

Yeates LV, Barmuta LA (1999) The effects of willow and eucalyptus leaves on feeding preference and growth of some Australian aquatic macroinvertebrates. Aust J Ecol 24:593–598

Zhou Q (2014) A review of sustainable urban drainage systems considering the climate change and urbanization impacts. Water 6:976–992

Zukowski S, Gawne B (2006) Potential effects of willow (*Salix* spp.) removal on freshwater ecosystem dynamics. A literature review. Murray-Darling Freshwater Research Centre, Wodonga

# Chapter 7
# Macroinvertebrates of Inland Waters

## 7.1 Introduction

The term 'macroinvertebrates' has both taxonomic and physical connotations. Conventionally, many biologists use the term to solely include arthropods, although other taxa—notably molluscs and larger annelids—are also well-represented in freshwater environments. All these groups are diverse, and vulnerable to changes. In contrast 'microinvertebrates' encompass many small-sized non-insect arthropods (such as Cladocera, Copepoda and Ostracoda) as well as Oligochaeta, Nematoda, Tardigrada and others, most of them not represented in any formal conservation planning or recognised as needy (except, perhaps by a few specialist advocates). In some studies, microinvertebrates also include small insect larvae—so that size is then the major criterion for categorising them. Thus, Robertson et al. (2014) characterised microinvertebrates as those organisms passing through a 1 mm sieve but retained on a 63 μm sieve, irrespective of taxonomic category. In that study in southern England, microinvertebrates included representatives of Ephemeroptera, Hemiptera, Diptera and Coleoptera. More conventionally, and as followed here, all insects are deemed macroinvertebrates, as the groups that have attracted most conservation attention as both tools in wider assessments and direct targets for action. Especially in the northern hemisphere, unionid molluscs (broadly, freshwater mussels) include many heavily threatened taxa, and about 70% of North American freshwater mussels are of conservation concern. Many have become extinct. Several Australian species are also of concern. They, and some crustaceans include many species exploited by people and increasingly vulnerable to environmental changes. Macroinvertebrates are the predominant animals in many lotic systems in many parts of the world (Resh and Rosenberg 1989).

Aquatic macroinvertebrates include a wide variety of arthropod taxa, with a collectively wide range of habitats and ecological roles. This short chapter introduces the variety of aquatic insects, larger crustaceans and molluscs in Australia, to establish a perspective on their biological variety and the scale of conservation needs.


T. R. New, *Insect Conservation and Australia's Inland Waters*,
https://doi.org/10.1007/978-3-030-57008-8_7

## 7.2 The Variety of Aquatic Insects

Aquatic insects comprise two very broad functional categories, both of them taxonomically and ecologically diverse and together including around a dozen insect orders. The categories are (1) those that are entirely, or almost entirely, restricted to aquatic habitats, with all or most species passing their early stages in water and adults—although embarking on a more familiar terrestrial existence—remaining associated with waterbodies and depending on water for reproduction; and (2) those that have secondarily invaded aquatic systems as isolated lineages within orders that are primarily terrestrial. Although the concept of 'aquatic insects' is usually clear, Chadd and Extence (2004) pointed out that some subjectivity may occur in the absence of any rigorous definition of their scope. Thus, some insects are associated specifically with water-dependent habitats such as exposed riverine sediments (ERS, p. 31): some Coleoptera (notably some ground beetles [Carabidae] and rove beetles [Staphylinidae]), for example, are restricted to such areas, but are not truly aquatic as they do not live in the water bodies themselves. Morse (2009) noted that, because most aquatic insects live below the water surface, their diversity and abundance is not readily apparent except through deliberate investigation. He considered this obscurity a contributor to the generally rather poor knowledge of their diversity. Perhaps no more than about 20% of species in some orders has been diagnosed or described, and most of these are likely to be within the under-explored tropics and subtropics (Morse 2009). For example, a projected figure of 50,000 species of Trichoptera (p. 198) includes about 40,000 species in southern Asia, where the density of species is about twice that in the next most speciose region, the neotropics.

Including Collembola (springtails, not true insects) Williams and Feltmate (1992) included 13 orders as to some extent 'aquatic' (Table 7.1), and much background biology is provided in their text and in numerous entomology books. For Australian fauna, the synoptic background in the major text ('The Insects of Australia': Naumann 1991) can be augmented from more recent accounts; well-illustrated accounts by Hawking and Smith (1977) and Gooderham and Tsyrlin (2002) facilitate recognition of most major taxa and recognition of many genera and species is aided further through series of technical or field guides (notably a series published through the Murray-Darling Freshwater Research Centre, Albury, from 1994), but new taxa and distribution records for all taxa continue to accumulate.

Other than for Hemiptera, in which all active stages may occur together in or near water, insects treated here have either (1) aquatic larvae and non-aquatic adults, or (2) aquatic larvae and aquatic adults of very different form. The often contrasting biology and environmental needs of larval and adult stages can thereby span aquatic and terrestrial regimes, and may necessitate rather different management portfolios in planning conservation—as well as the water body involved, management for a given species or community is likely to embrace riparian vegetation (for many species that remain close to water) and in some cases wider landscape features (for more widely ranging or strongly dispersive taxa). The transitional zones between water bodies and terrestrial habitats constitute a series of gradient environments, in some cases with

**Table 7.1** The aquatic orders of insects, as listed by Williams and Feltmate (1992) and others

| Order | Common names |
|---|---|
| *Collembola (Springtails, not true insects and not included here) Orders with incomplete metamorphosis*: | |
| Ephemeroptera | Mayflies |
| Odonata | Dragonflies, damselflies |
| Plecoptera | Stoneflies |
| Orthoptera | Grasshoppers and their allies (not truly aquatic but some species riparian) |
| Hemiptera | 'Bugs', aquatic Heteroptera include water boatmen, water scorpions, water bugs, backswimmers, pond skaters and their relatives |
| *Orders with complete metamorphosis*: | |
| Neuroptera | Lacewings |
| Megaloptera | Alderflies, Dobsonflies, fishflies |
| Coleoptera | Beetles |
| Diptera | Flies, including midges, mosquitos, blackflies, craneflies and others |
| Lepidoptera | Butterflies and moths |
| Trichoptera | Caddisflies |
| Hymenoptera | Wasps, bees and ants: very few wasps aquatic as parasitoids of aquatic insect hosts |

the level of connectivity influenced by changing water levels depending on seasonal rainfall patterns or usage through discharge. These zones can therefore present a dynamic variety, of hydrology, sediment features, vegetation extent and composition, and others with which insects such as many Heteroptera can be associated. Thus, for the aquatic Heteroptera of backwater systems that differed in their connectivity (measured as days of connection) to the Danube River in Austria, more species preferred the water covered with branches and floating vegetation (*Lemna*) and the shoreline vegetated, whilst few of the 20 species preferred open habitats with little shoreline vegetation (Skern et al. 2010). Five categories of habitats were recognised (Table 7.2), and a considerable variety of occupation patterns by Corixidae (eight species) and Gerromorpha (12 species) was evident—eight species occurred in only one habitat category, and only one spanned all five. Those with restricted distributions were rare, but each habitat was 'preferred' by particular species. Paucity of shoreline vegetation and associated erosion caused by periodic flooding can restrict values of such areas for the bugs. From this study, associations of Heteroptera segregated along gradients of vegetation cover and the patterns helped to understand influences of variations in floodplain and nearby areas. Importance of vegetation was exemplified by (1) some taxa being associated strongly with dense aquatic macrophytes; (2) most species preferring dense shoreline vegetation, possibly related to shelter from predators and other disturbances; and (3) branches or *Lemna* ('duckweed', Araceae) on the water surface benefit most species, as a substrate and increasing access to plant materials for food.

**Table 7.2** Five categories of habitat connectivity in the River Danube, Austria, used to assess distribution of aquatic Hemiptera in relation to extent of connectivity (Skern et al. 2010)

| Habitat category | Characteristics |
|---|---|
| 5. | No connectivity with main channel, temporary pools, water level primarily dependent on ground water levels, terrestrial vegetation |
| 4. | Water bodies connected for less than 100 days/year, terrestrialisation processes |
| 3. | Water bodies connected for more than 100 days/year, terrestrialisation processes, coverage of open water areas by macrophytes exceeds 50% of open water area |
| 2. | Water bodies connected for more than 100 days/year, coverage of open water areas by macrophytes does not exceed 50% of open water area, shoreline vegetation exceeds 20% of shoreline of sampling locations |
| 1. | Hydrologically dynamic water bodies connected with the main channel, only few macrophyte communities in the open water, open banks or few shoreline plants in the littoral area |

The ecological categorisation of the terrestrial adults of aquatic insects by Harrison and Dobson (2008) helps to indicate and clarify the variety of those needs. They distinguished (1) those that are most typically rather inconspicuous, short-lived, small, cryptic, poor dispersers and either non-feeding or feeding on non-protein materials—all features that can render them collectively far more difficult to study than their immature stages, and exemplified by Ephemeroptera, Plecoptera, Trichoptera and Diptera; (2) those with far more conspicuous adults (as in Odonata, and some larger predatory Hemiptera and Coleoptera) and that can be studied more easily and which have contributed substantially to knowledge of the taxa: they are most typically large, often brightly coloured, strong dispersers, and many with long adult life; and (3) blood-feeding Diptera, distinguished because, whilst actively feeding as adults (as in the previous category), larvae more resemble those of the first grouping and often feed on detrital material. The three categories thus differ functionally by flight/dispersal capability and adult feeding behaviour—which Harrison and Dobson regarded as adaptations to the two major drivers of insect life histories, namely habitat stability and larval food availability.

The variety of insect life history patterns in aquatic environments reflects the spectrum of species represented at any given time and, hence, their amenity to detection in sporadic sampling events, as opposed to more inclusive regular sampling programmes. In France, five different phenological patterns were distinguished among the 21 insect species assessed by Cayrou and Cereghino (2005). These are: (1) semivoltine, development extending over two years, with two overlapping generations; (2) univoltine with slow development—for example, eggs hatching in spring, and larvae taking 9–12 months to reach adulthood; (3) univoltine with fast development—mostly with overwintering eggs and a period of 2-4 months from hatching to adulthood; (4) multivoltine, with two or three generations a year; and (5) bivoltine, with two short generations a year. Variations in these patterns can reflect local

**Table 7.3** Major aquatic insect lineages and their values for diversification research (abbreviated from Dijkstra et al. 2014)

| Taxon | Comment |
|---|---|
| Ephemeroptera | Reasonably well-studied, ecologically different from Odonata, and opportunities for comparative studies |
| Odonata | Best researched aquatic insect group, only order with global perspective of threat and conservation needs |
| Plecoptera | Very understudied, ecologically sensitive and many restricted in ecological range |
| Hemiptera | Two major predatory radiations, moderately well studied and potential for study in historical biogeography |
| Coleoptera | Prominently studied group (after Odonata) with small proportion of aquatic taxa; aquatic taxa taxonomically varied; many riparian taxa |
| Diptera | Only major terrestrial order with large freshwater proportion; large body of work on taxa of medical importance, and order with great ecological diversity; many species poorly documented; only non-medical group well studied is Chironomidae |
| Trichoptera | Largest purely aquatic order, great diversity due to habitat and feeding specialisation; under-studied |

conditions and water temperatures (p. 79), the latter influencing growth rates, adult size, and fecundity. More generally, some variations in temporal patterns of development appear to be widespread, and sampling undertaken to inform management must be sufficient to detect and evaluate these. Pond management by dredging, for example, could easily reduce or eliminate populations if undertaken during their major hatching or growth periods (Cayrou and Cereghino 2005).

The global diversity of the major aquatic insect orders has been projected to number more than 200,000 species, of which about 100,000 described species were totalled by Dijkstra et al. (2014). The 12 major aquatic orders considered in that account reflected more than 50 separate invasions of aquatic environments and vary greatly in richness (Table 7.3). Five orders (Ephemeroptera, Odonata, Plecoptera, Megaloptera, Trichoptera) are wholly, or almost wholly, aquatic. Others represent lineages or more sporadic invasions from more predominantly terrestrial orders, and except for Megaloptera (with an apparently relict distribution) all orders are cosmopolitan. Of these, Diptera are by far the largest group, with each of their two largest aquatic lineages (Culicomorpha, Tipulimorpha) outnumbering the richest truly aquatic order, Trichoptera. However, Dijkstra et al. noted that although Trichoptera were then evaluated as containing a little over 14 000 species, they could in due course be recognised as the largest aquatic insect radiation, with up to 50,000 species. Both these most diverse orders exhibit a considerable variety of larval biology and feeding habits. Four orders are omitted from the table as having only very low proportions of their total species aquatic, and all are predominantly terrestrial. These, Orthoptera, Neuroptera, Lepidoptera, and Hymenoptera, generally comprise only minute proportions of aquatic insect richness and, again in general, do not participate in studies of aquatic insect conservation.

Six functional feeding groups are commonly recognised among aquatic insect larvae, following Merritt and Cummins (1996) and discussed further by Strand and Merritt (1999), and relative representation of these can imply condition or changes in the holding water body. The groups are: (1) 'shredders', which feed primarily by breaking down larger pieces (>1 mm diameter) of vascular plant material, together with associated microflora and fauna; (2) 'collector-gatherers', generally feeding on fine (<1 mm diameter) particulate organic matter; (3) 'collector-filterers', largely filtering out fine particles from the water column, and many having specialised structures to enable this; (4) 'scavengers', feeding by grazing or scraping materials such as algae (as 'periphyton') from a variety of substrates such as plants or stones; (5) 'plant piercers' that pierce cells or tissues and ingest fluids; and (6) 'predators, feeding on animal tissues. Although 'shredders' are essentially detritivores, they occupy a relatively unusual position in the process because, rather than depending on already fragmented debris, they deal primarily with whole leaves (Boyero et al. 2012).

Functional feeding groups among aquatic invertebrates have thus facilitated comparisons of community structure and fate of organic matter across stream ecosystems (Hearndon and Pearson 1991a, b). In some contexts, they can be viewed more broadly as 'ecosystem service providers' in the process of organic matter decomposition (Boulton et al. 2008). Declines of 'shredder biodiversity', in their example, could reduce removal and breakdown of leaf litter, affecting both local water quality and the supply of carbon to the wider stream food webs. However, many details of such roles are still largely undocumented, but emphasise the likely functional importance of insect conservation in streams and ground water systems.

Tentative figures for the richness of the aquatic insect orders comprising the approximately 7500 species signaled from the Australian region (including New Zealand and New Guinea) sum to about 10% of the global total (Balian et al. 2008). Amongst these, Diptera (43%), Coleoptera (18%) and Trichoptera (15%) predominate.

The major focus of this account is on the four largest wholly aquatic orders (Ephemeroptera, Odonata, Plecoptera, Trichoptera), but Megaloptera are also wholly aquatic. The other orders noted contain lineages that are sometimes diverse and abundant, but derived clearly from terrestrial ancestry: Hemiptera, Coleoptera and Diptera are the predominant representatives, but Neuroptera and Lepidoptera also have several aquatic groups and, in all, moves to freshwater environments have taken place on different occasions in different intraordinal groups. Thus, beetles have colonised aquatic habitats on many occasions: Bilton et al. (2019) suggested at least eight major transitions, four of which have led to radiations of more than 1000 aquatic species (Table 7.4) and noted also that only Diptera paralleled this frequency of aquatic invasions. In both orders, taxonomically disparate families are involved. Further, five families of beetles have colonised subterranean waters. Invasions by other orders are also complex: Hymenoptera have generated a number of small parasitoid wasps, in several families, that enter water to discover aquatic insect hosts. They oviposit, for examples, in eggs or larvae of aquatic insects that may be

**Table 7.4** Major transitions of aquatic Coleoptera: the six families of water beetles with more than 1000 species (after Jach and Balke 2008)

| Family | Estimated number of species |
|---|---|
| Dytiscidae | 4800 |
| Hydraenidae | 2500 |
| Hydrophilidae | 2320 |
| Elmidae | 1850 |
| Scirtidae | 1700 |
| Gyrinidae | 1000 |

related to the terrestrial hosts exploited by related wasps. Finally, Orthoptera are not truly aquatic, but a number of taxa are characteristically found on shorelines and are regarded as 'hygrophilous' as depending on those moist transitional regimes. Invasion of freshwater by some of the other terrestrial lineages noted may have been through hygrophily, and a number of true bugs and beetles are also characteristic of shoreline environments, including exposed sediments (p. 31) and do not usually occur elsewhere.

## 7.3 Other Macroinvertebrates

### *7.3.1 Crustaceans*

Some other groups of aquatic macroinvertebrates have attracted considerable conservation attention in Australia, and their needs clearly parallel, and augment, those of the various insect groups—but in some cases raise rather different priorities. Freshwater crayfish (Parastacidae) comprise well over 100 species in Australia, where they are among the largest and most ecologically influential invertebrates in many freshwater bodies. Horwitz (1995) noted that they are both 'keystone' and 'flagship' species, and at the time of his account he considered that more than a quarter of the species were in need of conservation attention. Perceived threatening processes listed by Horwitz are all widespread, and relevant also to insects: (1) agriculture, mainly trampling by cattle; (2) fire; (3) drainage of swamps; (4) clearing and loss of riparian vegetation; (5) recreational fishing; (6) channelisation and other instream changes such as desnagging, construction of culverts and weirs, and river regulation; (7) translocations, mainly for aquaculture, and casual discards; and (8) urbanisation. Recreational and aquaculture interests differentiate crayfish from most other aquatic arthropods, and may give their conservation 'extra urgency', in that those interests create an additional suite of concerns from exploitation, translocations, and spread of diseases from aquaculture stocks. Horwitz also discussed the vulnerability of different crayfish habitats—noting, for example, that unique cave communities could be threatened by crayfish introductions, as could those of the isolated mound springs of central Australia. Both those aquatic systems support unique endemic biota. The vulnerability of peatlands and coastal swamps to development and fire, suggests that the endemic burrowing crayfish of such systems, with their burrows sometimes harbouring localised endemic insects and other fauna, could also be threatened.

Lowland riverine habitats are susceptible to changes elsewhere in their corresponding catchment, and montane streams (as refugia for alpine and subalpine biota) are also threatened by immediate change as well as by longer-term climatic changes. As for many other aquatic biota, the precise effects of postulated threats to crayfish can often not be assessed in detail. Whilst the most obvious effects of direct habitat loss and change are relatively unambiguous, the multitude of more gradual or insidious changes leading to gradual sublethal changes in species' reproduction, behaviour and population dynamics can usually only be inferred. In addition to the relatively tangible array of current threats, more imponderable climate changes in the future are believed widely likely to affect crayfish.

Early conservation attention was largely to species of economic importance: both Marron (*Cherax tenuimanus*) in Western Australia and Murray crayfish (*Euastacus armatus*) in the south-east became of concern through excessive harvesting for food. They, and other commercially desirable species founding aquaculture interests have commonly been translocated to stock dams and other streams, with possible competitive interactions with resident species, and spread of various crayfish diseases that may not be wholly host-specific. The greater array of species of concern recognised more recently in part reflects taxonomic advances, whereby more and smaller species are now known to each occur in only very restricted ranges. Several are thus listed as threatened.

Perhaps the most notable crayfish is the spectacular Giant freshwater crayfish, *Astacopsis gouldi*, the world's largest freshwater crustacean and by far Australia's largest aquatic macroinvertebrate. *A. gouldi* is unusual, also, in that a full national recovery plan for its conservation has been prepared (Government of Australia 2017). The problems it faces, and measures proposed for conservation are paralleled for many less-infamous invertebrates—and *A. gouldi* can be considered an 'umbrella species' for those wider aquatic invertebrate needs. Listed as 'vulnerable', *A. gouldi* is endemic to northern Tasmania, where its needs include well-shaded streams with relatively low temperatures, good water quality, low sediment levels, snags for shelter, deeper pools and undercut banks.

Factors contributing to decline include over-fishing (with past records of individuals of up to 6 kg in weight, now largely replaced by individuals weighing 'only' 2–3 kg). A fishing ban was imposed in January 1998, and any take is prohibited—however, possible illegal take persists. Habitat loss and sedimentation are also significant threats, with relevant disturbances listed in the conservation plan as removal or loss of riparian vegetation, removal of snags, bank erosion, siltation, channelisation, pollution from nutrients and toxic chemicals, instream barriers, and changed flow and temperature regimes. Each of these is a familiar threat cited for far more aquatic invertebrates. Forestry is a major concern (Davies et al. 2016, p. 100), especially when streamside vegetation is removed by clearcutting, with the extent of change to the streams reflected in extent and duration of operations undertaken in any catchment. Agriculture may also lead to stream degradation, through input of sediment and chemicals, and point-source effluent and nutrient discharges have been implicated in some local losses of *A. gouldi*. The species' range has contracted from localised disturbances, to create a pattern of disjunct populations, and the crayfish has been translocated successfully to other sites on several occasions. The crayfish occurs in streams in both plantations and native production forest, and harmonising such production systems with conservation is a major management focus.

The five species of primitive Anaspididae (Malacostraca), all endemic to Tasmania, were recognised by Wells et al. (1983) as 'vulnerable'. At that time, however, they varied somewhat in status. The widespread *Anaspides tasmaniae* was not considered threatened, but others were possibly under threat from changes to their lake habitats and development of hydroelectric schemes. All those four have very limited distributions. Predation by introduced trout has been implicated in declines and, despite some of the species being within national parks, those threats persisted, and one conservation need was to avoid further spread of trout, through curtailing further deliberate releases.

### 7.3.2 Molluscs

Australian freshwater mussels all belong to the family Hyriidae. Most species are endemic and one Victorian species (*Hyridella glenelgensis*) is likely to be extinct. Many declines and extirpations of other species have been reported (Walker et al. 2001). Surveys in the Hawkesbury-Nepean River system in New South Wales confirmed such declines, attributed in part to disturbance of riparian vegetation and more intensified land uses, leading Brainwood et al. (2006) to consider some species under threat in that catchment. Physical disturbances such as the above have been implicated in declines elsewhere, and those trends may also reflect lack of recruitment to populations. Indeed, some residual mussel populations in the Hawkesbury system were considered 'geriatric', with juveniles absent from all the highly disturbed sites sampled by Brainwood et al.

Many freshwater mussels depend for development on the presence of particular host fish species. Their larvae, glochidia, are initially obligatorily parasitic on the gills or fins of a fish, to which they attach for a period before they become benthic. Reduced populations and dispersal opportunities for host fish augment mussel losses from physical changes that can also include impoundments, substrate changes and changed water depth. However, details of host fish specificity are not available for most Australian mussel species. The major exception is for the Western Australian *Westralunio carteri*, for which Klunzinger et al. (2012) compared glochidial numbers on eleven species of fish (seven native species and four alien species) to show that the mussel is a host generalist but that the various hosts examined varied in their ability to support glochidia to develop to the juvenile mussel stage. Their accompanying laboratory studies showed that four native and one alien fish species were 'competent hosts' in allowing this development, but two alien fish species were not.

## References

Balian EV, Segers H, Leveque C, Martens K (2008) The Freshwater Animal Diversity Assessment: an overview of the results. Hydrobiologia 595:627–637

Bilton DT, Ribera I, Short AEZ (2019) Water beetles as models in ecology and evolution. Annu Rev Entomol 64:359–377

Boulton AJ, Fenwick GD, Hancock PJ, Harvey MS (2008) Biodiversity, functional roles and ecosystem services of groundwater invertebrates. Invertebr Syst 22:103–116

Boyero L, Pearson RG, Dudgeon D, Ferreira V, Graca MAS (and 27 other authors) (2012) Global patterns of stream detritivore distribution: implications for biodiversity loss in changing climates. Glob Ecol Biogeogr 21:134–141

Brainwood M, Burgin S, Byrne M (2006) Is the decline of freshwater mussel populations in a regulated river in south-eastern Australia linked with human modification of habitat? Aquat Conserv Mar Freshw Ecosyst 16:501–516

Cayrou J, Cereghino R (2005) Life-cycle phenology of some aquatic insects: implications for pond conservation. Aquat Conserv Mar Freshw Ecosyst 15:559–571

Chadd R, Extence C (2004) The conservation of freshwater macroinvertebrate populations; a community-based classification scheme. Aquat Conserv Mar Freshw Ecosyst 14:597–624

Davies PE, Cook LSJ, Mallick SA, Munks SA (2016) Relating upstream forest management to stream ecosystem condition in middle catchment reaches in Tasmania. For Ecol Manage 362:142–155

Dijkstra K-DB, Monaghan MT, Pauls SU (2014) Freshwater biodiversity and aquatic insect diversification. Annu Rev Entomol 59:143–163

Gooderham J, Tsyrlin E (2002) The waterbug book. CSIRO Publishing, Collingwood

Government of Australia (2017) Recovery plan for the Giant freshwater crayfish (Astacopsis gouldi). Commonwealth of Australia, Canberra

Harrison SSC, Dobson M (2008) Evolutionary drivers and the ecological traits of adult insects. In: Briers RA (ed) Lancaster J. Aquatic insects.Challengers for populations. CAB International, Wallingford, pp 250–267

Hawking JH, Smith FJ (1977) Colour guide to invertebrates of Australian inland waters. Identification Guide no 8. Coop Res Cent Freshw Ecol, Thurgoona

Hearnden MN, Pearson RG (1991a) The diets of mayflies in a tropical Australian rainforest stream. Trop Freshw Biol 2:203–212

Hearnden MN, Pearson RG (1991b) Habitat partitioning among the mayfly species (Ephemeroptera) of Yuccabine Creek, a tropical Australian stream. Oecologia 87:91–101

Horwitz P (1995) The conservation status of Australian freshwater crayfish: review and update. Freshw Crayfish 10:70–80

Jach MA, Balke M (2008) Global diversity of water beetles (Coleoptera) in freshwater. Hydrobiologia 595:419-442

Klunzinger MW, Beatty SJ, Morgan DL, Thomson GJ, Lymbery AJ (2012) Glochidia ecology in wild fish populations and laboratory determination of competent host fishes for an endemic freshwater mussel of south-western Australia. Aust J Zool 60:26–36

Merritt RW, Cummins KW (1996) An introduction to the aquatic insects of North America, 3rd edn. Kendall/Hunt, Dubuque, Iowa

Morse JC (2009) Biodiversity of aquatic insects. In Foottit RG, Adler PH (eds) Insect biodiversity. Science and Society (1st edn) Wiley-Blackwell, Oxford, pp 165–184

Naumann ID (chief ed) (1991) The insects of Australia (2nd edn). Melbourne University Press, Carlton

Resh VH, Rosenberg DM (1989) Spatial-temporal variability and the study of aquatic insects. Canad Entomol 121:941–963

Robertson A, Dineen G, Baker R, Hancock B, Shaw P (2014) Microinvertebrate community colonization and succession in a new urban river: lessons for river restoration. Fundam Appl Limnol 185:31–41

Skern M, Zweimuller I, Schiemer F (2010) Aquatic Heteroptera as indicators for terrestrialisation of floodplain habitats. Limnologica 40:241–250
Strand M, Merritt RW (1999) Impacts of livestock grazing activities on stream insect communities and the riverine environment. Amer Entomol 45:13–29
Walker KF, Byrne M, Hickey CW, Roper DS (2001) Freshwater mussels (Hyriidae) of Australia. In: Bauer G, Wachtler K (eds) Ecology and evolution of the freshwater mussels, Unionoida. Springer, Berlin, pp 5–31
Wells SM, Pyle RM, Collins NM (comp) (1983) The IUCN Invertebrate Red Data Book. IUCN, Gland and Cambridge
Williams DD, Feltmate BW (1992) Aquatic insects. CAB International, Wallingford

# Chapter 8
# Insects of Australia's Inland Waters

## 8.1 Introduction

The enormous ecological and taxonomic variety of aquatic insects, many of them endemic and localised in occurrence in Australia, is still to be appreciated fully. Many of the 'rarer' freshwater insects in Australia occupy equally restricted habitats, whose loss and degradation pose the most severe threats to those insects, with many of those threats continuing to increase insect vulnerability. In addition, the changing status of any insect species—the 'Prestonian shortfall' of Cardoso et al. (2011)—is almost always unknown and (as suggested by Drinan et al. 2020, for the New Zealand fauna) largely reflects three factors. These are: (1) the difficulties of producing any realistic estimate of population size; (2) the impracticalities of gaining long-term data on population sizes and distributions; and (3) the natural fluctuations in species' abundance that necessitate such long-term investigations to determine. Such gaps in understanding persist even for species acknowledged as having high conservation importance, such as the flagship taxa noted in Chap. 9, due to lack of resources and expertise.

Many Australian aquatic insect lineages are regionally restricted, highly under-documented, and apparently susceptible to environmental changes. Several characteristic taxa are shared with New Zealand, and characterise a distinctive southern fauna of Gondwanan 'cool-adapted species' that almost exclusively frequent running waters. In general, these are best represented in Tasmania, the far south east of the Australian mainland, New Zealand and far southern South America (Cranston 2010a, b), with related species of several orders co-occurring in upland forested streams (Winterbourne 1980). For example, the different genera of Plecoptera found in Australia, New Zealand and South America show strong relationships, but usually circumscribed local distributions. All Australia/New Zealand stonefly genera are endemic, but a few occur in both countries. Many are flightless. Their distributions, together with those of Ephemeroptera (also predominantly a southern group but again with some more recent northern incursions) largely reflect vicariance, rather than active dispersal, in contrast to a second major faunal element which comprises


T. R. New, *Insect Conservation and Australia's Inland Waters*,
https://doi.org/10.1007/978-3-030-57008-8_8

taxa that have entered Australia from the north, with their closest relatives in the Oriental region, and their later-evolved descendants. Likewise, Trichoptera contain both Gondwanan and northern components in Australia. Bayley and Williams (1981) noted several characteristic features amongst the benthic insect fauna of running waters in Australia: (1) presence of several families of Plecoptera (Eustheniidae, Gripopterygidae, Austroperlidae) absent from the northern hemisphere; (2) absence of several northern hemisphere families of Ephemeroptera, but with Leptophlebiidae being predominant and ecologically diverse; (3) occurrence of aquatic larvae of Mecoptera (Nannochoristidae), unknown from the northern hemisphere; and (4) absence of several common northern hemisphere families of Trichoptera, and adaptive radiation within some families present, most notably Leptoceridae, Leptophlebiidae, Gripopterygidae and Rhyacophilidae. Others, such as the predaceous diving beetles (Dytiscidae), also have spectacular regional features, discussed in context later. Collectively, 11 orders of insects are represented strongly in Australia's inland water ecosystems, most with distinctive regional features. Brief outline perspective for each of these is followed by general appraisal of the Australian fauna.

The basic biology of each order summarised in general texts is not recapitulated in detail here, other than to summarise some broad features of life cycles and ecology relevant to assessing conservation status and needs, and to exemplify conservation advances (mostly from examples from other parts of the world) as foundations for advancing constructive progress. Rather, the peculiarities of Australian biota are emphasised to convey more relevant background to those themes in the wider contexts of wellbeing of aquatic environments and communities. For some insect orders, almost no such information is available, and extrapolation of information across even closely related species may prove misleading. Even the developmental duration of many species is unknown, and within some genera and orders is highly variable. Marchant's (1986) summary of family level information then available, for example, noted development of Austoperlidae (2–4 years over the four species included) and Siphloneuridae and Oligoneuridae both as 6 months to 2 years (across three and four species, respectively). For some other orders, examples from other parts of the world markedly enhance understanding of likely changes within Australia and provide avenues for their conservation assessment and management.

## 8.2 Ephemeroptera

Mayflies are one of the predominant groups of benthic insects and are advocated widely as indicators of water pollution and other environmental changes in both lotic and lentic waters. In contrast to the predatory Odonata (below) larvae of most mayfly species are grazers on plants or other substrates—relatively few (including all members of the Australian family Amelotopsidae) are carnivores. Whilst larvae can be long-lived, many species undergo several generations each year. Ephemeroptera is often considered to be the most primitive extant order of insects and may be the sister group to all other insects. All species are aquatic, and ecologically are very varied. Their diversity appears to be highest in the Holarctic region, although parts of the Oriental region are still very incompletely surveyed (Barber-James et al. 2008), and many families are widespread. The distinctive Australian fauna displays about 90% genus-level endemism, with many taxa confined to the eastern and south eastern parts of the continent. Campbell (1990) recognised three probable biogeographical groups of Australian mayflies, as (1) cosmopolitan genera (such as *Baetis*), (2) those with northern origins, and (3) those with Gondwanan affinities. Leptophlebiidae and Baetidae, both diverse and widespread in Australia, are the predominant families, but almost all species of Australian Ephemeroptera are endemic and many are restricted to the south-east. The few non-endemic taxa belong to cosmopolitan families, and most occur also in Indonesia. The northern tropical taxa appear to be largely relatively recent elements of the fauna.

Mayfly larvae are commonly the most abundant constituents of 'drift' (p. 132). However, conservation interpretations (as in some other aquatic insects, as a widespread restriction) can be thwarted by inability to associate larvae and adults reliably. Thus, the genus *Coloburiscoides* is of conservation interest in Australia (p. 84), but the species can be identified only from the short-lived and rarely seen adults, and not from the more conspicuous and more frequently-collected larvae. For these, Suter and McGuffie (1997) noted likely confusions, and considerable uncertainties over whether some species that appear to be common and widespread may in fact be potentially endangered, especially in the alpine region where fire is a clear threat to stream fauna. Larvae of *Coloburiscoides* became very scarce in foothill streams following runoff from the 2003 fires in the Australian Alps, where they were previously abundant, but numbers recovered somewhat over the next 18 months or so, but intense fire and ash-laden runoff could eliminate some populations (Suter and McGuffie 1997). In general, mayflies are susceptible to the many threats that affect aquatic invertebrates, notably pollution, changed flow regimes, and disturbances to riparian and nearby forest vegetation. Acidification led to disappearance of Ephemeroptera from some streams in Europe (Guerold et al. 2000). In the Vosges Mountains of northern France, those streams also lost molluscs and crustaceans. Several mayflies and stoneflies were subsequently considered regionally endangered, and it appeared that protection of streams against acidification was needed in order to maintain their high invertebrate diversity.

Changes in mayfly assemblages from such interventions can be profound, but they may also be subtle. Assemblages in the Czech Republic were compared through samples from 60 streams in 1955–1960 and 2006–2011, across a period of considerable anthropogenic changes from agricultural intensification and changes in water quality from organic pollution and siltation, leading to the expectation of substantial changes (Zedkova et al. 2015). Comparisons were from replicated larval samples from kick-sampling across a range of habitats (riffles, run, pool, woody debris, aquatic vegetation) in the streams, and sites categorised according to water quality and prevailing human pressures. Comparisons were restricted to the 59 sites from where at least five species in at least one pond were found. Gross alpha richness was very similar across the two periods: of the total 53 species obtained from both periods (42 in 1955–1960 and 49 in 2006–2011), the expected catastrophic declines were not evident. Nevertheless, some possible complex changes in assemblages occurred, with losses of some highly specialised species and loss or turnover of some common and widespread species that are generally resilient to imposed changes. Heavy pollution overrode some other influences, but both siltation and direct habitat loss contributed to assemblage changes, urging the need for a catchment-level scale of stream/river management both for restoration and for protecting the better reaches from further disturbances. Some of the trends in mayfly colonisation and loss from brooks, and small and large rivers (Fig. 8.1) illustrate the various factors and ecological groups of mayflies affected.

Habitat use by mayflies, compared by kick-sampling the benthos and accompanied by a series of implanted substrate trays in the stream bed, was assessed for Yurrabine Creek, in northern tropical Queensland (Hearnden and Pearson 1991a, b). All 12 detritivore species occurred in all microhabitats investigated—but different species were the most abundant in different microhabitats, implying some level of habitat partitioning. The distribution of fine debris and leaf litter was closely correlated with mayfly distribution patterns, and these parameters were affected by current velocity and substrate particle size. The tray study (using trays 314 × 254 mm in area and 48 mm deep, deployed for 30 days and each with substrate of one of the four substrate classes used and the series duplicated with addition of a mesh package of leaves) provided local more uniform habitat than the considerable small-scale heterogeneity of the creek. That heterogeneity would otherwise mask any such partitioning in the broader benthic samples.

The ecological variety of mayflies leads to conservation interests in species associated with particular kinds of waterbody. Thus, for Montana, United States, the nine species of conservation concern or potential concern (a further two species) are associated with large rivers (six species), small forested streams (two), large western rivers (one), wetlands/ponds (one), and sloughs/lotic backwaters/lakes (one), collectively spanning many available aquatic habitats (Stagliano 2016). More commonly, surveys of mayfly distribution patterns are used to associate richness with site or waterbody features to infer rarity, endemism or vulnerability, and suggest areas of greatest conservation significance (Brazil: Massariol et al. 2014).

Although mayflies have a low formal conservation profile, several North American species (as above) are thought to have become extinct regionally, and Barber-James et al. (2008), for Europe, commented that 'many more species need to be added to the list [my addition: the Bern Convention, p. 213)] as more and more habitat destruction takes place'.

## 8.3 Odonata

Unlike virtually all other aquatic insects, Odonata—as large, colourful, mostly diurnal and conspicuous insects, and of significant ecological and cultural importance—are a potent flagship group with considerable public appeal. In one sense, drawing from this appeal and their familiarity to many people who appreciate these

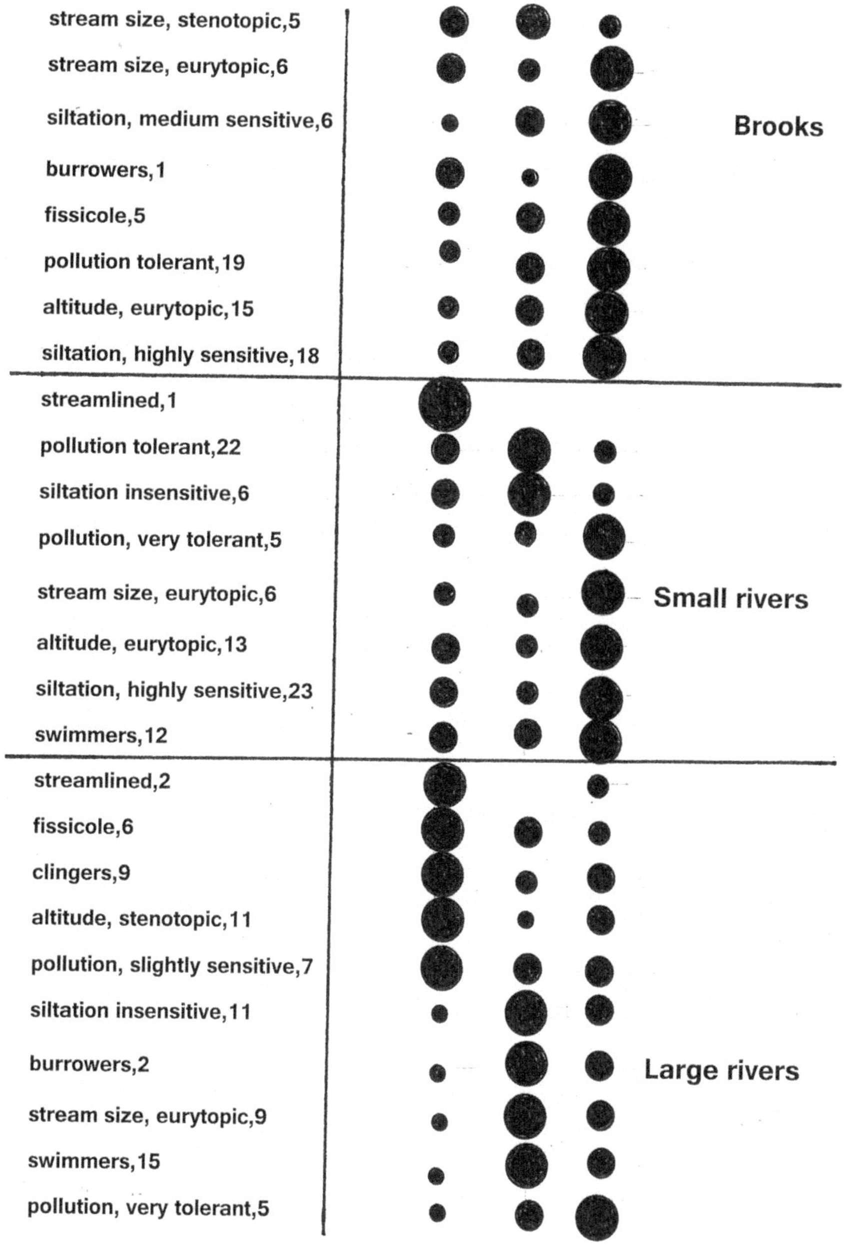


**Fig. 8.1** Representation of percentage of colonisation of new sites (middle column) by Ephemeroptera species of particular categories (list) at time of recent survey compared with earlier data, together with local loss (left column) and survival (right column). Numbers of species in each category shown, percentage expressed by size of circles (simplified from Zedkova et al. 2015)

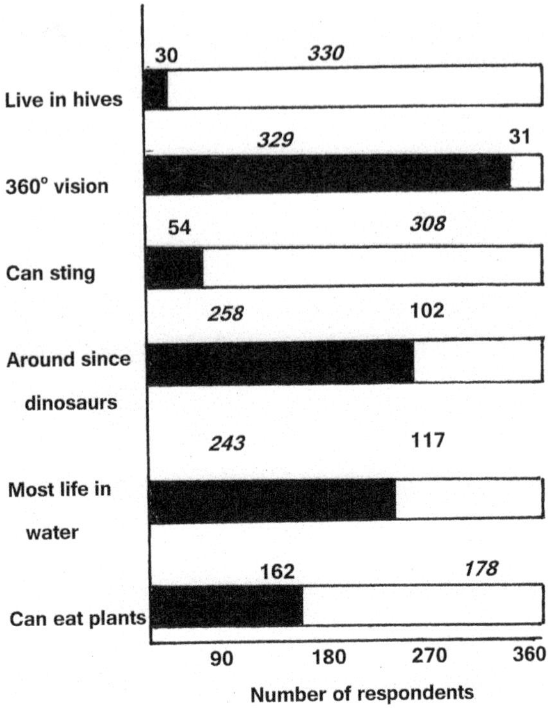


**Fig. 8.2** Results from a survey of public knowledge of Odonata: responses from 360 people to six 'true/false' statements as answered 'true' (black) and answered 'false' (open), with true answer in italics (Ngiam et al. 2017)

insects, dragonflies are the 'aquatic parallel' to butterflies in terrestrial insect conservation advocacy, and differ markedly from all other aquatic insects in the sympathies and interests they generate. As for butterflies, numerous local interests in conservation have arisen through drawing on this public acceptance. Thus, many people surveyed to adjudge their appreciation of urban ponds in greater London felt that seeing dragonflies was 'interesting', even though knowledge of their ecology was not always high (Fig. 8.2, Ngiam et al. 2017). Ngiam et al. proposed three principles to increase best practices for dragonfly conservation in urban ponds (p. 126), as (1) cultural interventions, with aesthetically pleasing ponds to enrich enjoyment of green spaces and cultivate a sense of 'stewardship'; (2) ecological interventions, with habitat creation and enhancement to promote diversity; and (3) educational interventions, such as signage, to improve knowledge. These principles extend to ponds beyond urban areas (Chien et al. 2019) and both lotic and lentic systems, despite their differing features, may be amenable to this approach (Prescott and Eason 2018). Odonata have been proclaimed widely as an excellent and significant focal taxon in advancing knowledge of freshwater biology and in addressing a variety of themes that advance conservation in freshwater environments, as ecological tools, indicators of disturbance and changes (Davis et al. 1987), and with both adult and larval stages easily sampled and, in many cases, identifiable to at least genus level—and often to species, through availability of well-illustrated handbooks. As summarised by Clausnitzer et al. (2012), their values encompass (1) Odonata are a key component of species assemblages in many freshwater systems; (2) they are sensitive to changes in both terrestrial and aquatic environments, through the differing requirements of adults and larvae; (3) they are abundant on all continents (except Antarctica), with many species restricted both geographically and ecologically; and (4) they are taxonomically tractable, with many local faunas identifiable to near species level both as adults and larvae. A slightly different emphasis is the parallel rationale presented by Villalobos-Jimenez et al. (2016), which broadens this perspective further. They emphasised the suitability of Odonata more specifically for the study of urban ecosystems because (1) they are sensitive to a range of stressors such as pollutants and temperature widespread in city environments; (2) species occupy terrestrial and aquatic environments, as above; (3) as important predators, Odonata have a wide range of ecological interactions in both these major ecosystems; (4) adults are useful in studying movements across landscapes; and (5) they are sufficiently well-documented for a robust foundation of knowledge to be laid.

That knowledge and awareness is fostered through a considerable number of specialist ‘interest groups’, which range from local to international in scope and have incorporated a wide range of naturalist observers and citizen scientists in their recording and other conservation activities. Many species are rare and noted individually as of conservation concern, and composition of local assemblages may depend substantially on environmental conditions. They have received far more species-focused conservation attention than any other aquatic insects. In short, Odonata are useful ‘model organisms’ for studying a wide range of topics in freshwater ecology, as well as insect behaviour and wider community features and changes—the titular phrase ‘Guardians of the watershed’ (Clausnitzer and Jodicke 2004) is indeed apt, and their importance, wherever damselflies and dragonflies occur, is evident throughout the magisterial account by Corbet (1999).

The wide potential for promoting Odonata in conservation (Bried et al. 2015) is reflected in a rapidly increasing literature that embraces most key themes of conservation biology, and demonstrates the relevance of both suborders, the dragonflies (Anisoptera) and damselflies (Zygoptera), as model organisms, indicators, tools for wider assessments of diversity and habitat condition, and conservation of individual threatened species and local to regional assemblages. The wealth of applications of information on Odonata in applied ecology and conservation was illustrated amply by Bried and Samways’ overview of 411 publications, with many papers focusing on ecology and conservation status of rare species, values as indicators of wider envronmental impacts, and improvements in methodology for varied applications and assessments within the theme of ‘applied odonatology’. That array accords Odonata a uniquely prominent place amongst aquatic insects—and, acompanied by their taxonomic tractability, one unlikely to be approached by any other aquatic insects in the forseeable future. Nevertheless, Bried et al. (2015) also suggested several further topics that could give them even greater relevance, with the prospect of ‘further linking odonatology with emerging issues in the ecology and conservation of fresh waters’ (Bried et al. 2015).

Recognition of the importance of Odonata in conservation planning was illustrated by their representation amongst 'Species of Greatest Conservation Need' in the United States, where nearly two-thirds of the then 441 species were accorded this status (Bried and Mazzacano 2010)—namely, as 'species with small or declining populations or other characteristics that make them vulnerable'. Listings for each State differed somewhat in scope; some were very selective, but Alaska included all documented Odonata and Hawaii included all endemic species (p. 117). The coverage, although substantial, was thus somewhat uneven but illustrated the attention that may be gained for this best-documented aquatic insect group and provided a wide basis for augmentation. Even for North America, increased recording effort is needed as current gaps in records hamper definition of the true distributions of many species (Hassall 2012) in a problem reflecting scales of area and interest, and which is even more evident in Australia. In a scenario that applies widely beyond dragonflies, 'recorder effort' may validate or contradict earlier work or opinion on distribution (as for *Hemiphlebia mirabilis* in Australia, p. 219) and local diversity, and lead to uncertainties over the relevance of environmental factors previously considered to be influential or critical.

Studies of the structural features of habitat suitability for different Odonata abound. Separate consideration of the needs of lotic and lentic Odonata can sometimes provide insights into the reasons for local loss of species, in conjunction with aspects of their distribution. In Finland, extinction risk for local dragonfly populations was particularly high for species with a narrow distribution (Korkeamaki and Suhonen 2002). Lotic species generally had lower extinction risk, and the highest risks were for species associated with specialised vulnerable wetlands such as peatlands (p. 221). Extinction risk was also higher in small upstream habitats than in the larger and relatively more stable downsteam reaches. Although they are not usually categorised as 'pests', the strongly dispersive abilities of some dragonflies can occasionally lead to large populations of larvae developing in open fish-rearing ponds, where their predation on large numbers of fish fry can lead to economic concerns.

Odonata dominate lists of threatened aquatic insect species, reflecting largely that they are indeed the best documented order amongst these. Thus, Kalkman et al. (2010) found that Odonata comprised 106 of the total 118 aquatic insects listed by the IUCN, and more recent comment has suggested that up to 15% of all Odonata species are likely to be threatened (Sanchez-Bayo and Wyckhuys 2019).

As Bried et al. (2015) observed, the values of Odonata in ecological assessments have been facilitated immensely by the legacy of biological and distributional information accumulated over the last century, rendering the order the only one among freshwater insects to become a paramount interest in aquatic conservation. Their intricate behaviour has also attracted considerable wider interest, and Odonata have been proclaimed 'an unparalleled insect model to integrate evolutionary genomics with ecology for the study of insect evolution' (Bybee et al. 2016). Moore (1997) commented that 'a count of dragonflies would provide a quick, and therefore low cost, indication of the health or richness of the lake or river'. Odonata thus play a central role in conservation of freshwater habitats and continue to be promoted strongly in conservation assessments through a global effort to assess species' conservation status (Clausnitzer et al. 2009). Especially in Europe, and progressively elsewhere, 'red lists' and distribution atlases have summarised enormous numbers of records and enabled well-informed definition of conservation status and needs. Thus, Kalkman et al. (2018) drew on such materials to describe patterns of endemism and diversity of the European species. Patterns of lotic and lentic species differed considerably—lotic species diversity (44 species assessed) was greatest in the Mediterranean region, whilst lentic diversity (99 species) was highest in western and central Europe. Lotic species may average smaller distributions, possibly from rather lower dispersal ability, but those ranges can still be substantial and allow for movements over at least several tens of Km. Many lentic species are absent from the Mediterranean region, possibly reflecting lower rainfall leading to scarcity of permanent marshlands, and the associated widespread absence of bogs and related habitats (Kalkman et al. 2018). Maintenance of suitable conditions for river specialist Odonata is a priority need in conservation, because these habitats are both increasingly restricted, and more natural conditions also help to foster natural species richness (Hoffmann and Mason 2005).

Odonata comprise a full range from 'generalist' to 'highly specialist' species, sensitive to changes in both aquatic and terrestrial environments. It is perhaps reasonable to suggest that, as for many other insects, generalist species can largely resist environmental changes, whilst specialist species may be far more susceptible to imposed changes. The range of threats to Odonata spans both aquatic and terrestrial changes—severe declines in populations have resulted from agricultural intensification, for example (Raebel et al. 2012). The importance of vegetation for dragonflies has for long been recognised, but the great variety of functions for dragonfly welfare make it easy to overlook some critical role. The list provided by Buchwald (1992) (Table 8.1) is a useful reminder of the possible values for any given species. Both individual plants or plant stands and the plant communities of surrounding landscapes must be considered in conservation, and Buchwald's list also emphasises the importance of understanding the often complex behaviour of dragonflies. Adult Odonata use 'wildcard visual cues' to assess habitat quality (Harabis and Dolny 2012) and in order to select new habitat areas. These may be a particular successional stage of macrophyte vegetation, and two alternative outcomes were postulated. First, such stages in anthropogenic ('secondary') habitats may act as an ecological trap (p. 138) because of their instability or, second, they may be short-term but important refuges for the species. Study of the endangered European *Leucorrhinia pectoralis* suggested that many secondary habitats might function as 'sinks' because their instability through frequent disturbances tends to induce local extinctions of specialist species for which optimal conditions are lost rapidly (Harabis and Dolny 2012). High diversity of freshwater habitats caused by frequent disturbances may be a requirement for dragonfly diversity to persist in secondary habitats. In their study of mine subsidence pools in post-mining areas, surface reclamations were regarded as the greatest threat to the diversity of secondary habitats, through leading to loss of heterogeneity.

Unusually among aquatic insects, for many parts of the world, samples of larvae and adults of Odonata can both be identified reliably, and extended sampling can

**Table 8.1** Possible functions of vegetation for dragonflies, as listed by Buchwald (1992)

| |
|---|
| *Single plants or plant stands as/for:* |
| Larval habitat: feeding, shelter from predators, etc. |
| Oviposition site or substrate |
| Substrate for hatching (mostly vertical element) |
| Perch site for territorial behaviour, hunting, waiting for females in order to mate, mating, sun-basking, roosting |
| Boundary demarcation of territory |
| Protection: from predators and unfavourable weather |
| Signal releasing habitat selection (specific plant species or community) |
| *Patterns of plant communities/surrounding landscape as/for:* |
| Feeding area |
| Maturation area for subadults |
| Influencing the microclimate |
| Signal releasing habitat selection ('aspect' of vegetation in the adjacent areas and the surrounding landscape) |

lead to effective inventory of the order of the wetland(s) studied. As in many such surveys, the more common analyses are based on single visit ('snapshot') surveys to define the assemblages, and those interpretations have relatively rarely been evaluated against more extensive sampling programmes. Adults and larvae were sampled separately on a series of occasions on a small tributary (Middle Creek) of the Kiewa River, Victoria, to reveal also the occurrence of three groups of species that cannot be separated convincingly from single-occasion samples of one or other life stage. Hawking and New (2002) collected 15 species as larvae, and 16 as adults, in a total of 18 species. The groups present were 'permanent resident' (13 species), 'transient or opportunist resident', and 'non-resident'—the second of these implying that vagile species may occasionally colonise temporarily, and become easy to confuse with permanent residents, while others are found only fortuitously, perhaps as vagrants from other reaches of a river where their critical resources are better represented. The fundamental needs for a dragonfly inventory as a reference point for comparisons, include a list of all taxa present, their relative abundance, and evaluation of their resident/non-resident status in relation to the habitats present (Schmidt 1995). In principle, inventories are easier to compile for lentic faunas. Many dragonfly inventories are for small ponds (Wissinger 1998), in which the longitudinal habitat variations of lotic systems do not occur—although the variety in lotic waters in the same regions may be a valuable source of information in defining local species richness and suggest the species that should be represented in any given more restricted survey.

The relative ease with which Odonata can be detected, as larvae and adults, and their abundance investigated has given them considerable advantages in conservation assessments over other aquatic insects. Whilst incidences of larvae and/or adults are the most commonly assessed measures of 'diversity', and clarify the number of 'core resident species' (as used by Niba and Samways 2006), Simaika and Samways (2009) pointed out that the presence of exuviae is perhaps the most useful measure as 'the only true demonstration that the habitat in question is suitable to odonates in both the aquatic and aerial parts of the life cycle'.

The 'Dragonfly Biotic Index' (DBI: Simaika and Samways 2009) has been proposed to evaluate the conservation value of odonate assemblages, by combining three parameters for each species: (1) the regional distribution of the species; (2) the national red list classification (or reference to any equivalent ranking assessment); and (3) the species' sensitivity to habitat changes. The total DBI for each species is the sum of these, each parameter given a value from '1' (maximum) to '3' (minimum), and the total value for a site is the sum of values of all species present, so that richness is incorporated. The index thereby acknowledges richness, species of conservation significance, and ecological specialists. It can help to indicate the relative values of particular sites or site categories for dragonflies well beyond South Africa, for which the index was initially designed. For example, the three categories among the 61 freshwater ponds as post-industrial sites on lignite spoil heaps in the Czech Republic each supported substantial dragonfly diversity, but each was also preferred by distinct threatened species (Harabis et al. 2013). A mosaic of all three restoration treatments was recommended, to incorporate the substrate and bank variables needed to support the more specialised species present. Strong correlations were found between DBI and vegetational heterogeneity. Further, DBI's value in monitoring activities includes that it gives weight to those species of greater concern as being geographically restricted, threatened and sensitive to change, rather than to species richness alone. In Simaika and Samways' (2009) example, of tracking habitat recovery from invasion by alien riparian vegetation, restoration led to both greater species richness and greater DBI at all sites assessed.

The wide utility of the DBI is demonstrated in a manual for its application to all species of Odonata in South Africa, in which detailed description of procedure and analysis are given (Samways and Simaika 2016), together with a conservation status evaluation for each species. The major categories of DBI are listed (Table 8.2), together with their associations with habitat condition.

In a somewhat similar approach, Kutcher and Bried (Kutcher and Bried 2014) devised a 'Coefficient of Conservatism' (CoC) to estimate sensitivity of adult dragonflies to disturbance. Their study embraced 135 species and their distributions among 510 wetlands, which were degraded to varying extents, in Rhode Island, United States. The list of parameters used to evaluate wetland condition (Table 8.3) indicates the wide variety that may need to be considered in parallel assessments elsewhere.

CoC was calculated as: $\left(N_{LD}/N\right) \pm \frac{(N_{MD}/N)}{2} \times 10$, where

$N_{LD}$ is the number of least-disturbed wetlands in which a species is recorded; $N_{MD}$ is the number of most-disturbed wetlands where it occurs, and N is the total number of wetlands on which it was detected, multiplied by 10 to scale the output. A species thereby has a value from '0' if found exclusively in the most-disturbed sites to '10'

**Table 8.2** Features of the 'Dragonfly Biotic Index' (DBI) (Simaika and Samways 2009; Khelifa 2019), and scoring of the components in allocating conservation status for African taxa

| Score | Distribution | Threat category | Sensitivity |
|---|---|---|---|
| 0 | Very widespread | LC | Apparently not sensitive to habitat disturbance; may thrive in natural and/or artificial waterbodies |
| 1 | Localised across wide area | NT | Low sensitivity to habitat disturbance such as bank degradation; may occur commonly in natural and/or artificial waterbodies |
| 2 | Quite localised | VU | Medium sensitivity to habitat disturbance; may recolonise areas where it went extinct |
| 3 | Patchy range and rare in region | EN, CR | Extremely sensitive to habitat change from alien plants; low tendency to recolonise areas where it went extinct |

**Table 8.3** Components of a scheme to evaluate freshwater wetland condition: the 'Rhode Island Rapid Assessment Method' as used by Kutcher and Bried (2014) to indicate possible variety of relevant metrics available

| Metric | Metric scoring criteria |
|---|---|
| Integrity of buffers | Estimates % cultural cover class within 30 m of unit |
| Integrity of surrounding landscape | Generates weighted average of four land-use intensity categories by relative proportion within 150 m of unit |
| Impoundment | Estimates water regime change and proportion of unit affected, and identifies barriers to resource movement |
| Draining or diversion of water | Estimates water regime change and proportion of the unit affected |
| Anthropogenic fluvial inputs | Estimates impacts of four types of fluvial inputs, including nutrients, sediments and solids, toxins and salts, and flashiness |
| Filling and dumping | Estimates the intensity of fill within or abutting the wetland and the proportion of the unit affected |
| Excavation and substrate disturbances | Estimates the intensity of substrate disturbances within the wetland and the proportion of the unit affected |
| Vegetation and detritus removal | Estimates the extent and proportion of vegetation and detritus removed from each of five vegetation strata |
| Invasive species within wetland | Estimates the collective cover class of all identified invasive plant species |
| Observed state | Rates the apparent integrity of five wetland functional characteristics, including hydrologic integrity, water and soil quality, habitat structure, vegetation composition, and habitat connectivity |

if found only in the least-disturbed sites. The mean CoC for all species at a site has been termed the 'Odonata Index of Wetland Integrity' (OIWI). In this survey CoC values indeed ranged from 0–10 (with a mean of 6.4 ± 2.2 SD), whist OIWI values across 51 sites ranged from 3.74–7.15 (mean 5.90 ± 0.77 SD). Kutcher and Bried (2014) considered that adult dragonfly CoC values based on their affinity to wetland disturbance classes indicated their worth as 'a meaningful and robust indicator of freshwater wetland condition'. Whilst the study used three broad groups of wetlands to assess the index—least-disturbed, intermediate-disturbed, and most-disturbed—Kutcher and Bried also suggested that using just the two extremes might be valid: although some intermediate information would be lost, more efficient and targeted monitoring effort and simpler computations would occur.

Conservation interests extend from the well-publicised values of Odonata as indicators of aquatic ecosystem condition to also indicating changes in some terrestrial ecosystems. Whilst the former is largely founded in the aquatic larval stages, the gradual changes in composition of adult assemblages in tropical rainforests of Kalimantan (Borneo) were associated with the level of anthropogenic disturbance (Dolny et al. 2012). Response of individual species to habitat disturbance reflects individual specialisations and needs, and Dolny et al. noted that the habitat divisions of Odonata are much more complex under closed canopy than in secondary forest with more open landscapes—and low insolation and interspecific competition can there segregate forest and non-forest species. In general, whereas forest species were strongly competitive within their closed forest environments, that competitiveness is much lower elsewhere, when more aggressive generalist species may prevail, and invade more open forest areas (Samways and Steytler 1996). That process may increase competition and lead to local extinctions of habitat specialists. Dolny et al.'s hypotheses, based on samples including a pool of 88 species across 14 families, were that disturbance would affect (1) the number of species found (no significant trend found); (2) the proportion of Zygoptera relative to Anisoptera; (3) the proportion of forest specialists; and (4) the proportion of Borneo endemic taxa. The last three were all higher in closed canopy habitats.

Responses to habitat changes have received considerable attention, with size and quality of the water bodies highly influential—but those influences are not always predictable. Thus, larger ponds do not always support richer dragonfly assemblages than smaller ponds. In South Africa, Osborn and Samways (1996) found that features such as vegetation, water quality and the variety of local habitat heterogeneity all influenced richness. Collectively, these factors were more influential than pond size alone, and dragonfly richness was highest at well-vegetated shallow ponds in which water was clear and well oxygenated. More specifically, Osborn and Samways found, from observations on the adult stage, that sunlight/shade regimes and marginal/submerged vegetation gradients strongly influenced assemblage patterns.

The broad habitat dichotomy between 'running waters' and 'still waters' can be used to designate relative habitat quality for species that can exploit both of these. Thus, in central Finland 15 species of Odonata occur in both water types but each species has a preference for breeding in only one category ('high quality habitat') so that the water bodies available to each species could be allocated contrastingly as either high quality or low quality habitat (Suhonen et al. 2010). Local extinction rates of Odonata were assessed from comparative surveys in 1930–1975 and 1995–2003, with an average of around 25 populations examined for each species. Also on average, 30% (range 0–100%) of populations of each species had been lost, this assumed from them not being found on previously occupied sites during the second round of surveys. Those lost were typically species with few localised populations, whilst common species were relatively unaffected—not least, because any extinctions could be compensated by recolonisations from the wider landscape and were thus unlikely to be detected. Extinction rates were higher from low quality habitats. However, contrary to expectation, specialist species in some cases had lower extinction rates than the generalists. Suhonen et al. (2014) suggested that this might be because generalists occur in both high quality and low quality habitats, fewer migrants might flow from the source populations as those habitats become degraded, their extent overall declined and the normal source populations lost or strongly diminished. Suhonen et al. suggested the latter context could apply particularly to peatland-associated Odonata because in central Finland the area of undisturbed peatland had been reduced very substantially as forestry activities increased.

Odonata exhibit strong responses to water quality and the broad structure of aquatic habitats (substrates and vegetation features, by larvae) and to nearby changes to terrestrial environments (such as forest clearance, by adults). The order is also widespread, both geographically and across numerous aquatic regimes. Together with their popularity, features such as general accessibility to study through relative ease of surveying both larval and adult stages, and the availability of reliable identification guides for many parts of the world have fostered global appraisal of their conservation status and needs. Clausnitzer et al. (2009) endorsed their values as an 'easy-to-use indicator group' for evaluating the status of wider biodiversity in freshwater ecosystems. They recommended a series of general conservation measures that would cover many of the threatened species, especially the most threatened species associated with lotic habitats in tropical forests (Table 8.4).

**Table 8.4** Some general conservation measures that cover needs for many threatened Odonata (after Clausnitzer et al. 2009, referring especially to threatened forest species in lotic habitats)

| |
|---|
| 1. Prevention of any further deforestation: forests should be regarded as a natural resource with high conservation priority |
| 2. Afforestation with indigenous trees: wide conservation value for biodiversity and also in resource management |
| 3. Fire control; reduced fires can aid natural succession |
| 4. Leaving a strip of natural riparian vegetation of at least 20 m on either side of streams in cultivated and mined landscapes |
| 5. Water pollution control measures, such as sewage works, mining runoff, factory effluent treatment, restricting pesticide uses |
| 6. Protection of watersheds, floodplains, water retention sites, swamps and others against adverse impacts of damming and other alterations |

A continental-scale overview of African dragonfly conservation status and needs (Clausnitzer et al. 2012) helped to reveal the diversity of threats and the scale and variety of needs that arise. In general, natural high dragonfly diversity was associated with the previously continuously forested lowland areas in western and central Africa, with richness gradients reflected in habitat heterogeneity as a combination of rainforest and highlands, and the large desert areas supporting rather few species. The 90 species ranked as threatened or near-threatened are mostly very restricted—many of them to highland areas and with the species in forested running waters most likely to be threatened. As elsewhere, including Australia, forest stream species adapt only poorly to changed conditions. Clausnitzer et al. found the single most embracing threat to be alterations to the natural landscape—involving agriculture, urbanisation and general deforestation, with riparian forests disproportionately highly susceptible to exploitation. Some species depend on wet forest litter without open water in forested areas. Damming and mining were recognised threats to specific habitats—with extreme cases such as the likelihood that the entire habitat of the endemic *Oreocnemis phoenix* in Malawi could be removed if mining of bauxite on Mount Mulanje was developed. Dams can have both upstream and downstream impacts through changed flow regimes (p. 119), such as by affecting rapids-specialist species by flooding, and by creating erratic flow regimes downstream.

The Australian odonate fauna is reasonably well known, and the species recognisable through national (Watson et al. 1991) and more regional guides and overviews. It includes strong representation of several small families largely confined to the region (Kalkman et al. 2008). Many Australian Odonata are southerly, Gondwanan, relict species (Watson and O'Farrell 1991) and recent explorations show a variety of threats to such species, some of which are very localised. Hawking (1997) recognised nine species of high conservation significance in the south east (Table 8.5) with the unifying threat being habitat loss, from a variety of influences. No regional species was then considered endangered, however. Their habitats are all restricted naturally, and preservation of those habitats was clearly the major conservation priority.

Later, Hawking (1999) informally assessed the conservation status of all Australian Odonata against IUCN Red List criteria, to imply that one species (*Archaeophya adamsi*, p. 220) was Critically endangered, 12 species were Endangered and 24 Vulnerable, leaving 39 as Near threatened, 154 (almost half the total fauna) as Lower Risk—Least Concern, and the remaining 84 species as Data Deficient. Those figures, however preliminary, raised three serious concerns: (1) a large number of endangered species; (2) a large number of priority species; and (3) the general lack of reliable information on which to focus definitive assessments. Hawking considered the Data Deficient species to be of major concern, recommending that they should be investigated immediately because some may be close to extinction—or even extinct.

The largest Australian dragonflies, the 'petaltail dragonflies' (Petaluridae), have attracted considerable attention as localised and scarce taxa (Hawking and Theischinger 2004).

The small family is regarded as a relic from the Jurassic era, and the Australian endemic *Petalura*, with five highly localised species, is the largest of the five included

**Table 8.5** nine species of Odonata regarded as of 'high conservation significance' in south-eastern Australia by Hawking (1997)

| | Species' status | Priorities |
|---|---|---|
| Hemiphlebiidae<br>*Hemiphlebia mirabilis* | Rare | Preserve known habitats |
| Megapodagrionidae<br>*Austroargiolestes isabellae* | Rare | Identify larva and habitat |
| Gomphidae<br>*Austrogomphus angeli* | Indeterminate | Identify larva and habitat; determine distribution |
| Austropetaliidae<br>*Archipetalia auriculata* | Rare | Preserve known habitats |
| *Austropetalia patricia* | Rare | Preserve known habitats |
| *Austropetalia tonyana* | Rare | Preserve known habitats |
| Petaluridae<br>*Petalura gigantea* | Rare | Preserve known habitats |
| Corduliidae<br>*Archaeophya adamsi* | Vulnerable | Establish distribution and Habitat |
| *Synthemiopsis gomphomacromiodes* | Rare | Identify larva and habitat |

genera. In Western Australia, *Petalura hesperia* occurs in scattered localities along the Darling Scarp, where it is associated with headwaters of permanent watercourses (Barrett and Williams 1998). Surveys are complicated by the small size of most habitat areas and the dragonfly's association with dense understorey vegetation rendering detection of exuviae very difficult, together with a short annual flight period. Barrett and Williams found it at six of the 30 sites they explored, but with few individuals at each. Several specific sites were recommended for protection from future disturbances. Although *P. hesperia* was not considered then to be threatened, changed conditions could easily increase its vulnerability.

However, more conservation concern has been paid to eastern taxa, notably the Giant petaltail, *Petalura gigantea* (p. 221). A broader review of threats to these taxa identified three major groups of interrelated threatening processes with a wide range of collective potential impacts to their groundwater-dependent mire and peat bog habitats (Table 8.6: Baird and Burgin 2016).

## 8.4 Plecoptera

Stoneflies have been suggested to be 'probably one of the most endangered groups of insects' (Fochetti and de Figueroa 2006). This comment arose from studies on the European fauna, for which some species had been recorded as extinct and many to have undergone severe reductions to leave only small remnant populations. There, the entire plecopteran fauna of lowland rivers could be considered threatened—several species once common in large rivers had not been seen for many years, and are

**Table 8.6** Summary of potential impacts of identified threatening processes relevant to petalurid dragonflies in Australia (Baird and Burgin 2016)

| |
|---|
| *Groundwater depletion* |
| Reduction/loss of seepage areas or saturated peaty substrates |
| Subsequent increased vulnerability of dry peaty soils to fires |
| Weed invasion |
| Increased photodegradation and oxidisation of peaty soils |
| Reduction/loss of suitable ovipositing and larval establishment sites |
| Mortality of later instar larvae when lowering water tables exceed burrow depths or capacity for adaptive burrow deepening |
| Loss of populations, reduced connectivity between mire patches and reduced geographic distribution |
| *More intense fire regimes* |
| Combustion of dry organic component of peaty soils |
| Increased erosion of mire soils during high rainfall events and increased sedimentation (of mineral soils) into mire from upslope following fire |
| Collapse of peat piping, channelisation, and lowering of adjacent water tables |
| Increased photodegradation and oxidisation of exposed peaty soils |
| Reduction/loss of suitable ovipositing and larval establishment sites |
| Increased mortality of recently laid eggs and early larval instars in early stages of burrow establishment and increased mortality of adults |
| Loss of populations, reduced connectivity between mire patches and reduced geographic distribution |
| *Climate change* |
| Increased erosion/channelisation of mire during extreme rainfall events |
| Increased variability in environmental conditions, higher temperatures, more dry days, and increased occurrence of drought |
| Weed invasion |
| Increased photodegradation and oxidisation of exposed peaty soils |
| Higher fire risk with potential impacts on eggs, larvae and adults |
| Reduction in groundwater inputs, increased surface drying and lowering water tables |
| Contraction in mire habitats and reduction/loss of areas of suitable oviposition and larval establishment sites |
| Loss of populations, reduced connectivity between mire patches and reduced geographic distribution |

presumed extinct. The only known examples of global extinctions of stoneflies known to Fochetti and de Figueroa were two such riverine species, *Taeniopteryx araneoides* and *Oemopteryx loewi*, but many others have been lost from parts of their European ranges. Yet others are known from small areas and few individuals, with the single or few known populations threatened. Within Europe and North America, it has been estimated that 35% of stonefly species are declining, 29% are 'threatened' and 19%

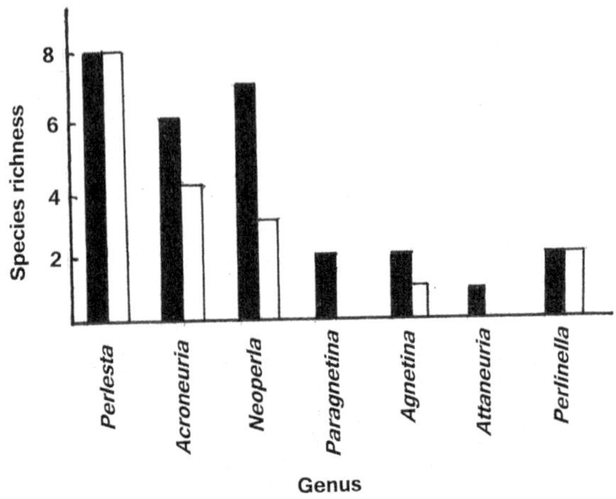


**Fig. 8.3** Richness of the genera of Perlidae (stoneflies) in Illinois, as total recorded species richness (black bars) and those remaining at the time of survey (open bars) (DeWalt et al. 2005)

are locally or regionally extinct (Sanchez-Bayo and Wyckhuys 2015). Many species have declined and, as well as extinctions and local extirpations, some ecological guilds have suffered substantially. DeWalt et al. (2005) noted the high losses amongst predatory and long-lived species in the United States. Most of these belong to the families Perlidae or Perlodidae, and information on the seven genera of Perlidae recorded from Illinois (Fig. 8.3) showed that only 18 of the 28 species recorded before 1950 were still present by the end of the century, as landscape changes proceeded. Two genera appear to have been lost completely. Losses were attributed mainly to hydrological modifications (such as damming and channelisation) and were greatest in the larger rivers. As a consequence of this, smaller species with shorter life cycles have replaced those long-lived species in some documented assemblages, although the functional ramifications of such changes are still unclear. However, DeWalt et al. (2019) noted earlier comments that losses of 'shredders' (p. 107) (11 of the 16 families of stoneflies, and the majority of species in the order, fall into this category) from headwaters could reduce availability of fine organic matter downstream, with cascading losses of numerous downstream fauna.

In general, stoneflies are among the most sensitive insect groups, and are among the first taxa to be lost from freshwater systems after even rather minor disturbances (DeWalt et al. 2019). Many Plecoptera thus have narrow environmental tolerances and are recognised as among the most sensitive responders when used as bioindicators. Rare or restricted taxa that are found in the upper reaches of streams and rivers, at higher elevations, and are restricted to cold waters will be amongst those most threatened (de Figueroa et al. 2010). Assessing possible impacts of climate change, at least 324 European stoneflies (63% of the taxa evaluated) were in some way considered vulnerable, with de Figueroa et al. noting that 'European Plecoptera as a whole are an endangered group'. The European regions likely to suffer the greatest impacts from changing climate are those with high standing species richness and which also contain almost all the most vulnerable taxa. Only a very small proportion of species, most of them widespread and with wider temperature and elevational ranges, are likely to benefit.

The major lineages of Plecoptera in Australasia are almost all Gondwanan groups shared, at most, only with New Zealand and/or South America. The most diverse family (Gripopterygidae) occurs in Australasia (214 species noted by DeWalt and Ower 2019) and South America (99 species), and the less diverse Eustheniidae (20 species in Australasia, two in South America) and Austroperlidae (11 species in Australasia, four in South America) are also in the Antarctoperlaria. The only Australasian family of Arctoperlaria (Notonemouridae) is more widely distributed, found in Australasia (62 species), Africa (39 species) and South America (20 species). Endemism is high, perhaps especially so in Tasmania (Cranston 2010a, b), and a number of characteristic species of conservation concern are restricted to higher elevations and cool headwater streams (p. 28). They are paralleled elsewhere, such as by the wingless genus *Scopura* (Scopuridae), which is restricted to small montane streams in Japan and the Korean Peninsula, where the detritivorous larvae take up to four years to develop and feed mainly on foliage from riparian deciduous trees (Jin and Bae 2005).

## 8.5 Hemiptera

Two rather different major radiations of largely predatory Heteroptera have become aquatic. The true water bugs (Nepomorpha) and water striders (Gerromorpha) are both widespread. In addition to these wholly aquatic groups, Polhemus and Polhemus (2008) recognised a third major taxon, the Leptopodomorpha (the shorebugs) as 'water-dependent species'. Fifteen families of Heteroptera have aquatic taxa or stages, all derived largely from terrestrial phytophagous lineages. All are characteristically carnivorous, with a few believed to be omnivores, and some are among the relatively few insects (together with some beetles) to occupy the river surface film. Larvae and adults usually co-occur, and have similar habits and, whereas most are entirely aquatic, some taxa are most characteristic of moist shoreline environments. Fourteen of the 54 genera in Australia are endemic, and the fauna includes representatives of 17 of the 20 regional families. Species-level endemism is high (Andersen and Weir 2004), but almost a quarter of the species are marine. Polhemus and Polhemus considered that most aquatic Hemiptera are not at risk, but also cited cases such as *Ambrysus amargosus* (p. 115), a saldid (*Salda morio*) disappearing from the Netherlands as its peat bog habitat diminished, and a Japaness aphelocheirid (*Aphelocheirus kawamurae*) believed extinct because of changes to a river system. The range of threats to waterbugs parallels those for other taxa, with their predatory habits perhaps influencing vulnerability of some species.

In Japan, the giant waterbug *Lethocerus deyrollei* (Belostomatidae) has been used as an indicator for unpolluted freshwater environments. It formerly depended heavily on traditional 'satoyama' landscapes and declined in parallel with these from around the mid-twentieth century (Nagaba et al. 2011; Ohba 2018). By 1978, the bug had become extinct in nine of the 47 prefectures, and was included in the Red Data Books for all prefectures, with losses (and declines elsewhere) attributed to pesticide applications, changing agricultural practices, and urbanisation, including street lighting (p. 239). Alien species implicated in declines of *L. deyrollei* in Japan include the Red swamp crayfish (*Procambarus clarkii*) and the Bullfrog (*Lithobates catesbeiana*), both from North America and both of which have expanded in distribution greatly since they were introduced to Japan in the early twentieth century.

Pollution-free systems are a likely key to conserving the bug in Japan. Rice paddies, with their regularly changing water regimes, are a major component of satoyama systems, as temporary waters with a seasonally stable management pattern. Study of *L.* (or *Kirkaldyia*) *deyrollei* across this variety showed how it could be maintained over its univoltine life cycle (Mukai et al. 2005). Adults are long-lived, with some reports of individuals living for more than a year. Adults move out of water, and over-winter in nearby coppice.

Belostomatidae, with about 150 species, are predominantly tropical or subtropical. Some of the largest species, in particular, are desirable as food for people (p. 97), but can also be predatory pests of fish hatcheries, whilst some smaller-bodied taxa are beneficial predators of pest insects such as mosquitos (Ohba 2018). Prey ranges from small arthropods to tadpoles and frogs, and small fish—with occasional records of larger species (Lethocerinae) attacking birds or snakes—which may reciprocate and attack the bugs! *L. deyrollei* reportedly feeds mainly on vertebrates, including diminishing components of the ricefield areas and that in some cases may threaten fish diversity and abundance, in turn affecting the bug. Because tadpoles are an important food (Ohba 2018, noted 'indispensable') for growth of earlier instar *L. deyrollei*, adequate populations of frogs are a key need. Water pollution has affected the bugs directly and their feeding on contaminated (by bioaccumulation) prey, with impacts from endocrine disruptors on reproductive physiology (Nagaba et al. 2011). Vegetation is also important as oviposition sites in water and for sheltering adults during their terrestrial phase. Thus, Kobayashi and Iwai (2007, cited by Ohba 2018) found that >75% vegetation cover in ponds, together with muddy substrate, water depth of 15–29.9 cm, and high temperature (around 30 °C) was a suitable combination of habitat features for the bugs.

## 8.6 Coleoptera

Three of the four suborders of Coleoptera include aquatic taxa, with incidence differing considerably across these groups: about 90% of the small suborder Myxophaga, with only 77 described species (Sharma et al. 2019) are aquatic, and this proportion falls greatly in others; Sharma et al. reported 18% of the 30 000 species of Adephaga and only 1.25% of the 370 000 species of Polyphaga aquatic. Despite their overall lower richness than Polyphaga, Adephaga are commonly the more diverse aquatic group. For example, a survey in South Africa showed aquatic Polyphaga (37 species) only about half as rich as Adephaga (68 species) (Bird et al. 2017).

Widespread aquatic beetles are thus dominated by species of Hydradephaga (diving beetles, whirligig beetles) and Hydrophiloidea (water scavenger beetles), with other significant aquatic Coleoptera including Elmidae (riffle beetles) and members of more predominantly terrestrial taxa such as Hydraenidae (moss beetles) and Scirtidae (marsh beetles). Collectively, aquatic beetles are 'perhaps second only to Odonata' in studies of diversification of aquatic insects (Dijkstra et al. 2014)—in part a legacy of long-term hobbyist interest in beetle collecting—but most of that attention has been outside Australia. The variety of transitions from terrestrial to aquatic life represented among the many beetle lineages, with more than 40 families involved, reflects considerable ecological complexity. Those transitions are related to four main factors, listed by Jach and Balke (2008) as (1) the amount of time passed in contact with water; (2) the degree of submergence; (3) the degree of dependence on water; and (4) the 'motivation' for contacting water—such as food, refuge, or other primary need. That variety was expressed in the six main ecological groups, distinguished by Jach (1998) (Table 8.7) and of which the first two are those regarded most generally as 'water beetles', with at least their larval stage fully submerged. The other four groups have been referred to as 'paraquatic' taxa but, as Jach and Balke commented, it is difficult to allocate many of the species to being truly aquatic or more truly riparian or terrestrial because their biology is not known in sufficient detail. Because many riparian zones are highly threatened (p. 100), those taxa may also be vulnerable—but they are easily overlooked by their exclusion from surveys of more characteristically aquatic fauna or of 'typical' terrestrial insects. Variations can occur within a family or other group. The Australian Elmidae, for example, comprise two subfamilies, of which Elminae (the more diverse) are wholly aquatic, and the Larinae in which adults are riparian or terrestrial (Glaister 1999). The small family Psephenidae ('waterpennies', after the shape of their flattened larvae) is represented in Australia by the single genus *Sclerocyphon*, largely concentrated in eastern Australia, but with one species (*S. fuscus*) reported also from central Australia (Davis 1998). Larvae cling to the

**Table 8.7** The six categories of freshwater beetles recognised by Jach (1998), as from Jach and Balcombe (2008)

| Category | Characteristic |
|---|---|
| True water beetles | At least partly submerged for most of the time of their adult stage |
| False water beetles | Submerged for most of the time of their larval stage, adults always predominantly terrestrial |
| Phytophilous water beetles | Living and feeding on water plants, submerged for at least some time in any developmental stage |
| Parasitic water beetles | Like phytophilous water beetles, but their hosts are aquatic mammals |
| Facultative water beetles | Actively submerged or actively dwelling on the water surface for a limited period of time during any of their developmental stages in at least one population |
| Shore beetles | Riparian, living close to the water's edge during all their developmental stages, not entering water voluntarily |

underside of stones and rocks in streams and rivers and are very distinctive. Adults are reported only rarely from surveys and are terrestrial. However, beetles occur within (or closely associated with) almost all kinds of freshwater habitats, some species being highly tolerant of salinity. Many have specialised habitat needs, and pollution, drought and other changes are likely to influence them strongly. As Jach and Balke (2008) commented 'any disturbance of the immediate surroundings of an aquatic habitat must be considered a major threat for its water beetle communities'.

The predaceous diving beetles (Dytiscidae) are one of the world's major aquatic insect radiations and are 'found in practically every form of inland water body on Earth' (Foster and Bilton 2014), where they may predominate amongst predatory taxa. The Australian fauna is diverse, with many very localised species. Recent studies on fauna of underground aquifers, for example, have led to discovery of numerous new taxa. Many of these are tiny—the smallest described species, *Limbodessus atypicalis*, is known from a single borehole in the Northern Territory, and is less than one mm in length—and many subterranean forms are flightless (Spangler 1986). Some other dytiscids are amongst the largest aquatic beetles, but the dytiscids of subterranean waters in Australia comprise perhaps the globally most diverse suite of such taxa. Australia's Dytiscidae are about 90% endemic, with the subterranean aquifer taxa especially notable.

The family has attracted considerable conservation attention in western Europe, in particular, from where experience may be extrapolated to other regions, and where their values as indicators have been explored in some detail. Foster and Bilton reiterated the values of dytiscids as surrogates, with the diversity and narrow ecological requirements of many species giving them status as indicators of habitat quality and conservation status of sites.

Fairchild et al. (2008) advanced the case for water beetle assemblages being 'unusually sensitive bioindicators' of habitat quality in ponds, in relation to most other aquatic groups. Three reasons for this were nominated: (1) the assemblages typically include both predators and a variety of other consumers, such as algal feeders and detritivores, so that assemblage composition may be influenced by changed food resources; (2) the beetles occupy a wide range of preferred microhabitats, both as larvae and adults, so that loss or change of particular habitat components might affect assemblage composition; and (3) the substantial size range (adult body mass from <1 mg to >100 mg) renders the assemblage likely to be sensitive to biotic influences such as fish predation. Assemblage structure has been linked to habitat features, including size of the water body, its acidity or salinity, level of connectivity within the local or regional landscape, its vegetation structure, and stresses such as flooding. The context for Fairchild et al.'s study was exploration of whether newly created ponds in Pennsylvania could mitigate losses from destruction of local wetlands. They compared the water beetles of 11 recently constructed ponds and seven older 'reference' ponds, collectively containing 77 beetle species across 47 genera. The two categories of ponds supported similar richness, but the reference sites yielded considerably more species found only at single localities. Ponds without fish or with only few fish had considerably greater (up to threefold higher) beetle biomass than ponds with larger fish assemblages. And, despite considerable variations among sites and sample years, Fairchild et al. endorsed the values of water beetles in reflecting habitat quality, but also emphasized the needs for extensive sampling to establish sound reference levels and to characterise the assemblages of different ponds. They also noted the importance of landscape-level perspective of the ponds, to complement single site implications.

## 8.7 Mecoptera

Most scorpionflies are terrestrial predators, but the small ancient Gondwanan family Nannochoristidae has aquatic larvae. *Nannochorista* occurs in Australia and South America, with other members of the family in New Zealand. Larvae of *Nannochorista* occur in silt in shallow, mostly slow-flowing, streams, where they fed on larval Chironomidae and perhaps other prey in the same habitats (Byers 1991). However, data are available for very few species—the above feeding habits are based on the New Zealand *Microchorista philpotti* and, without firm evidence, it has been assumed that habits of the Australian *Nannochorista* are similar (Ferrington 2008a). Adults have been found on vegetation close to water and are among the smallest mecopterans.

## 8.8 Megaloptera

All species of Megaloptera have aquatic larvae, but the two families appear to differ somewhat in their primary habitats. Larvae of Australian species of Corydalidae (all members of the subfamily Chauliodinae), by far the larger family, are found mainly in riffle areas, whereas those of Sialidae are associated more with slow-flowing stream sections. Reviewed by Theischinger (2000), the order was then known only from the southeast and southwest (the latter from a single species). Elsewhere, both families are widespread, and Cover and Resh (2008) recognised 328 described species, noting that further surveys are likely to increase this number to 'possibly more than 400 species'.

Megaloptera parallels several other aquatic groups, in that larvae of most species can be identified only by tentative association with adults, which are weak flyers and generally remain close to water: Theischinger noted that only four of the 22 Australian species of Corydalidae had been reared to establish certain identity. All members of the order in Australia are endemic, and their lineages Gondwanan with closest relatives in New Zealand and southern South America. In Australia, most species have very restricted distributions—or, at least, have been recorded from very few sites. The nature of that inference is exemplified by *Austrosialis ignicollis*, the only representative reported from Tasmania and known from a single adult female from Maria Island—where a single sialid larva (again the only one reported from Tasmania) was attributed by Theischinger to that species. Several mainland species are known only from single sites, but several species have been reported from the border areas of northern New South Wales and southern Queensland.

General degradation of freshwater and riparian habitats, including by pollution and eutrophication, is viewed as among the strongest threats to larval Megaloptera (Rivera-Gasperin et al. 2019). Little detail is available on their tolerances, although some species are known only from relatively pristine waters, free from pollution and introduced trout or predatory crayfish. The large size of both corydalid larvae and adults renders them conspicuous and, perhaps, attractive as candidates for use in management (Rivera-Gasperin et al. 2019), but their relatively low abundance renders this unlikely to occur in Australia.

## 8.9 Neuroptera

Whilst lacewings are predominantly terrestrial predators (many of them as both larvae and adults), representatives of three families are associated with fresh water. Two, Nevrorthidae and Sisyridae, are wholly aquatic and some members of the larger group Osmylidae also have aquatic (associated mainly with stream margins) or semi-aquatic larvae. Cover and Resh (2008) referred to them as 'water-dependent'. All three lineages occur in Australia, and most included species are poorly known and may be highly localised, as very few records or localities for them have been reported. The taxa were reviewed by New (2004).

Neither of the two truly aquatic lineages is diverse. Nevrorthidae contains very few species—two endemic species of *Austroneurorthus* have been described from Australia, and others occur in south-east Asia (*Nipponeurorthus*) and southern Europe (*Neurorthus*). The known larvae of these are very similar and are found in running waters. They can swim but are regarded as substrate dwellers and believed to feed on small benthic invertebrates. Sisyridae (spongeflies), still amongst the most poorly documented Australian lacewing groups, are amongst the few specialised predators of freshwater sponges, themselves some of the most poorly documented freshwater invertebrates. However, sponges 'occur in the majority of semi-permanent and permanent inland waters of Australia' (Racek 1969), thus transcending lotic and lentic systems and with some believed to be endemic. Spongefly larvae live in close association with sponges (and, possibly bryozoans), but their levels of feeding specificity are unknown. Sponges are susceptible to pollution and, perhaps, are most typical of well-aerated waters. However, one of the largest populations of a sisyrid known is from Lake Pedder, Tasmania (Forteath and Osborne 2012). *Sisyra pedderensis* is endemic to that lake impoundment area, where it is abundant in sheltered bays in association with the sponge *Radiospongilla pedderensis*. Racek (1969) also suggested that many Australian sponges have adapted to long exposure to muddy conditions, so that lotic habitats that are flooded seasonally or have suitable sediment loads may also be frequented. Two genera of Sisyridae occur in Australia, occurring mostly in the cooler regions of the southeast. *Sisyra* is by far the more diverse.

Osmylidae are far more complex, and most lineages are terrestrial—only one of five subfamilies, Kempyninae, is presumed to be wholly aquatic, but the habits of several species of Spilosmylinae and Stenosmylinae are still unknown. Kempyninae have aquatic or semi-aquatic larvae. The few reported incidences have not been identified to species level but their identity can sometimes be inferred from adult presence at the same sites, and larvae are from shallow or shoreline waters, where they are presumed to probe in mud and mosses for small invertebrate prey. Adults examined have yielded a variety of gut contents (New 1983) and may be rather generalist feeders rather than wholly predatory.

Kempyninae are a Gondwanan group, elsewhere found in New Zealand and southern South America. Most Australian species are very localised in the south east, some known from single sites or single water bodies, and are apparently univoltine, with adults appearing in early summer Adults of some species of the endemic genus *Australysmus* and *Kempynus* have been reported to aggregate under bridges or on waterside rocks at that time. The third included genus, *Clydosmylus*, is known only from its type locality in south-east New South Wales—a situation not unusual for members of most aquatic insect groups in Australia. Dedicated searches for such apparently restricted species have usually failed to discover further individuals.

## 8.10 Lepidoptera

Larvae of almost all Lepidoptera are terrestrial, and the order is dominated largely by their feeding on plant materials and as one of the major insect herbivore groups. However, within the large moth family Crambidae (or Pyralidae), and a few others, some taxa have invaded aquatic environments, where their larvae feed on macrophytes. Pabis (2018) distinguished between Lepidoptera that are 'truly aquatic' and the rather greater number that are 'semi-aquatic', as depending for larval food on emergent, amphibious or marsh-restricted plants. The latter encompass representatives of about 15 families, whilst truly aquatic species occur only within Cosmopterygidae, Crambidae and Erebidae. In particular, the crambid subfamilies Acentropinae (formerly known as Nymphulinae, with a very high proportion of aquatic moth species) and Pyraustinae contain species strongly adapted to freshwater life. Larvae respire through filamentous gills and those of many taxa construct cases from pieces of foliage; a few scrape algae from rocks and construct silken shelters or retreats on stones. As with numerous other Lepidoptera, the biology of the early stages of many of the species involved is poorly known, and assumption of aquatic status of some species is inferred from taxonomic position rather than biological evidence. Pabis (2018) concluded his overview by noting that these moths 'are probably one of the most poorly studied ecological groups with the Lepidoptera'.

Indeed, they are regarded widely as the most poorly known of all aquatic insect groups. Mey and Siedel (2008) speculated that this was because, in general, aquatic entomologists pay little attention to the predominantly terrestrial Lepidoptera, and the approaches to collecting and study used by lepidopterists (such as careful individual handling of specimens to preserve wing vestiture) are not usual among limnologists. They commented that 'the study of aquatic Lepidoptera has always been a subject of lepidopterists alone', with those lepidopterists only rarely having a limnological background. However, those Acentropinae associated with running water are considered highly vulnerable to water pollution and channelisation: Mey and Siedel commented 'They are among the first organisms which would disappear when facing a drop in water quality', but any such details are unavailable for Australian taxa.

Acentropinae are widespread, but a high proportion of Australian species appears to be endemic. Pending further surveys of the fauna of some nearby regions, Mey and Siedel listed 21 genera from the Australian region (including New Zealand, New Guinea and other nearby lands), seven of them not known elsewhere. Information on Australian species is sparse, but considerable taxonomic and ecological variety is present (Hawking 2001). In revising the group, Hawking (2014) later recognised 52 species across 17 genera in Australia, with the state of taxonomic knowledge reflected in 10 species and seven genera being undescribed. The eastern seabord in general was considered the area of greatest diversity, with the greatest numbers of species in northern Queensland but some others confined to the colder streams of the southeast mainland. Hawking tentatively allocated species among the IUCN Red List categories. Two species were considered 'vulnerable' because few populations were known, and five others regarded as 'near threatened' by having very limited distributions and/or unusual ecology. Two species were selected as in need of monitoring because they have not been rediscovered at historical Queensland sites, although occurring elsewhere in the state. Collectively, 28 species were widely distributed in standing waters and assessed as of 'least concern' and 23 species were 'Data Deficient' as only fragmentary biological information was available.

## 8.11 Trichoptera

Caddisflies are the predominant aquatic group of endopterygote insects, and the most diverse of all primarily aquatic insect orders, with the >16200 extant species greater than the global total of all other aquatic orders combined (Morse et al. 2019). Australia's rich fauna includes 26 families, and more than 100 described genera. Three families are endemic (Calocidae, Oeconesiidae, Plectrotarsidae). Many endemic caddisflies are largely confined to, or are most diverse in, the south east region, and larvae frequent many kinds of aquatic environments (Dean et al. 2004). The ecological diversity of the larvae renders them very informative in interpreting changes in their habitats. They are trophically diverse, and collectively include predators, detritus feeders, shredders, filterers and algal grazers. Guild changes have been reported amongst caddisflies, as in stoneflies, largely amongst those species with a long life cycle (Houghton and Holzenthal 2010), with most losses amongst shredders and predators so that present-day communities can become dominated by filterers. Many larvae are recognisable to family (in some instances, to genus) from their structure and habits, and different species are variously free-living, occupy fixed retreats on solid substrates, construct tubes or galleries along rock surfaces or in sediments, or—perhaps most familiar to many observers—construct portable cases in which the larva lives and moves around. The form of the case has considerable value in taxon recognition, and may be made of twigs, foliage, gravel, sand, or silk alone in different taxa.

Uncertainties over the conservation status of some taxa (including some that have received considerable publicity as putative flagships: p. 211) have emphasised the needs for more comprehensive surveys to detect many of the presumed scarcer taxa (both of caddisflies and other taxa)—the outcomes either endorsing conservation needs or revealing the species to be more secure than previously believed. Some surveys may be far less conclusive, because reasons for non-detection of poorly-known species may not be clear, and factors such as unknown phenology hampering their detection.

Investigations of the Trichoptera of Tasmania's Wilderness World Heritage Area included assessing the status of four caddisfly species listed under the Tasmanian Threatened Species Protection Act, among the 88 species recorded by Jackson (2000). Two of the four species were known only from Lake Pedder, and their discovery there had involved them in the case for opposing flooding of the original lake in 1972 (p. 123). The Lake Pedder caddisfly (*Taskiropsyche lacustris*) and McCubbin's caddisfly (*Taskiria mccubbini*) were both listed as endangered and had not been seen in the long interval since they were discovered, despite several dedicated searches. Both were re-discovered in Jackson's surveys, but both in very small numbers—a single specimen of *T. lacustris* and two individuals of *T. mccubbini*. The latter were among a total of about 950 individual caddisflies collected from that single site on the shore of the Lake Pedder impoundment, which could be affected by changing water levels and contains introduced brown trout.

The two other listed caddisfly species, both relatively inconspicuous 'microcaddis' species are both designated as 'rare'. The spotted microcaddisfly (*Orphninotrichia maculata*) was not re-collected from its original site but was found at another locality. It is, however, common on mainland Australia, but the Tasmanian site may be threatened by water pollution and vegetation clearance. The Miena microcaddisfly (*Oxyethira mienica*) was not re-discovered. It can be identified reliably only from adult males, so that larvae and adult females collected by Jackson at five sites could not be identified conclusively. The precise site of the original discovery of this species is unknown. Jackson believed that the earlier conservation status remained valid for the three species she re-collected, and her surveys collectively contained about half the Trichoptera species recorded from Tasmania.

The above species are among 17 Trichoptera species (13 of them endemic to the state) listed for protection in Tasmania by 2005. Most (13) of them are listed as 'rare' and are known from single localities. Two were listed as 'extinct': *Costora iena* and *Diplectrona castanea*. However, the latter is now a formal synonym of *Diplectrona lyelli*, clearly not extinct, and has been removed from the listing.

Elswhere in the world, substantial declines of Trichoptera have been reported or inferred. Houghton and Holzenthal (2010) commented that, should their findings on caddisfly declines in Minnesota be more widespread across the north-central United

States, 'a large amount of regional caddisfly biological diversity has already been lost'. Using criteria for 'critically imperiled' (that is, known from fewer than five sites within a state) and 'imperiled' (known from <20 sites in a state), about 30% of Minnesota Trichoptera were critically imperiled, and >60% of species imperiled. However, lack of historical ranking data for the great majority of Minnesota species extends also to other regional surveys on the continent—in which perhaps >90% of species can not be ranked reliably. This example illustrates the widespread difficulties in gaining credibility for declarations of levels of loss or threat for aquatic insects, and numerous parallels occur for Australia.

## 8.12 Diptera

Forty-one families of Diptera have aquatic representatives, and collectively include nearly 46 000 species—according to Adler and Courtney (2019), at least three times as many as Coleoptera or Trichoptera, as the other major endopterygote orders in aquatic environments. Dijkstra et al. (2014) listed five major lineages of true flies amongst their key taxa of aquatic insects. Two of these are numerically predominant. Culicomorpha (midges, mosquitos, blackflies) is the more speciose and the economic importance of many species, as biting or 'nuisance' flies and vectors of diseases, has led to far greater study on these than most other Diptera have received. Tipulomorpha (craneflies) are also very diverse but lack any equivalent impetus for study. However, Tipulidae (or Tipuloidea, with the four famiies of Tipulidae, Limoniidae, Pediciidae and Cylindrotomidae, collectively the craneflies) are one of the largest groups of Diptera and larvae of most species are aquatic or semi-aquatic, and the terrestrial adults of many are confined largely to moist waterside environments. Larval habitats are diverse, and tipulids occur in most kinds of waterbodies. Many species have rather restricted ecological ranges and distributions and have sometimes been suggested to be useful in biomonitoring (Kimura et al. 2011, Japan). Generic diversity in Australia is high and, although not commonly included in monitoring programmes at present, the abundance of tipulid larvae suggests considerable future potential (de Jong et al. 2008), as being sensitive or moderately sensitive to human disturbances. Craneflies are associated with numerous moist environments, where larvae of particular taxa may occupy rather restricted moisture gradients, ranging from fully aquatic to largely terrestrial microhabitats with larvae living in such moist sites as decaying wood or fungi. Although they are not represented in specific conservation endeavours, the numbers and richness of craneflies in mires and other waterbodies render them conspicuous and diverse components of that fauna. Larvae are retrieved commonly from both lentic and lotic systems, especially in samples close to water margins. Many of them feed on decaying plant material. Livestock grazing along shorelines is thought to threaten some semi-aquatic taxa (Yadamsuren et al. 2015). In Mongolia, declines were driven by decreased plant biomass and relative humidity, both affected by grazing, with vegetation regarded as both a food source and physical shelter to cranefly larvae. Yadamsuren et al. noted the correlation between decreased vegetation and decreased soil mosture, leading to increased desiccation of early stages.

Acknowledging that a number of aquatic Diptera are threatened or extinct, Adler and Courtney (2019) also commented that 'the full picture is woefully inadequate given the state of knowledge for most species'. There are few well-documented extinctions.

Flies occur in all kinds of freshwater bodies from ephemeral pools and numerous small water containers (ranging from natural phytotelmata to discarded food and drink vessels and old vehicle tyres) to large lakes and rivers. Many Diptera have been used as indicators of water quality. Chironomidae, in particular, are a common component of monitoring programmes, reflecting their ecological and taxonomic diversity.

Representatives of around 20 families of true flies in Australia have aquatic early stages. They represent many different lineages but, in contrast to almost all other aquatic insects, some are major pest taxa. Biting flies (such as blackflies, Simuliidae) or disease vectors (mosquitos, Culicidae), in particular, can pose severe health and welfare problems and large-scale control needs, including applications of pesticides to aquatic environments, arouse concerns over non-target impacts on other aquatic invertebrates.

As for Lepidoptera, above, the aquatic status of numerous Diptera is uncertain or rather speculative, because larval biology is unclear other than in very general or extrapolated terms. Likewise, lack of specialist attention to many families renders knowledge of their distribution and diversity highly incomplete. Thus, Wagner et al. (2008) omitted several families from their broad overview of the group simply because no specialist was available to deal with them. They reviewed 19 families, but for many of those the predominant larval habitat(s) are wet ground/mud bordering waterbodies, or rotting wet wood, and relatively few are 'truly aquatic' in having wholly aquatic early stages. Many of those families have been largely disregarded in aquatic insect assessments, which have focused far more on the few large fly families with more universal aquatic habits, and which are known to have valuable attributes as indicators (predominantly Chironomidae), economic and health importance (Culicidae, Simuliidae), or defined conservation interest (Australia: Blephariceridae, p. 227). In studies related to aquatic habitat condition and conservation, appraisals of Chironomidae continue to contribute significantly to understanding and monitoring programmes in many parts of the world.

Chironomidae, the non-biting midges, are often the most predominant freshwater insects. As Cranston (1995) put it, 'There are few water bodies anywhere in the world in which midges do not occur, and whatever the interest of the limnologist, chironomids are unavoidable'. Their environments span saline to fresh waters, and midges can be abundant in brackish waters. As well as being important as monitors of water condition, assemblages of chironomids have values in 'classifying' freshwater bodies. Lindegaard et al. (1995) reviewed much early work on these themes, and studies have continued to proliferate, and to dominate ecological appraisals of aquatic Diptera. Chironomidae have many and varied ecological roles—as predators, herbivores or detritivores, bioaccumulators of toxic materials such as mercury and chromium, key food items for fish (to which they can transfer toxic materials and render fish unsuitable for human consumption) and in some places for people.

Many chironomids respond in subtle ways to water quality, sediment loads and pollutants, giving them widespread uses as indicators of freshwater environment quality, in some cases without need to identify the aquatic larvae beyond subfamily or genus level. However, whilst Chironomidae are by far the most intensively studied aquatic Diptera in monitoring of water quality and environmental conditions, lack of definitive taxonomy is a significant impediment to non-specialist evaluations of assemblages and to expanding their use more generally (Nicacio and Juen 2015). Despite substantial advances in Australia, more detailed surveys and appraisal are needed (in common with other southern hemisphere regions) to approach the detail available for much of Europe and North America. Molecular taxonomy is likely to play an increasing role, as discussed by Sharley et al. (2004), in advancing capability to focus at levels beyond subfamily in diversity estimations.

An important application of chironomid larvae in pollution studies involves the functional relationships between heavy metals and pesticide levels and induced morphological deformities of the head capsules of their benthic larvae. Incidence and form of deformities (which can be assessed separately for different cranial structures: mentum, mandibles, premandibles, pecten epipharynx and antennae) are varied with some seemingly more related to heavy metal pollution (mandibles, premandibles) and others to other organic compounds (mentum, antennae) (Janssens de Bisthoven et al. 1994). However, 'deformity' encompasses both natural processes such as spontaneous mutations, as well as the irregularities caused by pollutants, so that some background information on the unpolluted rate at a site is needed to partition these. Madden et al. (1994) distinguished also between 'absolute deformities' (such as changed numbers of antennal segments or mandible 'teeth') and 'relative deformities' (such as changes in shape or size). They undertook laboratory rearings over two years to compare the levels of deformity produced in polluted water with those in natural water from streams.

Rather than simply the incidence of deformities, Clarke et al. (1994) discussed the applications of 'developmental stability analysis', manifest as departures from true bilateral symmetry in structures that are normally symmetrical, as 'fluctuating asymmetry'. Again, increased frequencies of asymmetry relative to background levels in 'control' populations provides an index of developmental stress—and Clarke et al. noted that a particular value of this measure is that it may furnish a more generally applicable early warning of environmental degradation, that can be used across different midge species.

However, more general use of larval Chironomidae in monitoring is to some extent thwarted by their often very high abundance, difficulties of species identifications (often necessitating analyses only to family or subfamily level, as noted above) and, in Australia, the high numbers of undescribed and poorly known species likely to be encountered. Species-level identifications or separations may be necessary in order to identify which chironomids are useful indicators of sediment pollution. Around Melbourne, for example, different species were positively or negatively associated with sediment pollution (Carew et al. 2007). Whilst categorisation of some benthic invertebrate groups only to higher taxonomic levels can enable their valid use as indicators (p. 49), the ecological and response variety within Chironomidae can mask such values without further taxonomic penetration, perhaps depending on DNA analyses for consistency in recognising key indicator species. Nevertheless, some valuable conclusions have come from Australian studies. In the Blue Mountains (inland from Sydney, New South Wales), undisturbed sites supported species that were absent or scarce at sites affected by sewage effluent discharges. The latter had fewer, but more abundant, species (Hardwick et al. 2004). In particular, the largely cold-adapted Orthocladiinae were sometimes numerically dominant at affected sites, with Chironominae becoming more prevalent at lower elevations, as more typical of warmer waters. Changes also occur in response to acidic and heavy metal-containing waters from mining activities. In the Northern Territory, Smith and Cranston (1994) reported clear changes in assemblage composition above and below acid discharge points, although richness remained similar, with species turnover reflecting the tolerances of individual species. Congeneric species can exhibit markedly different tolerances and responses.

Further attention to the flies arises from their 'nuisance values', when vast numbers of swarming adults create a variety of concerns—in tourist area water bodies in which the flies breed, visitors can be deterred by the midges, and additional cleaning and control costs become necessary together with remediation and maintenance (Ferrington 2008b). Other reported annoyances include large numbers entering buildings, attractions to light causing disruption to human activities (including sporting events), increased accident risk to people, respiratory allergens, contamination of food supplies, clogging vehicle radiator grills, and many similar disruptions.

### *8.12.1 Control of Aquatic Pest Flies*

In addition to uses of fish as predators of mosquito larvae (p. 114), concerns over mosquito-borne diseases, and changing human behaviour leading to proliferation of mosquito breeding sites, such as discarded containers, rainwater tanks and domestic ponds, have inevitably led to numerous other attempts to control these flies, especially the major vectors or other 'nuisance' species. The twin contexts for control of aquatic flies are thus (1) their roles as vectors of diseases, and (2) the nuisance caused by vast numbers, this ranging from annoyance to disruption of activities such as recreation and sporting events and enhanced in the more developed countries as many people elect to settle close to lakes or rivers. Both individual and economic welfare may be affected, the latter through impacts on tourism and hotel maintenance. Thus, Ali (1995) commented on the 'continuing increases of new midge-producing habitats' as human settlement patterns change, with consequent increase in some pollution-tolerant Chironomidae, and others. Many recent urban developments in coastal Australia have included 'canal developments' or linkages to lakes or the sea, and more inland developments often feature constructed wetlands.

Urban artificial wetlands systems to control and treat waste water and storm water (p. 277) are a symptom of increased urbanisation and human health relevance. In some cases it may be possible to construct such systems that are relatively unsuitable

**Table 8.8** Some aspects of physical modification of waterbodies that can aid in control of mosquitos by reducing opportunities or suitability for breeding (after SADH 2006)

| | |
|---|---|
| Environmental modification | Changes to characteristics such as pH or vegetation load can render site unsuitable |
| Water management | Runnelling (aiding tidal flushing of coastal sites), ditching or changing water depth can make water bodies unsuitable for mosquitos |
| Filling | Larval habitats filled in or covered with sand, earth or other material to eliminate the topographical depression that formerly allowed breeding |
| Draining | Drainage of the habitat so it can no longer support mosquito larvae; measures include open ditching, gravity drainage, and installation of tidal gates |

for mosquitos, and the list of considerations proposed for South Australia (SADH 2006), includes many such subtleties (Table 8.8).

Females of many species of aquatic flies must obtain a blood meal in order for egg maturation to occur, and giving them the potential to become vectors for viruses or other disease agents. Some non-biting flies, such as some Chironomidae, can cause allergenic reactions and irritations (Cranston 1995). The greatest health concerns for people are from arboviruses (Togaviridae), with both arbovirus and flavovirus diseases involved. The most significant examples, each with a significant record of investigation and documentation in Australia, include (1) Ross River virus; (2) Barmah Forest virus; (3) Murray Valley encephalitis; and (4) Dengue fever, each with a range of mosquito species as vectors, amongst which the freshwater *Culex annulirostris* is a widely distributed and dominant participant—and target for suppression. Reservoirs of each virus are maintained by other taxa: Ross River by native mammals and horses, Barmah Forest largely by domestic stock, Murray Valley mainly by waterbirds, and Dengue mainly in humans and vectored by *Aedes* mosquitos. Surveillance operations are sometimes feasible—sentinel flocks of chickens are deployed in the major range States and Northern Territory for detection of Murray valley encephalitis, for example.

A range of approaches to control of pest flies have been developed. Biological control by fish (p. 115) and chemical control by larvicides applied to water bodies have a long history of use and also of conservation concerns through non-target effects. They, and other approaches such as use of insect growth regulators and surface oiling to prevent emergence, focus on the aquatic early stages. Control of adult flies has rather less historical tradition, but chemical uses ('fly spray') and devices such as domestic 'bug zappers' can also pose non-target hazard which, if considered at all, is almost always secondary to pest supression.

Chemical larvicides and adulticides are used widely to suppress mosquito numbers, with the former used far more extensively and the latter employed mostly for domestic use or when disease outbreaks occur and rapid needs to curtail further transmission arise. Any such applications may need very careful consideration within the individual context of use, in order to minimise or avoid contaminations and impacts on non-target organisms.

However, all chemicals have some potential non-target impacts, and that recognition has—as in many other contexts of pesticide uses—led toward decreasing their use in favour of less-contaminating biological methods. For mosquitos, these have emphasised use of microbial pesticides, notably *Bacillus thuringiensis israelensis* and *B. sphaericus*. The latter, to quote from SADH (2006), 'is considered to be very specific, exhibits great toxicity against *Culex* spp. and *Anopheles* spp., provides effective control in polluted water systems and is unlikely to have an adverse impact on non-target organisms'.

Unexpected conservation complications can arise from pest fly control efforts. Blood-feeding blackflies (*Simulium* spp.) were associated with reduced nesting and reproductive success by the introduced (and endangered) Whooping crane (*Grus americana*) in Wisconsin, with swarms of flies (predominantly of *Simulium annulatus* and *S. johannseni*) leading to nest desertions and some egg breakages (Adler and Courtney 2019). Options available to attempt to reduce impacts raised the more general dilemma of how or whether to suppress populations of native flies in order to foster increased populations of endangered species. In this example, larval and adult *Simulium* may have significant ecological roles, but the key species are widespread and their vulnerability unlikely to be increased by their suppression. As in many other such decisions, some 'tradeoffs' are inevitable. The range of management actions embraced (1) do nothing; (2) suppress the blackflies; (3) manage crane nesting to occur outside the main flight periods of the simuliids; or (4) release the cranes into areas in which the problem flies are absent. The last has been attempted, with successful nesting, but destruction of the nests by predators thwarting any successful outcome (Adler and Courtney 2019).

## References

Adler PH, Courtney GW (2019) Ecological and societal services of aquatic Diptera. Insects 2109(10p):70. https://doi.org/10.3390/insects10030070

Ali A (1995) Nuisance, economic impact and possibilities for control. In: Armitage P, Cranston PS, Pinder LCV (eds) The Chironomidae. The biology and ecology of non-biting midges, Chapman and Hall, London, pp 339–364

Andersen NM, Weir TA (2004) Australian water bugs. Their biology and identification (Hemiptera -Heteroptera, Gerromopha and Nepomorpha. Entomonograph no 14. Apollo Books, Denmark/ CSIRO Publishing, Melbourne

Baird IRC, Burgin S (2016) Conservation of a groundwater-dependent mire-dwelling dragonfly: implications of multiple threatening processes. J Insect Conserv 20:165–178

Barber-James HM, Gattolliat J-L, Sartori M, Hubbard MD (2008) Global diversity of mayflies (Ephemeroptera, Insecta) in freshwater. Hydrobiologia 595:339–350

Barrett MD, Williams MR (1998) Distribution of the western petalura dragonfly *Petalura esperia* Watson in Western Australia. Pac Conserv Biol 4:149–154

Bayley IAE, Williams WD (1981) Inland waters and their ecology, 2nd edn. Longmans Cheshire, Melbourne

Bird MS, Bilton DT, Perissinotto R (2017) Diversity and distribution of polyphagan water beetles (Coleoptera) in the Lake St Lucia system, South Africa. ZooKeys 656:51–84
Bried JT, Hassall C, Simaika JP, Corser JD, Ware J (2015) Directions in dragonfly applied ecology and conservation science. Freshw Sci 34:1020–1022
Buchwald R (1992) Vegetation and dragonfly fauna—characteristics and examples of biocenological field studies. Vegetatio 101:99–107
Bybee S, Cordoba-Aguilar A, Duryea MC, Futahashi R, Hansson B (and nine other authors) (2016) Odonata (dragonflies and damselflies) as a bridge between ecology and evolutionary genomics. Front Zool 13:46. https://doi.org/10.1186/s12983-016-0176-7
Byers GW (1991) Mecoptera. In Nauman ID (chief ed) The Insects of Australia, 2nd edn. Melbourne University Press, Carlton. pp 696–704
Campbell IC (1990) The Australian mayfly fauna: composition, distribution and convergence. In: Campbell IC (ed) Mayflies and stoneflies. Kluwer Academic Publishers, Dordrecht, pp 149–153
Cardoso P, Erwin TL, Borges PA, New TR (2011) The seven impediments in invertebrate conservation and how to overcome them. Biol Conserv 144:2647–2655
Carew ME, Pettigrove V, Cox RL, Hoffmann AA (2007) The response of Chironomidae to sediment pollution and other environmental characteristics in urban wetlands. Freshw Biol 52:2444–2462
Chien W-C, Li M-H; Li H-J (2019) Appropriate management practices help enhance odonate species richness of small ponds in peri-urban landscapes. Urban Ecosyst 22:757–768
Clarke GM, Arthington AH, Pusey BJ (1994) Fluctuating asymmetry of chironomid larvae as an indicator of pesticide contamination in freshwater environments. In: Cranston P (ed) Chironomids: from genes to ecosystems. CSIRO Publications, Melbourne, pp 101–109
Clausnitzer V, Jodicke R (eds) (2004) Guardians of the watershed. Global status of dragonflies: critical species, threat and conservation. Int J Odonat 7:385–398
Clausnitzer V, Kalkman VJ, Ram M, Collen B, Baillie JEM (and 13 other authors) (2009) Odonata enter the biodiversity crisis debate: the first global assessment of an insect group. Biol Conserv 142: 1864-1869
Clausnitzer V, Dijkstra K-DB, Koch R, Boudot J-P, Darwall WRT (and five other authors) (2012) Focus on African freshwaters: hotspots of dragonfly diversity and conservation concern. Front Ecol Environ 10:129–134
Corbet PS (1999) Dragonflies. Behaviour and ecology of Odonata. Harley Books, Colchester
Cranston PS (1995) Introduction. In: Armitage P, Cranston PS, Pinder LCV (eds) the Chironomidae. The biology and ecology of non-biting midges, Chapman and Hall, London, pp 1–7
Cranston PS (2010) Insect biodiversity and conservation in Australasia. Annu Rev Entomol 55:55–75
Cover MR, Resh VH (2008) Global diversity of dobsonflies, fishflies and alderflies (Megaloptera: Insecta) and spongillaflies, nevrorthids and osmylids (Neuroptera: Insecta) in freshwater. Hydrobiologia 595:409–417
Davis J (1998) A guide to the identification of larval Psephenidae water pennies (Insecta, Coleoptera). Identification and ecology guide no. 17. Cooperative Research Centre for Freshwater Ecology, Thurgoona
Davis JA, Rolls SW, Balla SA (1987) The role of Odonata and aquatic Coleoptera as indicators of environmental quality in wetlands. (In Majer JD (ed) The role of invertebrates in conservation and biological survey. Department of Conservation and Land Managemen, Perth, Western Australaia, pp 31–42
Dean JC, St Clair RM, Cartright DI (2004) Identification keys to Australian families and genera of caddis-fly larvae (Trichoptera). Cooperative Research Centre for Freshwater Ecology, Identification Guide No 50, Albury-Wodonga
de Figueroa JMT, Lopez-Rodriguez MJ, Lorenz A, Graf W, Schmidt-Kloiber A, Hering D (2010) Vulnerable taxa of European Plecoptera (Insecta) in the context of climate change. Biodiv Conserv 19:1269–1277
de Jong H, Oosterbrock P, Gelhaus J, Reusch H, Young C (2008) Global diversity of craneflies (Insecta, Diptera: Tipulidea or Tipulidae *sensu lato*) in freshwater. Hydrobiologia 595:457–467

DeWalt RE, Ower GD (2019) Ecosystem services, global diversity, and rate of stonefly species descriptions (Insecta: Plecoptera). Insects 2019, 10, 99. https://doi.org/10.3390/insects10040099

DeWalt RE, Favret C, Webb DW (2005) Just how imperiled are aquatic insects? A case study of stoneflies (Plecoptera) in Illinois. Ann Entomol Soc Amer 98:941–950

Dijkstra K-DB, Monaghan MT, Pauls SU (2014) Freshwater biodiversity and aquatic insect diversification. Annu Rev Entomol 59:143–163

Dolny A, Harabis F, Barta D, Lhota S, Drozd P (2012) Aquatic insects indicate terrestrial habitat degradation: changes in taxonomical structure and functional diversity of dragonflies in tropical rainforest of East Kalimantan. Trop Zool 25:141–157

Drinan TJ, Grainger NPJ, Harding JS, Collier KJ, Smith BJ (and three other authors) (2020) Analysis of the conservation status of New Zealand freshwater invertebrates: temporal changes, knowledge gaps, impediments, and management implications. N Z J Zool https://doi.org/10.1080/03014223.2020.1778044

Fairchild GW, Faulds AM, Matta JF (2008) Beetle assemblages in ponds: effects of habitat and site age. Freshw Biol 44:523–534

Ferrington LC (2008a) Global biodiversity of scorpionflies and hangingflies (Mecoptera) in freshwater. Hydrobiologia 595:443–445

Ferrington LC (2008b) Global diversity of non-biting midges (Chironomidae—Insecta: Diptera) in freshwater. Hydrobiologia 595:447–455

Fochetti R, de Figueroa JMT (2006) Notes on diversity and conservation of the European fauna of Plecoptera (Insecta). J Nat Hist 40:2361–2369

Forteath GNR, Osborne AW (2012) Biology, ecology and voltinism of the Australian spongillafly *Sisyra pedderensis* (Neuroptera: Sisyridae). Pap Proc R Soc Tasm 146:25–35

Foster GN, Bilton DT (2014) The conservation of predaceous diving beetles: knowns, unknowns and anecdotes. In: Yee DA (ed) Ecology, systematics, and the natural history of predaceous diving beetles (Coleoptera: Dytiscidae). Springer Science + Business Media, Dordrecht, pp 437–462

Glaister A (1999) Guide to the identification of Australian Elmidae larvae (Insecta: Coleoptera). Identification and Ecology Guide no. 21. Cooperative Research Centre for Freshwater Ecology, Thurgoona

Guerold F, Boudot J-P, Jacquemin G, Vein D, Merlet D, Rouiller J (2000) Macroinvertebrate community loss as a result of headwater stream acidification in the Vosges (N-E France). Biodiv Conserv 9:767–783

Harabis F, Dolny A (2012) Human altered ecosystems: suitable habitats as well as ecological traps for dragonflies (Odonata): the matter of scale. J Insect Conserv 16:121–130

Harabis F, Tichanek F, Tropek R (2013) Dragonflies of freshwater pools in lignite spoil heaps: restoration management, habitat structure and conservation value. Ecol Engin 55:51–61

Hardwick RA, Wright IA, Jones HA, Chessman BC, Hollely DE (2004) Rapid biological assessment of streams in the Blue Mountains, Australia: characteristics of the Chironomidae fauna. In: Cranston P (ed) Chironomids: from genes to ecosystems. CSIRO Publications, Melbourne, pp 155–160

Hawking JH (1997) The conservation status of dragonflies (Odonata) from south-eastern Australia. Mem Mus Vict 56:537–542

Hawking JH (1999) An evaluation of the current conservation status of Australian dragonflies (Odonata). In Ponder W, Lunney D (eds) The other 99%. The conservation and biodiversity of invertebrates, Royal Zoological Society of New South Wales, Mosman, pp 354–360

Hawking JH (2001) An introduction to the identification of aquatic caterpillars (Lepidoptera) found in Australian inland waters. Identification guide no 37, Murray-Darling Freshwater Research Centre, Albury

Hawking JH (2014) Systematics and ecology of the Australian aquatic moths, Acentropinae (Insecta: Lepidoptera). Unpublished PhD Thesis, La Trobe University, Bundoora, Victoria

Hawking JH, New TR (2002) Interpreting dragonfly diversity to aid in conservation assessment: lessons from the Odonata assemblage at Middle Creek, north-eastern Victoria, Australia. J Insect Conserv 6:171–178

Hawking JH, Theischinger G (2004) Critical species of Odonata in Australia. Int J Odonat 7:113–132
Hearnden MN, Pearson RG (1991a) The diets of mayflies in a tropical Australian rainforest stream. Trop Freshw Biol 2: 203–212
Hearnden MN, Pearson RG (1991b) Habitat partitioning among the mayfly species (Ephemeroptera) of Yuccabine Creek, a tropical Australian stream. Oecologia 87: 91–101
Hofmann TA, Mason CF (2005) Habitat characteristics and the distribution of Odonata in a lowland river catchment in eastern England. Hydrobiologia 539:137–147
Houghton DC, Holzenthal RW (2010) Historical and contemporary biological diversity of Minnesota caddisflies; a case study of landscape-level species loss and trophic composition shift. J N Am Benthol Soc 29:480–495
Jach MA (1998) Annotated checklist of aquatic and riparian/littoral beetle families of the world (Coleoptera). In Jach MA, Li I (eds) Water beetles of China, Vol 2, Vienna, pp 25–42
Jach MA, Balke M (2008) Global diversity of water beetles (Coleoptera) in freshwater. Hydrobiologia 595:419–442
Jackson J (2000) Threatened Trichoptera (caddisflies) from the Tasmanian Wilderness World Heritage Area. Proc R Soc Tas 134:55–62
Janssens de Bisthoven L, Huysmans C, Ollevier F (1994) The *in situ* relationships between sediment concentrations of micropollutants and morphological deformities in *Chironomus* gr. *Thummi* larvae from lowland rivers (Belgium): a spatial comparison. In: Cranston P (ed) Chironomids: from genes to ecosystems. CSIRO Publications, Melbourne, pp 63–80
Jin YH, Bae YJ (2005) The wingless stonefly family Scopuridae (Plecoptera) in Korea. Aquat Insects 27:21–34
Kalkman VJ, Clausnitzer V, Dijkstra K-DB, Orr AG, Paulson DR, van Tol J (2008) Global diversity of dragonflies (Odonata) in fresh water. Hydrobiologia 595:351–363
Kalkman VJ, Boudot J-P, Bernard R, Conze KJ, De Knijf G (and six other authors) (2010) European Red List of Dragonflies. European Union, Luxembourg
Kalkman VJ, Boudot J-P, Bernard R, De Knijf G, Suhling F, Termaat T (2018) Diversity and conservation of European dragonflies and damselflies (Odonata). Hydrobiologia 811:269–282
Khelifa R (2019) Sensitivity of biodiversity indices to the life history stage, habitat type and landscape in Odonata community. Biol Conserv 237:63–69
Kimura G, Mishima T, Hirabayashi K (2011) Species composition and abundance of craneflies (Diptera: Tipulidae) in the highland lakes of Japan. J Freshw Ecol 26:91–97
Kobayashi M, Iwai D (2007) Environmental conditions of the microhabitat of *Lethocerus deyrollei* (Vuillefroy) during its activity period in the northern part of the Kanto Plain, Japan. Jap J Environ Entomol Zool 18:133–136 (in Japanese, cited by Ohba 2018)
Kutcher TE, Bried JT (2014) Adult Odonata conservatism as an indicator of freshwater wetland condition. Ecol Indic 38:31–39
Lindegaard C (1995) Classification of water-bodies and pollution. In: Armitage P, Cranston PS, Pinder LCV (eds) the Chironomidae. The biology and ecology of non-biting midges, Chapman and Hall, London, pp 385–404
Madden CP, Austin AD, Suter PJ (1994) Pollution monitoring using chironomid larvae: what is a deformity? In: Cranston P (ed) Chironomids: from genes to ecosystems. CSIRO Publications, Melbourne, pp 89–100
Marchant R (1986) Estimates of annual production for some aquatic insects from the La Trobe River, Victoria. Aust J Mar Freshw Res 37:113–120
Massariol FC, Soares EDG, Salles FF (2014) Conservation of mayflies (Insecta, Ephemeroptera) in Espirito Santo, southeastern Brazil. Rev Bras Entomol 58:356–370
Mey W, Speidel W (2008) Global diversity of butterflies (Lepidoptera) in freshwater. Hydrobiologia 59(5):521–528
Moore NW (1997) Dragonflies: status survey and conservation action plan. IUCN/SSC Odonata Specialist Group. IUCN,Gland and Cambridge
Morse JC, Frandsen PB, Graf W, Thomas JA (2019) Diversity and ecosystem services of Trichoptera. Insects 10:125. https://doi.org/10.3390/insects10050125

Mukai Y, Baba N, Ishii M (2005) The water system of traditional rice paddies as an important habitat of the giant water bug, *Lethocerus deyrollei* (Heteroptera: Belostomatidae). J Insect Conserv 9:121–129

Nagaba Y, Tufail M, Inui H, Takeda M (2011) Hormonal regulation and effects of four environmental pollutants on vitellogin gene transcription in the giant water bug *Lethocerus deyrollei* (Hemiptera: Belostomatidae). J Insect Conserv 15:421–431

New TR (1983) The food of adult Osmylidae: Kempyninae (Neuroptera). Aust Entomol Mag 10:45–46

New TR (2004) Introduction to the systematics and biology of Australian aquatic Neuroptera. Identification and Ecology Guide no 49. Cooperative Research Centre for Freshwater Ecology, Thurgoona

Ngiam RWJ, Lim WL, Collins CM (2017) A balancing act in urban social-ecology: human appreciation, ponds and dragonflies. Urban Ecosyst 20: 743–758

Niba A, Samways MJ (2006) Development of the concept of 'core resident species' for quality assurance of an insect reserve. Biodiv Conserv 15:4181–4196

Nicacio G, Juen L (2015) Chironomids as indicators in freshwater ecosystems: an assessment of literature. Insect Conserv Divers 8:393–403

Ohba S (2018) Ecology of giant water bugs (Hemiptera: Heteroptera: Belostomatidae). Entomol Sci (2018) https://doi.org/10.1111/ens.12334

Osborn R, Samways MJ (1996) Determinants of adult dragonfly assemblage patterns at new ponds in South Africa. Odonatologica 25:49–58

Pabis K (2018) What is a moth doing under water? Ecology of aquatic and semi-aquatic Lepidoptera. Knowl Manag Aquat Ecosyst 419:42. https://doi.org/10.1051/kmae/2018030

Polhemus JT, Polhemus DA (2008) Global diversity of true bugs (Heteroptera: Insecta) in freshwater. Hydrobiologia 595:379–391

Prescott VA, Eason PK (2018) Lentic and lotic odonate communities and the factors that influence them in urban versus rural landscapes. Urban Ecosyst 21: 737–750

Racek AA (1969) The freshwater sponges of Australia. Aust J Mar Freshw Res 20:267–310

Raebel EM, Merckx T, Feber RE, Riordan P, Macdonald DW, Thompson DJ (2012) Identifying high-quality pond habitats for Odonata in lowland England: implications for agri-environment schemes. Insect Conserv Divers 5:422–432

Rivera-Gasperin SL, Ardila-Camacho A, Contreras-Ramos A (2019) Bionomics and ecological services of Megaloptera larvae (dobsonflies, fishflies, alderflies). Insects 2109, 10,86: https://doi.org/10.3390/insects10040086

Samways MJ, Simaika JP (2016) Manual of freshwater assessment for South Africa: Dragonfly Biotic Index. Suricata 2. South African National Biodiversity Institute, Pretoria

Samways MJ, Steytler NS (1996) Dragonfly (Odonata) distribution patterns in urban and forest landscapes, and recommendations for riparian management. Biol Conserv 78:279–288

Sanchez-Bayo F, Wyckhuys KAG (2019) Worldwide decline of the entomofauna: a review of its drivers. Biol Conserv 232:8–27

Schmidt E (1995) A survey of threatened dragonfly habitats in Central Europe, especially bogs and bog management. In: Corbet PS, Dunkle SW, Ubukata H (eds) Proc Int Symposium on Conservation of dragonflies and their habitats, Kushiro, Japan. Japanese Society for Protection of Birds, Kushiro, pp 45–68

Sharley DJ, Pettigrove V, Parsons YM (2004) Molecular identification of *Chironomus* spp. (Diptera) for biomonitoring of aquatic ecosystems. Aust J Entomol 43:359–365

Sharma S, Sharma G, Pir FA (2019) Diversity and habitat selection of aquatic beetles (Coleoptera). IOSR J Pharm Biol Sci 14:31–37

Simaika JP, Samways MJ (2009) An easy-to-use index of ecological integrity for prioritizing freshwater sites and for assessing habitat quality. Biodiv Conserv 18:1171–1185

Smith M, Cranston PS (1994) "Recovery' of an acid mine-waste impacted tropical stream—the chironomid story. In: Cranston P (ed) Chironomids: from genes to ecosystems. CSIRO Publications, Melbourne, pp 161–173

Spangler J (1986) Insecta: Coleoptera. In: Botosaneunu L, Stock JH (eds) Stygofauna: a faunistic, distributional and ecological synthesis of the world fauna inhabiting subterranean waters (including the marine interstitial). Brill, Leiden, pp 622–631

Stagliano DM (2016) Mayflies (Insecta: Ephemeroptera) of conservation concern in Montana: status updates and management needs. Western N Am Nat 76:441–451

Suhonen J, Hilli-Lukkarinen M, Korkeamaki E, Kuittunen M, Kullas J, Pentinnen J, Salmela J (2010) Local extinction of dragonfly and damselfly populations in low- and high-quality habitat patches. Conserv Biol 24:1148–1153

Suhonen J, Korkeamaki E, Salmela J, Kuitunen M (2014) Risk of local extinction of Odonata freshwater habitat generalists and specialists. Conserv Biol 28:783–789

Suter PJ, McGuffie P (1997) Conservation of mayflies (Ephemeroptera) especially *Coloburiscoides* in the Victorian Alps: impediments and threats. Vict Nat 124: 273–277

Theischinger G (2000) Australian alderfly larva and adults (Insecta: Megaloptera). Preliminary guide to the identification of larvae and survey of adults of Australian alderflies. Cooperative Research Centre for Freshwater Ecology, Identification Guide No 29, Albury-Wodonga

Villalobos-Jimenez G, Dunn AM, Hassall C (2016) Dragonflies and damselflies (Odonata) in urban ecosystems: a review. Eur J Entomol 113: 217–232

Wagner R, Bartak M, Borkent A, Courtney G, Goddeeris B (and nine other authors) (2008) Global diversity of dipteran families (Insecta Diptera) in freshwater (excluding Simuliidae, Culicidae, Chironomidae, Tipulidae and Tabanidae). Hydrobiologia 595: 489–519

Watson JAL, O'Farrell AF (1991) Odonata (dragonflies and damselflies). In Naumann ID (chief ed) The insects of Australia, 2nd edn, pp 294–310, Melbourne University Press, Carlton

Watson JAL, Theischinger G, Abbey HM (1991) The Australian dragonflies. A guide to the identification, distribution and habits of Australian Odonata. CSIRO Publishing, Melbourne

Winterbourn MJ (1980) The freshwater insects of Australasia and their affinities. Palaeogeog, Palaeoclim, Palaeoecol 31:235–249

Wissinger SA (1998) Spatial distributions, life history and estimates of survivorship in a fourteen-species assemblage of larval dragonflies (Odonata: Anisoptera). Freshw Biol 20:329–340

Yadamsuren O, Hayford B, Gelhaus J, Ariuntsetseg L, Goulen C, Podemas S, Podeniene V (2015) Declines in diversity of crane flies (Diptera: Tipuloidea) indicate impact from grazing by livestock in the Hovsgol region of Mongolia. J Insect Conserv 19:465–477

Zedkova B, Radkova V, Bojkova J, Soldan T, Zahradkova S (2015) Mayflies (Ephemeroptera) as indicators of environmental changes in the past five decades; a case study from the Morava and Odra River Basins (Czech Republic). Aquat Conserv Mar Freshw Ecosyst 25: 622–638

# Chapter 9
# Australia's Flagship Freshwater Insects

## 9.1 Introduction

The following series of comments on individual species reveals the variety of taxa that have attracted conservation concern in Australia, mostly through being 'listed' under some relevant legislation. Many of those listings have eventuated from rather flimsy evidence, but with the wise precaution that poorly known species that are supposedly or demonstrably scarce, those restricted to or known from only single localities, and those that have not been seen for considerable periods may indeed be in need of focused attention to clarify their status and assure their wellbeing. Selecting those species in a poorly documented fauna is problematical and allocating conservation status and ranking species for practical management needs is a dynamic exercise open to severe revision and reallocation of resources as new information is accumulated, new populations are discovered, older ones lost, and new threats arise. In many cases, designation of purportedly needy species is somewhat serendipitous and, however well-intentioned, may need further investigation to justify.

The problems of reliably allocating conservation status to poorly known aquatic invertebrates emerged clearly from two examples discussed by Doeg (1997). Both the Otway stonefly (p. 225) and the Dandenong amphipod (*Austrogammarus australis*) had been considered extinct, but this inference clearly resulted from inadequate surveys. From Doeg's surveys, *Austrogammarus* was found at only two of the 47 sites sampled so, although it was clearly not extinct, the amphipod was still of high conservation concern. *Eusthenia nothofagi*, in contrast, was widespread in the Otway Ranges and appeared not to be in danger. Doeg emphasised that surveys of equivalent intensity for aquatic species are rarely undertaken, and that their true distributions may not be revealed through more sporadic (and more usually undertaken) investigations, or in cases in which taxonomic confusion is possible. Descriptions of new species (often from single localities), or clarification of species complexes, may also complicate interpretations and invalidate historical information. A classic European example is of the caddisfly *Hydropsyche tobiasi*, for long considered the only apparent extinction of an aquatic insect, and with no specimens reported since


T. R. New, *Insect Conservation and Australia's Inland Waters*,
https://doi.org/10.1007/978-3-030-57008-8_9

the 1920s from its only recorded range, in the River Rhine in Germany (Polhemus 1993). However, the species was described only in 1977 and was previously confused with the common *H. exocellata*—so that Polhemus considered the possibility that some populations of *H. tobiasi* could persist unrecognised as masquerading within the more common taxon. This has not been confirmed. Taxonomic uncertainty may extend to undetected errors in synonymy—and may become especially relevant when species are nominated by non-specialists for 'listing'. Another instructive example, from North America, is the Platte River caddisfly (*Ironoquia plattensis*) which was described from Nebraska in 2000 but because it was not found at the type locality in 2004 was apparently rare and declining (Vivian et al. 2013). However, surveys for larvae and adults prompted by this assumption revealed 23 populations along the Platte River, and others on nearby rivers; this larger documented range was the direct result of targeted sampling effort to clarify the species' status and exemplifies the approach needed urgently for many Australian taxa. Because the caddisfly was absent from the most degraded regions of the Platte River, with other sites vulnerable, it may be an important indicator of healthy regions, and conservation efforts should include attempts to slow and manage future degradation (Vivian et al. 2013).

Extant lists of insect species under conservation legislations are rarely definitive, and differences in numbers across different taxa tend to reflect the activities of specialists nominating them, rather than their real vulnerability, together with the realistic major gaps in knowledge so that many species are still data deficient. The strong bias toward Odonata amongst aquatic insects on protected species lists reflects this relative interest and attention.

Formal 'listing' of a species as threatened in principle supports commitment to conservation of that species. However, in most cases little practical action flows from such priority designation. Typically, listed vertebrates and vascular plants are more likely to be attended than even the most needy or 'popular' invertebrates. In turn, aquatic invertebrates constitute a small proportion of listed invertebrates, so that the bulk of the very limited resources devoted to insect conservation falls to terrestrial taxa. For many, the complications arising from the dual lifestyle of aquatic larvae and terrestrial adult insects add to the complexity of conservation management and assessing needs.

Detailed life histories and seasonal patterns, distributions and resource needs of most of Australia's aquatic insects are incompletely known—so that the factors contributing to declines, and their seasonal patterns of vulnerability are also unclear and only limited predictions of impacts of imposed disturbances may be possible. Assumptions about a species' biology—such as that a species living in ephemeral streams is adapted to withstand periods of stream drying—may be difficult to validate (Robson et al. 2011). As Robson et al. put it 'The difficulty is that we know relatively little of the adaptations of stream macroinvertebrate fauna to intermittent flows', so that impacts of additional water extraction or other change can rarely be predicted with any certainty. Likewise, seasonal use or availability of putative refuges may vary. These, and many other topics, are much under-investigated for most species of conservation concern.

As Abellan et al. (2005) noted, for many insects *any* increase in knowledge could lead to changed conservation category. Leading documents, such as the World Conservation Union's 'Red Lists', thus need continual updating to accommodate changes, but this does not preclude needs for further targeted surveys to provide the foundations for selecting and managing those species of greatest concern, namely those allocated currently as having high conservation need or known (or suspected) to be narrow range endemics in restricted threatened environments. Some form of priority Red List of Australian insects, expanded to include representatives of all major biological regions of Australia was an integral component of the national strategy proposed by Taylor et al. (2018), and must build on the scanty framework suggested from the few species currently listed as of concern. Following the path advocated by Haslett (1997) in suggesting additions to the Bern Convention in Europe to also expand habitat representation, some equivalent to satisfying the needs of one of three target criteria of concern ('threats to species of running waters') could be approached in Australia. The Bern Convention (fully, the Convention on the Conservation of European Wildlife and Natural Habitats) was signed in 1979. A later literature review discussed by Foster (1991) reported that 192 species of water beetles were considered threatened in at least one country of Europe, following three criteria used to formulate priorities, namely (1) the species should be under serious threat in Europe as a whole, but not necessarily in every place; (2) the species should be reasonably easy to identify; and (3) the species' distribution should be mainly in Europe. The second of these conditions, combined with the very small size of many such taxa, rendered some species very difficult to appraise. Haslett (1997) recognised the wide range of threats to European Odonata and noted that 15 species of Odonata and two species of aquatic Coleoptera (*Dytiscus latissimus*, *Graphoderus lineatus*, both in standing waters) were by then listed, and reported on a further eight species of Odonata that merited inclusion. Most of these were inadequately documented. Thus, comment on *Coenagrion caerulescens* included that threats 'are hard to define'—a sentiment that applies easily to many aquatic taxa in Australia—but with some other taxa more clearly succumbing to changes to their major habitats. The list in Appendix A (p. 243) contains many such relatively poorly documented species.

Aquatic insect 'flagships' are relatively few in any part of the world and most of the species designated as threatened or in need of conservation have not received focused management attention or detailed study, and lack conservation recovery plans. Whilst some conspicuous or 'attractive' feature can help to capture or augment public appreciation of a potential flagship insect, less generally appealing species are also evident. The first aquatic insect proposed for addition to the United States List of Endangered and Threatened Wildlife and Plants was a rather inconspicuous shore bug, described by Resh and Sorg (1983) as 'small (4 mm) and nondescript'. The Wilbur Springs shore bug (*Saldula usingeri*, Saldidae) occurred in the Hot Springs area of California. Initially known only from a single spring, and then nominated for listing, searches for the bug confirmed that it occurred also at several other springs in the area and was locally abundant in some areas where other *Saldula* spp. were absent. Water chemistry, including salinity, was an important correlate of the bug's occurrence, together with incidence of the predominant prey, larvae of the brine fly

*Paracoenia calida* (Ephydridae). Some suitable springs with high concentrations of sodium, chlorine and lithium have been changed to eliminate surface flows—such as by diversion of the water into baths for health resorts. Somewhat ironically, as Resh and Sorg pointed out, their surveys confirmed that *S. usingeri* did not qualify for the proposed listing but the prey brine fly was much more restricted by temperature and water chemistry and 'would be a far better candidate for protected status'.

Most aquatic insects are the subjects of 'passive conservation' rather than being the primary 'active' focus as specific targets of an individually designed conservation plan. Passive conservation implies that the species become the legatees of more general conservation measures or attention that primarily seek to protect habitat features, site condition or other focal species present. For Dytiscidae, Foster and Bilton (2014) remarked 'Sadly, most Dytiscidae will always depend on 'passive' conservation for their survival', and parallel sentiments apply to most other insect taxa—notwithstanding that many species are signalled as of conservation concern and need. That passive approach can easily lead to key sites occupied by individual notable species being overlooked—for example, Foster and Bilton noted the absence of inland saline water bodies in European priorities. In all insect groups, cryptic taxa or closely related species may easily be overlooked without closer focus. In many cases it is unknown whether these taxa depend on such restricted biotopes or have less specialised needs, so that defining 'suitable' or 'unsuitable' habitats can become a very uncertain exercise.

Nevertheless, in some cases, particular threatened aquatic insects have become important flagship species in conservation advocacy. Two North American taxa, contrasting in conspicuousness and public interest, demonstrate how conservation of such species may be pursued following formal designation as 'endangered' and so accorded conservation priority. Such status, in Australia under the Commonwealth's Environmental Protection and Biodiversity Conservation Act 1999 (EPBC Act), brings notice of their plight and needs to wider attention—but still does not guarantee that further investigation and practical protection will occur. Without such recognition, however, any such progress remains highly unlikely.

The two North American examples both occur in the central region of the continent and in both Canada and the United States. The small (adult body length about 3.7–4.4 mm) Hungerford's crawling water beetle (*Brychius hungerfordi*, Haliplidae) is endemic to the Great Lakes region. It is listed as 'endangered' in both range countries, in which it is restricted to Ontario and Michigan, respectively. The beetle is restricted to small to medium-sized streams with moderate to fast flow, good aeration, inorganic substrate and alkaline cool (15–25 °C) water (COSEWIG 2011). It has been recorded only from three rivers in Ontario, and five streams in northern Michigan, with an estimated global extent of occurrence of <6 Km$^2$ (COSEWIG 2011) and is a presumed postglacial relict species—so that incidence elsewhere is considered highly unlikely. Larvae are believed to be herbivores and probable specialist feeders on the filamentous alga *Dichotomosiphon tuberosus* in Canada, with a wider diet implied in the United States (USFWS 2006). Concerns continue over future security of the very restricted habitats. All the Canadian localities, for example, are likely to undergo increased threats from changed water quality and structural changes, as well as alterations to the shoreline sites used for pupation. However, threats are not well understood (USFWS 2006), and it is assumed that changes such as the above that affect the aquatic environment, together with the small and highly localised populations, will increase vulnerability and susceptibility to stochastic events. However, a statement by USFWS (2006) is salutary also for many examples of insect flagship species conservation—namely 'the greatest threat to advancing recovery of this species may be the lack of information on its ecology and natural history', with more information needed on microhabitat and resource needs and limitations. At that time the critical habitat for *B. hungerfordi* could not be defined properly, but it was considered important to protect the few occupied sites by keeping them free from disturbance. The USFWS Recovery Plan also listed the need to develop and implement site conservation plans for each, in order to address local threats, and also to encourage a wider watershed approach to conservation. As in many parallel cases of insect species conservation, the general measures emanating from sound biological knowledge should wherever possible be combined with the specific measures needed for each site of concern. The diversity of actions needed for *B. hungerfordi* is exemplified in Table 9.1 (USFWS 2006).

The second example is Hine's emerald dragonfly (*Somatochlora hineana*, Corduliidae), one of the rarest North American Odonata and listed as endangered (in the United States in 1995, the first dragonfly acknowledged under the national Endangered Species Act). It is known from parts of southern Ontario, and from sites in Illinois, Wisconsin, Michigan and Missouri and is believed to have become extinct in Ohio and Indiana. In all its occupied sites, the dragonfly is associated with marshes and sedge meadows with slowly flowing water fed from calcareous groundwater seepages and having forested edges. The major cause of decline is habitat loss from agricultural and other development (USFWS 2001), with surviving populations continuing to be under threat from this loss and from succession processes.

Genetic studies (Monroe and Britten 2014) on this long-lived species (larvae may take up to five years to develop) implied need to treat the discrete northern and southern range units separately for management. The low number of populations, most of them small and perhaps representing a metapopulation structure, dictated that their protection within suitable aquatic and adjacent terrestrial environments could help to counter some declines. 'Critical habitat' for *S. hineana* was designated in 2010 (USFWS 2001). The considerable complexities and controversies of achieving this for an aquatic insect with widely-dispersing adults across lands with differing tenures and planning options/uses were displayed in that designation. As one unusual example, train speeds across two sites in Illinois were restricted to only 4–6 miles/h during the adult flight season in an effort to reduce mortality from direct impacts, but sediment released from rail ballast may also affect the larval environment. The main specific components ('Primary Constituent Elements') of critical habitat for the dragonfly are listed in Table 9.2, to indicate the subtleties of providing a reasonably complete description of the needs for conserving the species, and which may be paralleled in many other taxa.

**Table 9.1** The broad scope of recovery actions proposed for the conservation of Hungerford's crawling water beetle, *Brychius hungerfordi*, in North America (USFWS 2006, in which many further details are explained)

| Conserve known sites |
| --- |
| Define and conserve areas of essential habitat |
| Develop and implement site conservation plans for each site to address threats |
| Review Federal, State and private actions |
| Review and conservation, under provisions of Sect. 9.7 of the Endangered Species Act |
| Permits: under Chap. 10 of the Endangered Species Act 'enhancement of survival' permits are needed for conservation research, and 'incidental take' permits needed, contingent on development of a habitat conservation plan |
| Land acquisition and conservation |
| Encourage watershed level conservation |
| Conserve riparian buffers |
| Conduct restoration activities that result in overall benefits to the watershed after ensuring that benefits to *B. hungerfordi* outweigh risks |
| Implement erosion control best management practices for road/stream crossings and other projects to minimise sedimentation as appropriate |
| Conduct in-stream projects such as bank stabilisation projects as appropriate |
| Conduct other stream and watershed restoration activities that result in benefits to occupied watersheds, as appropriate |
| Investigate the potential for transportation of hazardous materials (e.g. oil and other chemicals) on roads within occupied watersheds and potential for spills |
| Coordinate with Canadian officials (regarding a particular Canadian site) |
| Conduct scientific research to facilitate recovery efforts |
| Conduct studies to examine life history and ecology of *B. hungerfordi* |
| Examine habitat requirements |
| Confirm threats to the species |
| Conduct studies to examine population dynamics and demography |
| Investigate genetic heterogeneity and population viability |
| Investigate utility of captive propagation and translocation |
| Investigate the hydrological needs of the species |
| Convene scientific meeting/conference to share information |
| Conduct additional surveys and monitor existing sites |
| Develop standard surveys and monitoring protocols |
| Continue to survey new locations to identify new populations or areas of suitable habitat |
| Develop and implement a monitoring plan for all known sites |
| Develop and implement public education and outreach |
| Design educational materials and presentations for the public |
| Develop educational materials on *B. hungerfordi* and endangered species conservation |

(continued)

**Table 9.1** (continued)

| |
|---|
| Conserve known sites |
| Develop informational materials on the importance of local and watershed-level conservation |
| Conduct landowner contact and educational programmes to increase awareness of *B. hungerfordi* |
| Contact local organisations to inform them of the beetle |
| Revise recovery criteria and recovery actions, as appropriate, based on research and new information |
| Develop a plan to monitor *B. hungerfordi* after it is delisted |

**Table 9.2** The 'Primary Constituent Elements' (PCEs) aiding definition of critical habitat for Hine's emerald dragonfly, as physical and biological features that are essential to the conservation of the species and which may require special management considerations or protection (USFWS 2001)

| |
|---|
| *General: PCEs include, but are not limited to*: |
| 1. Space for individual and population growth and for normal behavior |
| 2. Food, water, air, light, minerals, or other nutritional or physiological requirements |
| 3. Cover or shelter |
| 4. Sites for breeding, reproduction and rearing (or development) of offspring |
| 5. Habitats that are protected from disturbance or are representative of the historical, geographical, and ecological distributions of a species |
| *Specific: PCEs designated for Hine's emerald dragonfly:* |
| *Habitat requirements are:* |
| 1. For egg deposition and larval growth and development |
| a. Organic soils overlying calcareous substrate |
| b. Calcareous water from intermittent seeps and springs and associated shallow, small, slow-flowing streamlet channels, rivulets, and/or sheet flow within fens |
| c. Emergent herbaceous and woody vegetation for emergence facilitation and refugia |
| d. Occupied burrows maintained by crayfish for refugia |
| e. Prey base of aquatic macroinvertebrates, including mayflies, aquatic isopods, caddisflies, midge larvae, and aquatic worms |
| 2. For adult foraging; reproduction; dispersal; and refugia necessary for roosting, resting, refuge for adult females to escape from male harassment, and predator avoidance (especially during the vulnerable teneral stage): |
| a. Natural plant communities near the breeding/larval habitat which may include fen, marsh, sedge meadow, dolomite prairie and the fringe (up to 100 m) of bordering shrubby and forested areas with open corridors for movement and dispersal |
| Prey base of small flying insect species (e.g. dipterans) |

## 9.2 Selected Flagship Taxa

A list of the aquatic insect species reported individually as of conservation concern in Australia (Taylor et al. 2018) included many that are known to few people beyond a small group of specialists in that group and which, therefore, scarcely intrude on wider awareness or concern. The list is inevitably a 'first pass', with some taxa listed on the basis of very incomplete knowledge of their distributions and biology. It does, however, include a variety of insects that are regarded as rare and/or signaled, some by formal 'listing' under national or regional legislation, as in some way 'threatened' and in need of conservation. The list is dominated numerically by Odonata (Appendix A)—by far the best known and most intensively studied order—followed by Trichoptera and Plecoptera, with very low representation of other orders. Several orders are not represented, but this is by no means an assurance of their security, rather of the amount of assessment that remains to be done in this still relatively poorly known fauna. However, despite its highly incomplete coverage, the list is sufficient to imply considerable variety and uncertainty of conservation needs. It includes the selected 'flagship species' noted here, and others are possible future candidates for that status.

The taxa noted below represent many of Australia's freshwater ecosystems, but most of the species are from the south-eastern regions. A number of rare alpine stream insects support the general assertion by Gomi et al. (2002) of the importance of headwaters for aquatic insects, that 'new and endangered species are often found in headwater streams because such streams are relatively unexplored'. Those waters are also usually isolated, so that they support genetically isolated species or populations as important components of local biodiversity. As Gomi et al. emphasised, headwater features are linked intricately with downstream riparian ecosystems, and can influence their wellbeing. Macroinvertebrates of first- and second-order headwater streams are dominated by 'shredders' that break larger pieces of organic matter into smaller particles, but those streams may also become dry for part of the year, and their invertebrate residents must then depend on refuges such as remnant wet pools or having desiccation-resistant stages in their life history. Otherwise, replenishment of losses must depend largely on upstream movements and recolonisation—in many cases impossible simply because the species of concern do not occur in downstream locations, and colonisation by others might constitute a further threat. The wide ecological importance of headwaters (p. 16) as 'guardians of many downstream resources and ecosystem services on which humans rely' (Lowe and Likens 2005), as well as their being critical habitat for numerous specialist species not found elsewhere—and many of which are of significant conservation interest—emphasises the need for their effective long-term protection.

## 9.3 Ephemeroptera

### *9.3.1 The Large Blue Lake Mayfly,* Tasmanophlebia lacuscoerulei *(Oniscigastridae)*

One of Australia's largest mayflies, fully grown larvae of *T. lacuscoerulei* can reach about 40 mm in length, and adult wingspan is >30 mm. It occurs near the highest point of Australia, Mount Kosciuszko (New South Wales), where the five lakes from which it has been reported are all within the national park. No other similar alpine meadow habitats are known in the region, and *T. lacuscoerulei* takes its name from the type locality, Blue Lake. Concerns have been expressed over water quality there, as Blue Lake is a popular campsite area, and the neighbouring Lake Albina receives effluent from a nearby septic tank, as noted by Wells et al. (1983).

Suter (2014) discussed the few records of the mayfly and considered the verified distribution to comprise only Blue Lake and the Blue Lake inlet stream, with possible incidence in Lake Albina and Lake Cootapatamba, with some doubts over synonymy within the genus rendering other putative records needing further investigation. Assessment as 'Endangered', following a series of earlier appraisals of 'Vulnerable' or 'Rare', reflected the likely area of occupancy of 1 $Km^2$ and association with a climatically sensitive habitat (within a very restricted elevational range of only 1890–1900 m) that is likely to decline with raised temperature and rainfall fluctuations.

## 9.4 Odonata

### *9.4.1 The Ancient Greenling,* Hemiphlebia mirabilis *(Hemiphlebiidae)*

This tiny damselfly, for long known simply as the Hemiphlebia damselfly, has commonly been considered a 'living fossil', most similar to species from the Permian period, and accorded a superfamily (Hemiphlebioidea) of its own. Endemic to south-eastern Australia, this unique taxonomic placement, and the belief that it was actually or nearly extinct due to habitat losses to agriculture accorded it global interest. *H. mirabilis* was for some time a global priority for Odonata conservation. Then recorded only from a few small seasonally flooded swamps in Victoria, burning of those areas was also considered a threat (New 1993, 2007).

That notoriety rendered it of major interest to specialists, and much of the impetus for wider investigation flowed from that stimulus. Formal listing in Australia led to broader surveys that have now revealed the damselfly to occur in north eastern Tasmania (and the Bass Strait Flinders Island) and elsewhere in western Victoria and confirmed that it is less threatened than earlier supposed. Discovery of a very large population in south western Victoria enabled far greater evaluation of the damselfly's complex display behaviour, and Cordero-Ribaga (2013, 2016) estimated that the population could comprise more than a million individuals a season—far from the few tens, at most, of adults revealed in most of the earlier-known populations. The damselfly was thus not then considered to be threatened, and Cordero-Ribaga (2016) also noted incidence of large populations in the Grampians region of Victoria and recommended that the conservation status of *Hemiphlebia* should be revised. Those recently discovered populations, together with those in abutting parts of South Australia are the currently known 'stronghold' of *H. mirabilis* (Crowther 2011). In contrast, succession and agriculture have led to loss of more easterly populations, where targeted surveys have failed to reveal the damselfly or, even, apparently suitable areas in which it could persist.

### *9.4.2 The Sydney Hawk Dragonfly,* **Austrocordulia leonardi** *(Austrocorduliidae)*

This species was brought to the attention of conservationists by Moore (1997) listing it as 'critically endangered', because of its scarcity and very restricted distribution in central eastern New South Wales. Hawking and Theischinger (2004) noted that only 11 adults had by then been recorded, with intensive surveys by Theischinger failing to reveal the dragonfly, although Chessman and Williams (1999) reported larvae from several sites in southern Sydney. The larval habitat is natural deep and shaded river pools with cooler water. *A. leonardi* disappeared from its type locality (along the Woronora River) following removal of a weir, and Hawking and Theischinger commented that suitable deep pool habitats probably 'disappeared some time ago with the creation of large dams'.

### *9.4.3 The Horned Urfly or Adams Emerald Dragonfly,* **Archaeophya adamsi** *(Gomphomacromiidae)*

This rare dragonfly is legally protected and listed as 'vulnerable' in New South Wales. Extensive searches have revealed fewer than 10 adults and a few larvae from few localities within about 100 km of Sydney. Conservation interest is stimulated by the rapid expansion of Sydney, considered likely to threaten the six localised populations so far known (Hawking and Theischinger 2004).

### 9.4.4 *The Giant Petaltail (or South-Eastern Petaltail) Dragonfly,* **Petalura gigantea** *(Petaluridae)*

Notable as among the largest Australian Odonata, the Giant petaltail (actually *not* the largest species in the genus, hence the alternative common name!) has aroused considerable conservation interest, and several surveys have confirmed its very restricted distribution in New South Wales, mostly in the Blue Mountains area. Nevertheless, *P. gigantea* males are about 11 cm in wingspan, and females somewhat larger. *P. gigantea* occurs in permanent seepages, bogs and swamps, essentially in groundwater-dependent ecosystems and is restricted to areas in which the water table is high enough to saturate the peaty substrates. Females lay their eggs deeply into the substrate, and larvae excavate burrows of up to 75 cm deep. The burrows are occupied over the long larval life, which Baird (2012) estimated as 'at least six and possibly at least 10 years' (Clarke and Spier-Ashcroft 2003, indeed, quoted an estimate of 10–30 years, based on the related New Zealand genus *Uropetala*). Over that time, larvae become large, leaving exuviae of up to 5 cm in length. They are believed to be semiaquatic and do not occur in open waters. Their tunnels open to both water and land, with implications that larvae can leave burrows at night and hunt on land.

Conservation of *P. gigantea* can flow only from fuller understanding of pressures on the unique mire habitats, and their protection over the full ecological range they occupy. One site (Wingecarribee Swamp, in the Southern Highlands region of New South Wales) was the largest montane peatland swamp in Australia until peat mining and a large landslide after heavy rains reduced the suitable area considerably. Many other populations have become extinct as their habitats are lost to development. The dragonfly is listed as 'endangered' and the restricted mire habitats are also listed as 'Endangered Ecological Communities' under the Commonwealth EPBC Act.

Early assessments of *P. gigantea* included populations from the far north of coastal New South Wales and nearby south-eastern Queensland, but these are now recognised as a distinct species, the Coastal petaltail, *Petalura litorea*, which is also listed as endangered and was surveyed in detail by Baird (2017). Both species occur only to the east of the Great Dividing Range. The two taxa are rather similar in appearance and may be sister species. Whilst most records of *P. gigantea* are from national parks, state forests or water supply catchments, the variety of tenures necessitates equally varied protection for these habitats. The daunting array of threats projected to Australian Petaluridae (Table 8.6, p. 189: Baird and Burgin 2016) vary considerably across different sites. Some are amenable to mitigation, but others are far more complex and intractable.

*P. gigantea* is a valuable 'umbrella species' for mire habitats and their associated groundwater-dependent inhabitants and has become a prominent flagship for this in the Blue Mountains.

## 9.5 Plecoptera

### 9.5.1 *The Mount Donna Buang Wingless Stonefly,* **Riekoperla darlingtoni** *(Gripopterygidae)*

*R. darlingtoni* is known only from a few small sites near the summit of a single mountain in central Victoria. Dedicated searches elsewhere have so far been fruitless, and this very small stonefly (the females, although considerably larger than the males, are only about 12 mm in length) appears to be a very restricted relict narrow range endemic species with the area of occupancy around 2–4 $Km^2$. Larvae occur in the gravel substrate of small temporary streams and over a life cycle that lasts approximately three years their burrowing habit facilitates survival through periods of drought. Adults shelter in rolled fragments of eucalypt bark on the ground. The major concentration of stoneflies is on Mt Donna Buang, and two individuals (an adult in 1993, a larva in 1999: Ahern et al. 2003) have been found about 3 km away. The mountain is within the Yarra Ranges National Park, and the stonefly's presence led to Donna Buang being added to Australia's Register of the National Estate.

*R. darlingtoni* was one of the first Australian insects to be signaled as of conservation concern and was included in the early IUCN Invertebrate Red Data Book (Wells et al. 1983) as 'rare'. More recently, Clarke and Spier-Ashcroft (2003) noted that 'it may be critically endangered'. However, legislative acknowledgement of its status is complex, with the disparate evaluations under Commonwealth and State Acts discussed by New (2008). A major cause of concern is the progressive development of the summit area of Mt Donna Buang for tourism, with likely changes to stream water quality from road and carpark run-off (the stream harbouring the major stonefly population abuts a sealed carpark area). Ahern et al. (2003) noted, for example, that 'a single fuel spillage in a carpark could destroy a substantial proportion of the known stonefly population'. Direct trampling from visitors, and damage to the riparian vegetation are also regarded as threats.

### 9.5.2 *The Kallista Flightless Stonefly,* **Leptoperla kallistae** *(Gripopterygidae)*

Previously regarded as a subspecies of *Leptoperla kimminsi*, this small stonefly is known only from part of the Dandenong Ranges, Victoria, where it occurs in the headwaters of creeks in the Olinda, Monbulk and Sassafras catchments. As for other flightless stoneflies, it is likely to disperse only poorly, and the species is found only in stream lengths with dense vegetation and natural flow regimes, where larvae occur in patches of organic matter. At that scale, removal of riparian vegetation has been suggested to be the greatest threat, but at the catchment level directly connected drains from impervious surfaces are the major threat (Danger and Walsh 2008). Strict regulation of stormwater runoff into catchments supporting *L. kallistae* is a key need, together with retrofitting means to lower stormwater delivery to the streams. The stonefly may also be vulnerable to bushfires, and the measures used to control them.

### *9.5.3 The Mount Kosciuszko Wingless Stonefly,* **Leptoperla cucuminis** *(Gripopterygidae)*

At the time of designating this species as 'rare' in the IUCN Invertebrate Red Data book, Wells et al. (1983) noted that it was known only from a small stream near the mountain summit. The wingless adult is relatively small (about 8 mm long), and occurs among herbage on the water bank. Larvae live under stones on the stream beds.

### *9.5.4 The Alpine Stonefly,* **Thaumatoperla alpina** *and Mount Stirling Alpine Stonefly,* **Thaumatoperla flaveola** *(Eustheniidae)*

These closely related species are the largest Australian stoneflies, and both have become significant flagships for alpine invertebrate communities badly in need of more assured conservation through protecting the tiny alpine stream habitats from disturbances. *T. alpina* is endemic to the Bogong High Plains area of Victoria, within the catchment of the Kiewa River, where it has been found in about a dozen small first order alpine streams above about 750 m. Historical records and more recent surveys (2005) following bushfires in 2003 confirmed *T. alpina* at eight sites in streams spanning an elevational range of 660–1720 m, two of the sites within the Alpine National Park (McKay et al. 2005): that number has now been increased (below). The more common *Eusthenia venosa* can occur in the same streams as *T. alpina* but has a far wider distribution in Victoria's alpine region, where it may comprise a complex of species.

Larvae, up to about 5 cm in length, occur within the streams and the flightless adults apparently disperse little from streamside vegetation. Water temperature may be a critical factor of habitat suitability, with normal water temperatures there within the range of 5–15 °C. Although not confirmed, Crowther (in EPBC 2011) suggested that introduced trout (p. 114) may be a threat to the aquatic larvae at some sites, with spread of trout reducing predator-free habitat. However, the major threat is the expansion of winter sports activities, especially ski resorts and associated roads increasing contamination of previously pristine alpine tarns. Increased sediment and nutrient loads are of concern (and may change as climate changes and local developments increase), as well as chemical pollution. Snow-making, for example, may lead to chemical contamination. Whilst fire was not confirmed to have affected larvae of *T. alpina* adversely, adults on burned vegetation are highly exposed, and are likely to suffer increased predation. Collectively, a wide array of direct and indirect threats can influence alpine stoneflies and their restricted environments. Additional to those noted above, increased stream temperature from greater exposure following vegetation loss and climate change may be significant.

Grounds for assigning *T. alpina* 'endangered' status reflected the very restricted distribution (area of occupancy about 55 $Km^2$, with very fragmented incidence within this) and the continuing threats. Six of the known sites are within the Alpine National Park, and five others are in the Falls Creek Alpine Resort. Series of measures to augment conservation of *T. alpina* were proposed by McKay et al. (2005) and involved protection and accumulating better knowledge to hone management needs (Table 9.3). The range of research topics listed are paralleled in the needs for other alpine aquatic species.

*T. flaveola* is also very restricted, endemic to the Mount Stirling/ Mount Buller area, and is also poorly known. Extensive surveys yielded the stonefly at 28 localities around Mount Buller (Doeg 1999), all within an area of about 12 × 10 Km, and almost all within the upper tributaries of the Delatite River, above 1000 m elevation. The species' distribution appeared confined to that single massif. Major threats encompass alpine resort development on Mount Buller, and timber harvesting and grazing on Mount Stirling, and Doeg considered that no populations were free from potential disturbance and loss: he assessed *T. flaveola* as 'endangered'.

Life histories of both species appear to be long, with suggestion that a generation may take three years.

Crowther et al. (2008) surveyed both species following a further fire in 2006 (following the previous fire in 2003), the earlier incinerating >60% of the Alpine

**Table 9.3** Conservation recommendations proposed for the stonefly *Thaumatoperla alpina* in the Bogong alpine region of Victoria (McKay et al. 2005)

| |
|---|
| 1. Protecting *T. alpina* from threats: (1) ensure that vegetation within 10 m of streams and the distribution area of the stonefly is protected and that skiing and other ski resort activities do not encroach on the streams; (2) maintain water quality in streams within the distribution areas of *T. alpina*; (3) ensure trout are absent from streams identified as having *T. alpina* populations, removing them where necessary |
| 2. The conservation status of *T. alpina*: nominate for listing under the Environment Protection and Biodiversity Conservation Act 1999 |
| 3. Improving knowledge on the distribution of *T. alpina* and other stonefly species of conservation significance: (1) further surveys of streams outside the Bogong alpine area, to confidently map distribution; (2) annual assessments of currently known populations; (3) in conjunction, improve knowledge of distribution of *Riekoperla intermedia* |
| 4. Research to gain a better understanding of the biology and threats to *T. alpina*: (1) assess sediment loads in streams identified as habitat; (2) examine impacts of increased nutrient load in streams where grey water has been used for snow making, and effect on stonefly populations; (3) examine adaptive significance of adult colouration (warning or camouflage); (4) laboratory diet studies to determine what natural diet may be; (5) dietary studies on Mountain pygmy possum (*Burramys parvus*) to determine if it is a predator on the stonefly; (6) identify threats of predation on adults from lizards and birds |

National Park, whilst the 2006–2007 'Great Divide Fire' burned 1.1 million hectares, including much of the park. From their 2008 surveys, Crowther et al. concluded that neither stonefly had been severely affected by the fires—numbers of both were little different in tarns burned by both fires than in unburned sites.

Two other montane species of *Thaumatoperla* also appear to be highly restricted in occurrence, and have been designated as 'Data Deficient'. *T. robusta* is known from Mount Donna Buang to Mount Baw Baw, and *T. timmsi* from a single location on Mount Wellington, near Lake Tali Karng.

### 9.5.5 *The Otway Stonefly,* Eusthenia nothofagi *(Eustheniidae)*

This forest-dwelling stonefly was for long known only from a series of specimens collected in the Otway Ranges of southern Victoria. It was listed as 'endangered' by Wells et al. (1983) and as not having been found for more than 40 years, was later regarded as 'presumed extinct' under the Flora and Fauna Guarantee Act listing. However, in extensive surveys in Otway streams by Doeg and Reed (1995), larvae of *Eusthenia* were recorded in 19 of the 52 sites sampled. Rearing from last instar larvae confirmed the presence of *E. nothofagi* at nine sites. The wider distribution of the stonefly in the Otway Ranges, with some occupied sites in several catchments in secure areas, was considered sufficient to conserve the species, and it was subsequently delisted from the Victorian act, as foreshadowed by Doeg (1997).

The putative co-occurring species, *Eusthenia venosa*, was not present amongst the above rearings, and Doeg considered it likely that all nymphs from the surveys could be *E. nothofagi*, so that its actual range and abundance might be even greater than that fully documented.

## 9.6 Hemiptera

### 9.6.1 Tenogogonus australiensis *(Gerridae)*

This waterstrider is known from scattered populations in the near- coastal zone of north Queensland, where it occurs in streams within a closed rainforest canopy. It is believed to be one of the most specialised representatives of the family in Australia, and may be susceptible to changes in streamside vegetation, with loss of shade possibly detrimental (Andersen and Weir 1997). The species is likely to be 'Data Deficient' and has not been listed formally as needing conservation. Clarke and Spier-Ashcroft (2003) noted that protection of the closed rainforest areas against clearing or disruption are key management objectives, together with further surveys and study of the bug's biology.

## 9.7 Coleoptera

### 9.7.1 Hygrobia australasiae *(Hygrobiidae)*

All of the few Australian species of *Hygrobia* appear to be rare, and *H. australasiae* was selected for treatment by Clarke and Spier-Ashcroft (2003) as a representative of this poorly-documented group, sometimes termed 'screech beetles'. It is, however, quite widespread over much of south eastern Australia, but with populations very localised. Its main habitats, ponds with little or no water movement and often with an open gravel substrate, are mainly ephemeral and dry out in summer. Loss of those sites to agricultural and urban development is a clear threat to the beetle, and Clarke and Spier-Ashcroft also cited eutrophication as 'thought to be a major threat', so that control of nutrient inputs (such as by restricting access by stock) during winter may be sound management.

As with several other species noted in this chapter, far more detailed biological knowledge is needed to formulate any detailed management plan.

## 9.8 Trichoptera

### 9.8.1 Taskiria otwayensis *(Kokiriidae)*

This highly localised caddisfly exemplifies the many freshwater insects that have aroused comment for their apparent scarcity and possible conservation need but for which sound information is very limited. It, and others, are necessarily treated as 'Data Deficient' at present. As Clarke and Spier-Ashcroft (2003) commented, *T. otwayensis* is regarded as 'Endangered' in Victoria, but has not been listed formally under the state act. At that time, it was known only from three sites in the Otway Ranges, where it had been found near small streams in eucalypt forest or pine plantations—the latter is sometimes associated with heavy siltation levels and overgrowth of the stream by blackberries but influence of these disturbances is unknown.

*T. otwayensis* was described only in 1984, and the genus is otherwise known only from Tasmania (Neboiss 1984). The major conservation need identified by Clarke and Spier-Ashcroft is for further survey to define the caddisfly's distribution. The three sites they noted are in different catchments, this possibly helping to protect the caddisfly from localised pollution or other threat.

The Tasmanian Kokiriidae (*Taskiria mccubbini*, *Taskiropsyche lacustris*) have been reported only from within about 100 m of the Lake Pedder impoundment (p. 200), but speculation that they may be 'possibly more widespread' (Threatened Species Unit 2003) parallels hopes for *T. otwayensis*. However, intensity of surveys of Trichoptera in southern Tasmania (Jackson 2000) implies that this may not be so. Protection of the known sites and nearby potential sites is needed, together with further surveys.

## 9.9 Diptera

### 9.9.1 *The Giant Torrent Midge,* Edwardsina gigantea, *and the Tasmanian Torrent Midge,* Edwardsina tasmaniensis *(Blephariceridae)*

Both these Blephariceridae were included as 'endangered' in the initial Invertebrate Red Data Book (Wells et al. 1983). Larvae of torrent midges occur in rapidly flowing water and have a series of six strong ventral suckers that enable them to attach firmly to stones and resist being swept away. Adult torrent midges are weak fliers, with very little likelihood of colonising new sites.

*E. gigantea* is among the largest Australian blephariceridss, with wingspan of about 25 mm. It is known only from the Snowy Mountains region, where larvae occur in fast-flowing streams with high rainfall, and its disappearance from much of the Snowy, Cotter and Geehi Rivers has been attributed to impacts such as dam construction, changed flow regimes and water levels, substrate changes and disturbance, all emanating from the Snowy Mountains Hydroelectric Scheme and related activities (Clarke and Spier-Ashcroft 2003). Wells et al. (1983) noted that both known localities then known to support the midge are within the Mount Kosciuszko National Park, where they could be affected by river pollution and effluent contamination. The midge had apparently been lost from several formerly inhabited sites from the latter cause. Both inhabited streams received sewage effluent. Maintenance of those currently occupied habitats is a conservation priority for the midge, with control of pollution and prevention of further hydrological changes to the streams both of primary concern.

*E. tasmaniensis*, described from Cataract Gorge on the Esk River near Launceston, Tasmania, apparently became extinct there following water diversion for hydroelectric power generation from 1956. However, it was rediscovered in the Dennison River in 1976, but impoundment for the Lower Gordon hydroelectric scheme was anticipated to lead to extinction there.

The specialised nature of preferred habitat for these flies renders it unlikely that many further populations will be found.

## References

Abellan P, Sanchez-Fernadez D, Velasco J, Millan A (2005) Assessing conservation priorities for insects: status of water beetles in southeast Spain. Biol Conserv 121:79–90

Ahern LD, Tsyrlin E, Myers R (2003) Mount Donna Buang wingless stonefly, *Riekoperla darlingtoni*. Action Statement no 125, Flora and Fauna Guarantee Act. Department of Sustainability and Environment Victoria, Melbourne

Andersen NM, Weir TA (1997) The gerrine water striders of Australia (Hemiptera: Gerridae): taxonomy, distribution and ecology. Invert Taxon 11:203–299

Baird IRC (2012) The wetland habitats, biogeography and population dynamics of *Petalura gigantea* (Odonata: Petaluridae) in the Blue Mountains of New South Wales. PhD thesis, University of Western Sydney. http://handle.uws.edu.au:8081/1959.7/509925

Baird IRC (2017) A review of current knowledge of the coastal petaltail dragonfly, *Petalura litorea* (Odonata: Petaluridae). New South Wales Office of Environment and Heritage Save our Species Program, North-East Regjon. Sydney

Baird IRC, Burgin S (2016) Conservation of a groundwater-dependent mire-dwelling dragonfly: implications of multiple threatening processes. J Insect Conserv 20:165–178

Chessman BC, Williams SA (1999) Biodiversity and conservation of river macroinvertebrates on an expanding urban fringe: western Sydney, New South Wales, Australia. Pacif Conserv Biol 5:36–55

Clarke GM, Spier-Ashcroft F (2003) A review of the conservation status of selected Australian non-marine invertebrates. Environment Australia/ National Heritage Trust, Canberra

Cordero-Rivera A (2013) Behaviour and ecology of *Hemiphlebia mirabilis* (Odonata: Hemiphlebiidae). Report on work undertaken. Department of Environment and Primary Industries, Victoria, Melbourne

Cordero-Ribaga A (2016) Demographics and adult activity of *Hemiphlebia mirabilis*: a short-lived species with a huge population size (Odonata: Hemiphlebiidae). Insect Conserv Divers 9:108–117

COSEWIG (Committee on the Status of Endangered Wildlife in Canada) (2011) COSEWIG assessment and status report on the Hungerford's crawling water beetle *Brychius hungerfordi* in Canada. COSEWIG, Ottawa

Crowther D (2011) The Ancient Greenling: synthesis of new information to improve conservation outcomes. Arthur Rylah Institute for Environmental Research, Department of Sustainability and Environment, Melbourne

Crowther D, Lyon S, Papas P (2008) The response of threatened aquatic invertebrates to the 2006 fire in north-eastern Victoria. Tech Rep Series no 179, Arthur Rylah Institute for Environmental Research, Department of Sustainability and Environment, Melbourne

Danger A, Walsh CJ (2008) Management options for conserving and restoring fauna and other ecological values of urban streams in the Melbourne Water region. Report to Melbourne Water, Victoria

Doeg TJ (1997) Gone today, here tomorrow—extinct aquatic macroinvertebrates in Victoria. Mem Mus Vict 56:531–535

Doeg TJ (1999) Distribution and conservation status of the stonefly *Thaumatoperla flaveola* Burns and Neboiss in the Mt Buller-Stirling area. Proc R Soc Vict 111:87–92

Doeg T, Reed J (1995) Distribution of the endangered Otway stonefly *Eusthenia nothofagi* Zwick (Plecoptera: Eustheniidae) in the Otway ranges. Proc R Soc Vict 197:45–50

EPBC (Environment Protection and Biodiversity Conservation Act, scientific committee) (2011) Approved conservation advice for Thaumatoperla alpina (Alpine stonefly), Canberra

Foster GN (1991) Conserving insects of aquatic and wetland habitats, with special reference to beetles. In: Collins NM, Thomas JA (eds) The conservation of insects and their habitats. Academic Press, London, pp 237–262

Foster GN, Bilton DT (2014) The conservation of predaceous diving beetles: knowns, unknowns and anecdotes. In: Yee DA (ed) Ecology, systematics, and the natural history of predaceous diving beetles (Coleoptera: Dytiscidae). Springer Science + Business Media, Dordrecht, pp 437–462

Gomi T, Sidle RC, Richardson JS (2002) Understanding processes and downstream linkages of headwater systems. Bioscience 52:905–916

Haslett JR (1997) Suggested additions to the invertebrate species listed on Appendix II of the Bern Convention. Final report to the Council of Europe, Strasbourg

Hawking JH, Theischinger G (2004) Critical species of Odonata in Australia. Int J Odonat 7:113–132

Jackson J (2000) Threatened Trichoptera (caddisflies) from the Tasmanian Wilderness World Heritage Area. Proc R Soc Tas 134:55–62

Lowe WH, Likens GE (2005) Moving headwater streams to the head of the class. BioScience 55:196–197

McKay S, Bryce C, Papas P (2005) Impacts of fire on the distribution of a predatory stonefly (Eustheniidae: *Thaumatoperla alpina*) in the Bogong alpine region. Tech Report 155, Arthur Rylah Institute for Environmental Research, Melbourne

Monroe EM, Britten HB (2014) Conservation in Hine's sight: the conservation genetics of the federally endangered Hine's emerald dragonfly, *Somatochlora hineana*. J Insect Conserv 18:353–363

Moore NW (1997) Dragonflies: status survey and conservation action plan. IUCN/SSC Odonata Specialist Group, IUCN, Gland and Cambridge

Neboiss A (1984) Four new caddis fly species from Victoria (Trichoptera: Insecta). Vict Nat 101:86–91

New TR (1993) *Hemiphlebia mirabilis* Selys: recovery from habitat destruction at Wilsons Promontory, Victoria, Australia and implications for conservation management (Zygoptera: Hemiphlebiidae). Odonatologica 22:495–502

New TR (2007) The Hemiphlebia damselfly, *Hemiphlebia mirabilis* Selys (Odonata: Zygoptera), as a flagship species for aquatic insect conservation in south-eastern Australia. Vict Nat 124:269–272

New TR (2008) Legislative inconsistencies and species conservation status: understanding or confusion? The case of *Riekoperla darlingtoni* (Plecoptera) in Australia. J Insect Conserv 12:1–2

Polhemus DA (1993) Conservation of aquatic insects: worldwide crisis or localized threats? Amer Zool 33:588–598

Resh VH, Sorg KL (1983) Distribution of the Wilbur Springs shore bug (Hemiptera: Saldidae): predicting occurrence using water chemistry parameters. Environ Entomol 12:1628–1635

Robson BJ, Chester ET, Austin CM (2011) Why life history information matters: drought refuges and macroinvertebrate persistence in non-perennial streams subject to a drier climate. Mar Freshw Res 62:801–810

Suter P (2014) Tasmanophlebi(a) lacuscoerulei. The IUCN Red List of Threatened Species 2014: e T40728A21425993: https://doi.org/10.2305/iucn.uk.2014-1.rlts.t40728a21425993

Taylor GS, Braby MF, Moir ML, Harvey MS, Sands DPA (and 10 other authors) (2018) Strategic national approach for improving the conservation management of insects and allied invertebrates in Australia. Austral Entomol 57: 124-149

USFWS (United States Fish and Wildlife Service) (2001) Hine's emerald dragonfly (*Somatochlora hineana* Williamson) recovery plan. U S Department of the Interior, Fort Snelling, MN

USFWS (United States Fish and Wildlife Service) (2006) Hungerford's crawling water beetle (*Brychius hungerfordi*) recovery plan. U S Department of the Interior, Fort Snelling, MN

Vivian LA, Cavallaro M, Kneeland K, Lindroth E, Horack WW (and three other authors) (2013) Current known range of the Platte River caddisfly, *Ironoquia plattensis*, and genetic variability among populations from three Nebraska rivers, J Insect Conserv 17:885–895

Wells SM, Pyle RM, Collins NM (comp) (1983) The IUCN Invertebrate Red Data Book. IUCN, Gland and Cambridge

# Chapter 10
# Ecology and Management

## 10.1 Introduction

Several aspects of species' ecology are especially important in appraising conservation status and needs, in addition to assessing the direct critical resources such as food need and habitat condition and protection. Some relate to themes such as population structure and isolation and for many aquatic insects dispersal capability and changes in habitat connectivity as landscapes are altered are important considerations in assessing taxonomic integrity and understanding population dynamics. The processes of dispersal and consequences of habitat changes are introduced here to endorse their importance in conservation management.

Potential to colonise or re-colonise increasingly isolated and, perhaps, increasingly inaccessible habitats can be changed markedly by human interventions but is likely to depend on both dispersal capability and population or metapopulation structure. The latter is apparently not unusual among aquatic insects but for many taxa details of population structure are unknown. However, functional isolation of habitat sites may engender both differentiation of populations through loss of gene flow, and vulnerability through genetic restriction and increased stochastic impacts.

## 10.2 Dispersal

Assessing human impacts on ecological connectivity for denizens of freshwater ecosystems, and the functional roles of insect dispersal in assuring their integrity, necessitates understanding the great array of structural features and isolating processes that can occur. Within the complex branching systems of streams and rivers, isolation of populations can occur at relatively small scales. Crook et al. (2015) discussed four general models that can help to understand the varying contexts. These (Fig. 10.1) are shown through populations that exist in four subcatchments with headwaters in higher elevation regions. Each is explained in the figure legend,



T. R. New, *Insect Conservation and Australia's Inland Waters*,
https://doi.org/10.1007/978-3-030-57008-8_10

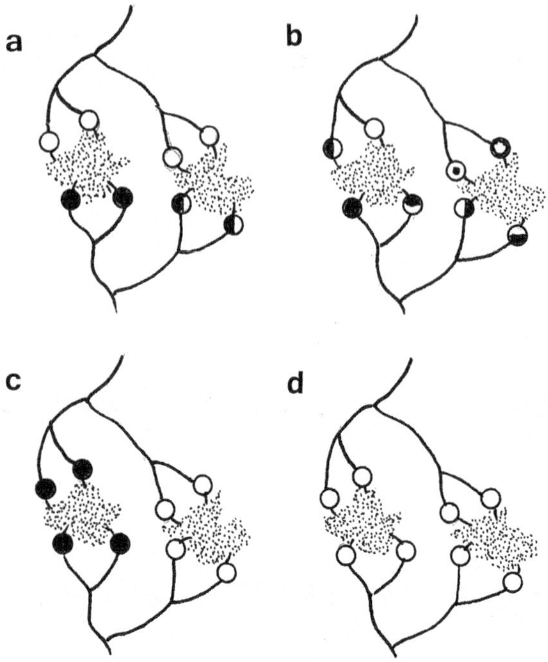


**Fig. 10.1** Four models of ecological connectivity in stream systems shown by populations (dots, with those of same colour being connected populations) in four subcatchments, with headwaters at higher elevations separated (dotted areas): **a** Stream hierarchy model, with species correlated in way reflecting stream dendritic nature; **b** Death Valley model, with extensive isolation among populations, each confined to small disconnected habitat patches; **c** Headwater model, a pattern of ecological connectivity essentially opposed to 'a', with populations in particular headwater regions but with some capacity to disperse between them; **d** Widespread gene flow, representing species with highly mobile terrestrial phase or adapted to temporary associations with highly mobile animals, and for which the pattern of the stream network has little influence on ecological connectivity (Crook et al. 2015)

with the patterns reflecting levels of functional connectivity and so linking with those biological and life history characteristics of the insects that affect population structure.

Very little is known of the population structure and dispersal capability of most aquatic insects. Other than for Odonata, adults of many are generally considered to be weak fliers, but some studies have indeed revealed that considerable distances may be traversed. However, as Beebee (2007) commented there is 'much to learn about the mechanisms and extent of the processes involved'. Even apparently poorly dispersing taxa such as the caddisflies *Plectrocnemia conspersa* (United Kingdom) and *Tasimia palpata* (Australia) may undergo widespread adult movements. In short, adults of many aquatic insects are found far from their natal waterbodies and may then become important contributors to terrestrial food webs as prey for insectivores that are not in any way fundamentally aquatic. This role has rarely been assessed, but in Kakadu National Park, Northern Territory, catches of insects were compared at distances from midstream through riparian zones (10–15 m) to 160 m from the stream (Lynch et al. 2002). Most aquatic groups were far more abundant close to the water, with greater overall abundance (of both terrestrial and aquatic taxa) implying the importance of the riparian zones. Some Chironomidae and Trichoptera were captured at the 160 m station, but small size of the former suggested that they would be eaten by predatory arthropods rather than vertebrates. Land use can markedly affect such 'lateral dispersal': in Sweden, abundance of adult aquatic insects was higher at agricultural sites than at forested sites, but most individuals occurred close to the stream edge, in contrast to rather little decline with increasing distance from forested streams (Carlson et al. 2016).

Some other taxa readily fly strongly. The large and rare Great silver beetle (*Hydrophilus piceus*, Dytiscidae) has declined considerably in Britain as its habitats have been lost over the last century, and the few remaining occupied sites are isolated. However, the beetle may even migrate between Britain and mainland Europe and could be a very capable colonist of new or restored habitats (Beebee 2007). The beetle is of broad conservation interest in Europe and is listed as a threatened species in several countries (Karaouzas et al. 2014). In Britain, most historical and currently occupied sites are low-lying marshy areas, with some apparent preference for open sites with abundant macrophytes. Major threats include pollution and direct habitat loss, with drainage of fens a predominant historical influence. As a relatively specialised feeder on aquatic molluscs, maintenance of these is critical for *H. piceus* and small-scale clearance of sites using hand tools is far preferable to use of less discriminating heavy machinery.

What may appear initially to be relatively isolated populations of some species may in reality represent more complex structures. Many aquatic insects occur as metapopulations—discrete population units that are linked by sequences of local extinctions and (perhaps infrequent) recolonisations amongst a mosaic of sites across the landscape that collectively assure their overall population integrity. Effects of human activities on connectivity are, in Crook et al.'s words 'numerous, complex, and often highly specific to the species and environment of concern', so that any single approach to mitigation is unlikely to succeed. However, embracing that variety involves four key areas of knowledge and understanding, as (1) autecology—understanding the biology of the focal species and how it/they may respond to connectivity; (2) population structure—levels of connectivity in space and time, and the impacts of changes in connectivity (such as in changed gene flow); (3) movement characteristics—how and why individuals move, as the mechanism that can drive connectivity at the population/metapopulation scales; and (4) environmental tolerances and phenotypic plasticity—information on tolerance limits and the extent to which the insects are able to modify their behaviour, physiology and morphology in response to environmental change can help to predict responses to changed connectivity. Those species with limited dispersal capability and low resistance to environmental changes are especially vulnerable, and prone to local extinctions. Crook et al. (2015) recommended that research in these fields could provide the basis for strategies to mitigate human disturbances to connectivity but integrating data from the very varied relevant studies may be difficult.

The interpretations of pond ecology throughout a landscape are thus perhaps better considered through the approach of a metapopulation and metacommunity structure sustained through continuing dispersal amongst habitat units, rather than treating ponds and allied water bodies as isolated units. Abilities and opportunities for insects to disperse or otherwise counter variations across habitat units thereby determine survival. Most relevant studies have been on single taxa, but these are augmented by several broader surveys across different invertebrate groups. For example, Juracka et al. (2019) considered Odonata, aquatic Hemiptera and aquatic Coleoptera in small fishless ponds in the Czech Republic and compared their incidence in 42 pools over two consecutive years, in each of which the insects were sampled in spring, summer and autumn. Whilst pond size (surface area and depth) was a key determinant of species richness, followed by the extent and composition of aquatic macrophytes, dispersal and metacommunity assembly were affected strongly by landscape structure. The region included steep ridges separated by deep valleys, in which spatial behaviour of the insects is likely to be a critical aspect of site occupation.

Most aquatic insects have some potential for dispersal over their lifetime—in water as immature stages and in the air as (most commonly) winged adults. The two major insect phases of larvae and adults usually have very different needs, in many cases with different feeding habits and consequent different needs for movement to track resources and colonise other sites. Adults of many aquatic insects are, perhaps uncritically, thought of as weak or 'reluctant' fliers that typically remain close to their natal water bodies—for example, on riparian vegetation. Many adults do indeed remain close to the water's edge, these commonly being taxa with short adult lifespans and with their primary 'function' being to mate and initiate the next generation. Their dispersal is limited and a major conservation consideration is to maintain or restore suitable conditions for shelter, feeding (if necessary) and mating behaviour in the riparian and near-riparian zones. Some other taxa, especially some larger and more vigorous species of Odonata, disperse far more widely. Thus, management for the threatened *Ischnura pumilio* in the United Kingdom should include the hinterland adjacent to natal waterbodies. Vegetation heights and structure were influential for at least 20 m from the water (Allen et al. 2010), with dense vegetation providing shelter from climate extremes and predators. A further complication in conserving *I. pumilio* is that it inhabits two rather different categories of habitats. One is high quality already maintained habitats where other species of conservation concern may also occur and where some management options—such as overgrazing to keep vegetation low—may be difficult. The second is polluted anthropogenic sites with high levels of human intervention and which, in general, are considered undesirable and discarded from conservation considerations. Losses of *I. pumilio* from some sites in the past were attributed to changes such as succession, over-shading and scrub invasions, all of which must be managed effectively. Without that management, those sites are likely to remain unsuitable (Allen et al. 2010).

Even small insects can characteristically disperse and, although many mayflies and non-biting midges most commonly congregate within a few metres of water, some can swarm and/or move away from water, in some cases phototactically attracted to other areas (p. 239). Dreyer and Gratton (2014) attributed the high concentration of adult aquatic insects at the edges of water bodies to their being habitat specialists, with possible implications that rather restricted conditions may be needed, and loss of those conditions increase vulnerability. They noted, also, the contrasts between lakes and streams, relevant to interpreting insect incidences, notably (1) that lakes represent large areas but have low perimeter length per unit water area, whist (2) streams represent high edge habitats with relatively small productive water area. Land adjacent to lakes is exposed to higher concentrations of aquatic insects than land adjacent to streams—but also, more land is closer to streams than to lakes, reflecting the linear form and large perimeter length.

The structure of waterside vegetation may influence dispersal of those insects and, again in general, far less is known of movements away from water and the distribution of adults in and across terrestrial environments, than of 'longitudinal' movements along rivers or short distance flights close to ponds or lakes. Odonata are a major exception, as large, conspicuous, diurnal insects, with elaborate adult behaviour patterns that have for long attracted attention from naturalists. They also commonly undertake 'lateral dispersal', some moving far from water during relatively long life times. Females of some species move away from water for several weeks to months of egg maturation, and then return to oviposit. Adults of *Sympetrum depressiusculum* in Europe (p. 139) occupy terrestrial environments for up to more than three months—at least as long as the period of aquatic larval development (Dolny et al. 2014). Adults of many other taxa are typically short-lived, almost by definition limiting opportunities for long distance active flights. Many of those adults do not feed or feed little, in contrast to Odonata and longer-lived Diptera and Coleoptera. The values of terrestrial habitats for Odonata have received rather little attention, but some species clearly have very specific microhabitat requirements in those environments. However, in general 'very little is known about the effects of individual characteristics of terrestrial environments on the distribution of odonates' (Dolny et al. 2014), despite clear awareness that heterogeneity, in the form of vegetational variety and land use, is highly relevant in their conservation. Agricultural intensification can reduce accessibility of shelter and forage, and some dragonflies prefer more natural vegetation such as abandoned fields and ruderal patches. The absence of continuing disturbances during adult flight periods can be important.

Phoretic dispersal, whereby the early stages, in particular, are dispersed by animal vectors—either attached to the exterior or as resistant stages in food—occurs in many aquatic molluscs and crustaceans. Flying insects can occasionally act as vectors, but most carriers are vertebrates. However, the predominant dispersal modes for aquatic insects are 'passive' (by water currents or wind) and 'active' (essentially, by flight). For many taxa, upstream migrations by flight are viewed as compensating for the early stages having been earlier displaced downstream by 'drift' in moving water. As discussed by Bilton et al. (2001), both components are complex, and recent genetic studies have helped in their functional interpretation and understanding evolutionary impacts. Incidence of 'cryptic species' may also be revealed. The caddisfly *Lectrides varians* (Leptoceridae) is distributed widely in eastern Australia, with genetic analyses revealing extensive gene flow and implying that it is a strongly dispersive species (Wickson et al. 2014) that (as with some other large caddisflies, such as Tasimiidae) embraces lateral dispersal as usual behaviour. Those analyses also revealed that the 'species' is paraphyletic with two well-defined forms present. One is a possible endemic species in the Grampians National Park of Victoria, where it is highly isolated from other suitable habitats. The two forms are morphologically similar but separable on larval colour pattern.

Many lotic insects are liable to drift with water currents, with their removal from natal areas and transportation to downstream habitats that might then be colonised. The larval stages of benthic insects most characterise this process, which is an integral part of the development and ecology of numerous taxa. Several different functional categories of 'drift' have been suggested, but these have led to 'much confusion' (Brittain and Eikeland1988). Nevertheless, high incidence of drift occurs, and various interpretations are unified through the most obvious consequences of (1) local removal of insects as they enter drift and (2) local colonisation and community entry as they later regain the substrate. Both processes may be influenced by local flow conditions, and many factors have been implicated in initiating or increasing drift—perhaps the most obvious is simply increased water flow, as from flooding, but Brittain and Eikeland noted also photoperiod, water temperature changes, increased density in populations, predator pressure, pollution, and physical disturbance(such as from dredging or electrofishing, p. 99). Early field experiments showed that up to 82% of invertebrates that colonised denuded stream beds in the River Medway, England, arrived by drift (Townsend and Hildrew 1976), and drift is regarded as the predominant mode of movement amongst microhabitats and helps to account for the patchy distributions of many benthic insects, for which substrate features may be critical. Leaving the drift to regain the substrate may be an active behavioural process.

The relatively overlooked process of 'benthic crawling' by larvae may also contribute to recolonisation of upstream sites following seasonal drying or isolation. Studies on a range of EPT taxa in New Zealand undertaken in the context of insect recolonisation of intermittent streams from a more permanent downstream area showed that upstream crawling distances commonly did not exceed 200 m for most genera (Graham et al. 2017). A few individuals occurred at the maximum sampled distance of 400 m upstream—one mayfly species (*Delatidium* sp.) was represented at that point by 17 individuals, together with a single stonefly larva (*Zelandobius* sp.) and nine species of Trichoptera. Some of these incidences were strongly isolated from the trends of far shorter distances traversed by most conspecific individuals, and mean travel distance for many taxa was less than a metre, supporting earlier studies suggesting that short distance dispersal by crawling was usual. Graham et al. had anticipated finding a 'moving front' of crawling larvae. This did not eventuate, but they also noted earlier accounts suggesting that crawling dispersal may be a more important mode of recolonisation in sites with continuous flow and unrestricted time for movements to occur.

Dispersal distance of adult mayflies, amongst others, can be influenced strongly by land cover and local topography. Landscape structure may impede or facilitate dispersal of insects between patches of suitable habitat, so that knowledge of dispersal behaviour and efficiency has strong relevance to assessing impacts of population isolation and vulnerability. Under arid conditions, effective trans-site dispersal by insects may be very restricted, and that isolation can drive diversification among taxa (Razeng et al. 2017). That analysis involved two arid regions, both considered important biogeographical refugia, each with many perennial pools, amongst which genetic variation was assessed among selected dragonflies and mayflies. In both the Pilbara (Western Australia) and the West MacDonnell and George Gill Ranges (Northern Territory), larvae were collected from series of pools and adult dragonflies captured from the latter area for DNA analysis. Two strongly flying dragonflies (*Orthetrum caledonicum*, *Diplacodes haematodes*) showed only minimal genetic differences between localities. In contrast, the mayflies (*Cloeon*, *Tasmanocoenis*) respectively showed seven and nine lineages sufficiently distinctive to be likely species-level characteristics, implying that low dispersal in such arid regions may indeed promote isolation and differentiation. Loss of aquatic connections between populations, with reduced gene flow, are likely drivers of differentiation in such environments.

Many lotic insects have a duality of dispersal modes (drift and flight, above), with directed upstream flight believed widely to compensate for impacts of drift on upstream populations. The predominant dispersal form for any given species can sometimes be assessed by comparing genetic variation within single streams and across streams over a wider local landscape. This approach has been pursued in a number of studies on different taxa in south-east Queensland: a rainforest mayfly (*Bungona narilla*, Baetidae: McLean et al. 2008), a caddisfly (*Tasimia palpata*, Tasimiidae: Murria and Hughes 2008), and a hydropsychid caddisfly (*Cheumatopsyche* sp.: Baker et al. 2003). All have provided greater perspective on contemporary or historical distribution patterns (Hughes et al. 2011), but for both orders, it appeared that adult flight was the major dispersal mode, with impacts of larval drift being relatively minor.

In parts of south-east Queensland, groups of rainforest streams are separated by lowland areas, apparently dividing the known range of some insects, and that could comprise barriers to dispersal. Genetic analysis of larvae of *B. narilla* from series of 'north' and 'south' streams separated by about 100 Km of heavily cleared lowland implied that the lowland indeed reduced gene flow and was associated with reduced dispersal and habitat fragmentation (McLean et al. 2008). However, populations were not extensively differentiated to either extreme. Likewise, the region was an ancient dispersal barrier to *T. palpata*.

Surveys on a California stream, using sticky traps at distances up to 150 m from the stream into mixed evergreen forest (Jackson and Resch 1989a, b) showed insect numbers to decrease with increased distances, but interspecific differences in incidence occurred. Thus, two species of Trichoptera were abundant on trees next to the stream, but *Helicopsyche borealis* was almost absent at 40 m away whilst *Gumaga nigricula* was still common at 150 m into the forest. Differences were also evident with height of the suspended traps (operated at 2, 5, and 8 m above the ground, the last near the tops of the trees). This variety implies needs to manage the riparian zones to accommodate needs of adult insects with varying behaviour patterns, as well as the more usual emphasis on managing the waterbodies themselves.

For some relatively long-lived aquatic insects, assemblage composition also changes with distance from water (Fig. 10.2). Trapping of adult Chironomidae at different distances from streams in Brittany (France) yielded 42 species, of which 31 had aquatic larvae (Delettre and Morvan 2000). Each site included species not found elsewhere, and only seven species were trapped at all sites. Numbers declined considerably with distance from water. More locally, and reflecting the hedgerow networks on the sites, the density of hedgerows and the 'openness' of the landscape influenced chironomid dispersal from water bodies—with removal of hedges at one site associated with lower species richness there. Differences were predominantly in the loss of rarer (low abundance) species reported from the other sites. Riparian vegetation and features such as hedges can comprise either corridors or barriers to chironomids, and most rare species occurred only close to water. The 'hedge effect' (Delettre and Morvan 2000) was evident at two spatial scales, with both quality and quantity of hedgerows influencing spatial distribution of the midges. Delettre and Morvan concluded that they must be considered simultaneously in explaining dispersal, as (1) landscape scale in which both distance to water and landscape openness influence species' distributions and (2) more local scales in which hedge quality and structure determine the sheltering effects of the hedgerows. The first of these may induce isolation of aquatic habitats and limit connectivity, whilst the second is likely to be mediated through choice of resting sites by the flies.

Not all insect individuals or species that successfully reach a new waterbody as flying adults can breed there. As McCaulay (2006) found for the Odonata of artificial ponds (cattle tanks) in Michigan, both dispersal and subsequent recruitment are 'filters' of the eventual richness of the communities present. In addition, adult

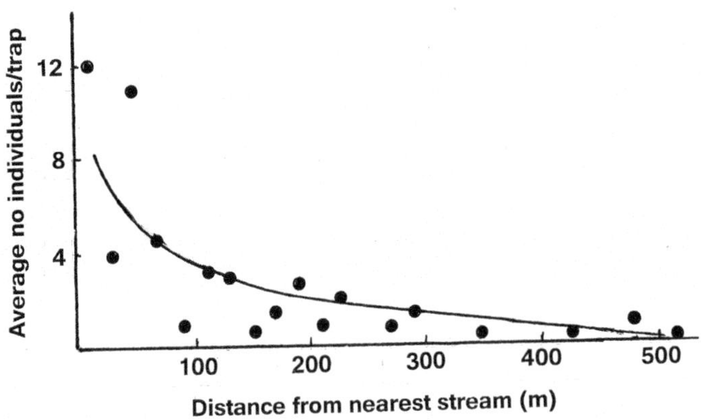


**Fig. 10.2** Average number of adult Chironomidae/trap at increasing distances from the nearest stream. Points are for pooled data from three sites (Delettre and Morvan 2000)

dispersal may be male-biased so that effective dispersal may differ from that estimated initially by observers, with features of individual water bodies also influencing outcomes. More broadly, any potential coloniser must negotiate several variable local 'filters' in order to establish, as (1) its own biological capabilities; (2) its range of tolerances to the local chemical and physical conditions encountered; (3) the morphological features (such as depth, area, hydroperiod) of the water body; and (4) the structure of the existing receptor community. These also relate to levels of isolation and water body density within the region (Incagnone et al. 2015), so that dispersal potential and actual colonisation are not necessarily linked closely. Even under rather uniform conditions, colonisation rates can vary considerably (zooplankton: Caceres and Soluk 2002). Dispersal may also have a strong seasonal component and, as Boda and Csabai (2013) observed 'understanding the kind of dispersal behaviour of aquatic insects is an old goal for ecologists'. Their studies in Hungary revealed three main seasonal patterns (spring, summer, autumn) among Coleoptera (69 taxa) and Hemiptera (21 taxa), with a variety of 'sub-patterns' across different taxa with differing peaks, flight season duration and other variations across April-October.

### 10.2.1 *Impacts of Urban Lighting*

The negative impacts of urban lighting, through attracting large numbers of phototactic flying insects are cited commonly as 'ecological traps' (p. 138), essentially removing those insects from breeding populations and in many cases increasing their vulnerability to insectivorous predators. Most commonly reported from Lepidoptera, many other insect groups—including aquatic taxa—succumb. Thus, a number of cases involving Ephemeroptera were noted by Egri et al. (2017), so 'huge swarms' of mayflies have been observed around street lamps in Europe and North America. The ecological trap impacts are compounded by the effects of polarised light. Reflected from dark shiny surfaces (such as asphalt roads, glass and black plastic sheeting), polarised light leads female mayflies and others to oviposit there, rather than in water. Szaz et al. (2015) reported an unusual (or, at least, rarely observed) such impact on *Ephoron virgo* from lamplit bridges. The unpolarised bridge lighting attracted female mayflies, interrupting their normal upstream compensatory flight (p. 239) as the influence of the strongly polarised river surface is interrupted. The females lay on the polarised asphalt of the bridge, leading to what Egri et al. described as 'huge piles of female carcasses and eggs' covering the bridge.

As a potential remediation measure, Egri et al. (2017) experimented with hanging LED beacon lights from bridges on tributaries of the Danube in Hungary. These lights were suspended below bridge level and facing the oncoming migrating mayflies. Their high attractiveness prevented mayflies from diverting from the river before they laid eggs so had a very positive effect in survival. The lights would be needed over only very limited periods each year, during the short flight season, and their use is potentially manipulable to help conserve different species. These novel trials on *E. virgo*, which had for long been absent from many rivers in central Europe but returned as water quality improved over the 1990s, revealed a potentially simple approach to preventing substantial reproductive mortality.

Installations of artificial lights close to streams should preferably be avoided or, if necessary, be spaced well apart. Perkin et al. (2014) suggested spacing of >80 m, based on their surveys of insects attracted to lights along the Spree River, Germany. This interval contrasts with the more usual spacing of around 30 m between lights.

## *10.2.2 Pond Colonisation*

With proliferation of constructed ponds (p. 270), monitoring to detect the progress of invertebrate communities and their resemblances and convergences toward those of more natural ponds involves appraisals of dispersal, although not always explicitly, and values of reservoir sources within the local landscape. One such example of colonisation is of the Pinkhill Wetland Project, Oxfordshire, in which a series of approximately 40 pools were monitored over seven years (Williams et al. 2008), during which the approximately 3.2 ha mosaic accumulated about 20% of the United Kingdom's wetland plant and macroinvertebrate species. Eight of the invertebrates were 'Nationally Scarce'. The four major monitored pools were colonised relatively slowly, so that 4–8 species were present a few months after they were dug. Richness then increased rapidly to average 52 species (46–57) three years later, after which richness levelled out, despite differences between individual pools in different years. Over the sampling period the site as a whole supported at least 167 macroinvertebrate species. The high richness might reflect the semi-natural surrounds of the Project and the complex of different ponds, providing considerable opportunities for colonisation from a variety of recruiting environments.

Both dispersal and habitat features act as 'filters' that affect assemblage composition of aquatic invertebrates, and the latter may have stronger influence (Smith et al. 2015). Nevertheless, Smith et al. emphasised the importance of maintaining or improving potential dispersal pathways, both along streams and rivers and overland between streams in neighbouring catchments. Understanding how aquatic insects interact with, and are restricted or facilitated by their enveloping non-aquatic landscapes is a key to planning effective conservation.

## References

Allen KA, Le Duc MG, Thompson DJ (2010) Habitat and conservation of the enigmatic damselfly *Ischnura pumilio*. J Insect Conserv 14:689–700

Baker AM, Williams SA, Hughes JM (2003) Patterns of spatial genetic structuring in a hydropsychid caddisfly (*Cheumatopsyche* sp. AV1) from southeastern Australia. Molec Ecol 12:3313–3324

Beebee TJC (2007) Population structure and its implications for conservation of the great silver beetle *Hydrophilus piceus* in Britain. Freshw Biol 52:2101–2111

Bilton DT, Freeland JR, Okamura B (2001) Dispersal in freshwater invertebrates. Ann Rev Ecol Syst 32:159–181

Boda P, Csabai Z (2013) When do beetles and bugs fly? A unified scheme for describing seasonal flight of highly dispersing primary aquatic insects. Hydrobiologia 703:133–144

Brittain JE, Eikeland TJ (1988) Invertebrate drift: a review. Hydrobiologia 166:77–93

Caceres CE, Soluk DA (2002) Blowing in the wind: a field test of overland dispersal and colonization by aquatic invertebrates. Oecologia 131:402–408

Carlson PE, McKie BG, Sandin L, Johnson RK (2016) Strong land-use effects on the dispersal patterns of adult stream insects: implications for transfers of aquatic subsidies to terrestrial consumers. Freshw Biol 61:848–861

Crook DA, Lowe WH, Allendorf FW, Eros T, Finn DS (and 12 other authors) (2015) Human effects on ecological connectivity in aquatic ecosystems: integrating scientific approaches to support management and mitigation. Sci Tot Environ 534:52–64

Delettre YR, Morvan N (2000) Dispersal of adult aquatic Chironomidae (Diptera) in agricultural landscapes. Freshw Biol 44:399–411

Dolny AQ, Harabis F, Mizicova H (2014) Home range, movement and distribution patterns of the threatened dragonfly *Sympetrum depressiusculum* (Odonata: Libellulidae): a thousand times greater territory to protect? PLoS ONE 9(7):e100408. https://doi.org/10.1371/journal.pone.0100408

Dreyer J, Gratton C (2014) Habitat linkages in conservation biological control: lessons from the land-water interface. Biol Control 75:68–76

Egri A, Szaz D, Farkas A, Pereszlenyi A, Horvarth G, Kriska G (2017) Method to improve the survival of night-swarming mayflies near bridges in areas of distracting light pollution. R Soc open sci 4:171166

Graham SE, Storey R, Smith B (2017) Dispersal distances of aquatic insects: upstream crawling by benthic EPT larvae and flight of adult Trichoptera along valley floors. N Z J Mar Freshw Res 51:146–164

Hughes JM, Huey JA, McLean AJ, Baggiano O (2011) Aquatic insects in eastern Australia: a window on ecology and evolution of dispersal in streams. Insects 2:447–461

Incagnone G, Marrone F, Barone R, Robba L, Naselli-Flores L (2015) How do freshwater organisms cross the"dry ocean"? A review on passive dispersal and colonization processes with a special focus on temporary ponds. Hydrobiologia 750:103–123

Jackson JK, Resh VH (1989a) Activities and ecological role of adult aquatic insects in the riparian zone of streams. USDA Forest Service Gen Tech Rep 110:342–344

Jackson JK, Resh VH (1989b) Distribution and abundance of adult aquatic insects in the forest adjacent to a northern California stream. Environ Entomol 18:278–283

Juracka PJ, Dobias J, Boukat DS, Sorf M, Beran L, Cerny M, Petrusek A (2019) Spatial context strongly affects community composition of both passively and actively dispersing pool invertebrates in a highly heterogenous landscape. Freshw Biol 53:2093–2106

Karaouzas I, Andriopoulou A, Gritzalis K (2014) Contribution to knowledge of the distribution of the rare great silver water beetle *Hydrophilus piceus* (Linnaeus, 1758) (Coleoptera, Hydrophilidae) in Greece. Pol J Entomol 83:99–107

Lynch RJ, Bunn SE, Catterall CP (2002) Adult aquatic insects: potential contributors to riparian food webs in Australia's wet-dry tropics. Austral Ecol 27:515–526

McCaulay SJ (2006) The effects of dispersal and recruitment limitation on community structure of odonates in artificial ponds. Ecography 29:585–595

McLean AJ, Schmidt DJ, Hughes JM (2008) Do lowland habitats represent barriers to dispersal for a rainforest mayfly, *Bungona narilla*, in south-east Queensland? Mar Freshw Res 59:761–771

Murria C, Hughes JM (2008) Cyclic habitat displacements during Pleistocene glaciations have induced independent evolution of *Tasimia palpata* populations (Trichoptera: Tasimiidae) in isolated subtropical rain forest patches. J Biogeogr 35:1727–1737

Perkin EK, Holker F, Tockner K (2014) The effects of artificial lighting on adult aquatic and terrestrial insects. Freshw Biol 59:368–377

Razeng E, Smith AE, Harrisson KA, Pavlova A, Nguyen T (and eight other authors) (2017) Evolutionary divergence in freshwater insects with contrasting dispersal capacity across a sea of desert. Freshw Biol 62:1443–1459
Smith RF, Venugopal PD, Baker ME, Lamp WO (2015) Habitat filtering and adult dispersal determine the taxonomic composition of stream insects in an urbanising landscape. Freshw Biol 60:1740–1754
Szaz D, Horvath G, Barta A, Robertson BA, Farkas A (and four other authors) (2015) Lamp-lit bridges as dual light-traps for the night-swarming mayfly *Ephoron virgo*: interaction of polarized and unpolarized light pollution. PloS one 10(3):e0121194
Townsend CR, Hildrew AG (1976) Field experiments on the drifting, colonization and continuous redistribution of stream benthos. J Anim Ecol 45:759–772
Wickson SJ, Chester ET, Valenzuela I, Halliday B, Lester RE, Matthews TG, Miller AD (2014) Population genetic structure of the Australian caddisfly *Lectrides varians* Mosely (Trichoptera: Leptoceridae) and the identification of cryptic species in south-eastern Australia. J Insect Conserv 18: 1037-1046
Williams P, Whitfield M, Biggs J (2008) How can we make new ponds biodiverse? A case study monitored over 7 years. Hydrobiologia 597:137–148

# Chapter 11
# Conservation

## 11.1 Introduction

The vital importance of conserving freshwater ecosystems is acknowledged globally, with the central relevance of environmental flow regimes in running waters emphasised as a core driver of much of their functioning and sustainability of their biota. The major findings and recommendations of the Brisbane Declaration (2007, see also Arthington et al. 2010) (Table 11.1) illustrate the variety of concerns that arise. That Declaration was revisited in 2017, following belief that it 'was significant in setting a common vision and directions for environmental flows internationally' (Arthington et al. 2018), and that meeting led to further urgent call for actions to protect and restore environmental flows and aquatic ecosystems. The earlier declaration had also been described as 'a pivotal statement and synthesis' (Poff and Matthews 2013), and the expanded later version included greater reference to cultural and social dimensions of environmental flows. However, lack of progress in the intervening decade meant that environmental flow requirements had not been assessed adequately for most aquatic ecosystems. Arthington et al. (2018) commented that major obstacles to implementation, included (1) lack of political will and public support; (2) institutional barriers and conflicts of interest; and (3) limited resources, capacity and knowledge—but also that, with ever-increasing demands for water, the need for implementing environmental flows is also increasingly urgent.

However, systematic conservation planning—used widely in terrestrial and marine environments—is a relatively recent advance for freshwater ecosystems (Barmuta et al. 2011). That planning incorporates identifying the 'assets' that need conservation, setting priorities among these, designing and implementing conservation measures, and monitoring to appraise the outcomes against the predetermined targets. Barmuta et al., and others, have expressed concerns that emphases on methods and validity of conservation assessment have not generally been followed through to implementation or effective management. This gap has been referred to as the 'research-implementation gap' (Knight et al. 2008), in which stakeholder involvement and understanding is a key remedial element. Nevertheless, far more is needed


T. R. New, *Insect Conservation and Australia's Inland Waters*,
https://doi.org/10.1007/978-3-030-57008-8_11

**Table 11.1** The 'Brisbane Declaration, 2007' as points for a global action agenda to protect rivers, developed from a conference of more than 750 participants: key findings and major actions called for (from Arthington et al. 2010)

| |
|---|
| Key findings: |
| 1. Freshwater ecosystems are the foundation of our social, cultural, and economic well-being |
| 2. Freshwater ecosystems are seriously impaired and continue to degrade at alarming rates |
| 3. Water flowing to the sea is not wasted |
| 4. Flow alteration imperils freshwater and estuarine ecosystems |
| 5. Environmental flow management provides the water flows needed to sustain freshwater and estuarine ecosystems in coexistence with agriculture, industry and cities |
| 6. Climate change intensifies the urgency |
| 7. Progress has been made but much more attention is needed |
| Global action agenda: |
| 1. Estimate environmental flow needs everywhere immediately |
| 2. Integrate environmental flow management into every aspect of land and water management |
| 3. Establish institutional frameworks |
| 4. Integrate water quality management |
| 5. Actively engage all stakeholders |
| 6. Implement and enforce environmental flow standards |
| 7. Identify and conserve a global network of free-flowing rivers |
| 8. Build capacity |
| 9. Learn by doing |

to overcome the wide accusation that 'the science of conservation assessment has lost its way and become a displacement behaviour for academia' (Whitten et al. 2001, as discussed by Knight et al. 2008).

The effective conservation of 'biodiversity' in freshwater environments reflects practices founded on widely applicable general principles, with site- or range-specific measures superimposed to hone the desired outcomes for particular contexts. All occur within the wider needs to promote connectivity across the wider landscape, such as by networks of pond habitat that are accessible to biota. General protocols or recommendations for conservation of freshwater invertebrates have been produced for many places, and each contains ideas or measures that may merit emulation elsewhere, sometimes with considerable refinements to local needs and priorities. Thus, Wells et al. (1983) cited recommendations made by Stansberry and Stein (1971) for North American pearly mussels (Unionidae) in broad terms to alleviate known threats or causes of loss, as (1) improve pollution control; (2) greatly reduce or stop the rate of stream and river impoundment; (3) greatly reduce or stop stream channelisation; (4) greatly reduce or stop commercial harvesting until populations recover their former abundance; (5) determine the ecological requirements for rare and endangered species; and (6) establish sanctuaries in rivers known to harbour relict populations of such species. Wells et al. listed 26 taxa of *Epioblasma* unionids of conservation concern amongst these endemic North American molluscs. Such general themes appear daunting, but embrace the broad conditions that a practical conservation programme must consider.

Six major actions, also all complex to accomplish, have been identified to aid conservation and protection of New Zealand's freshwater biodiversity (Table 11.2,

**Table 11.2** The six major actions proposed to enhance protection of New Zealand's freshwater biodiversity and prevent further declines (Weeks et al. 2016)

| |
|---|
| 1. Change legislation to adequately protect native fish and invertebrate species, including those harvested commercially and recreationally |
| 2. Protect habitat critical to the survival of New Zealand's rare and range- restricted fish, plant and invertebrate freshwater species |
| 3. Include river habitat to protect ecosystem health in the National Objectives Framework for the National Policy Statement on Freshwater Management |
| 4. Establish monitoring and recovery plans for New Zealand's threatened freshwater invertebrate fauna |
| 5. Develop policy and best management practices for freshwater catchments, which include wetlands, estuaries, and groundwater ecosystems |
| 6. Establish, improve, and maintain appropriately wide riparian zones that connect across entire water catchments |

Weeks et al. 2016). Collectively, they display the universal needs for political goodwill, wide stakeholder participation, informed science, and logistic support, accompanied by general acceptance of the needs to initiate, support and sustain those actions. As in Australia, interest in actions for freshwater biodiversity conservation in New Zealand can benefit heavily from the wider interests in native fish, and—less sound ecologically but politically persuasive—support for angling, with pressures to repeatedly release alien fish species as a basis for recreational activities. Weeks et al. (2016) noted that up to 74% of New Zealand's native freshwater fish are considered endangered or at risk. New Zealand's freshwater invertebrate fauna includes 'at least 638 species not found elsewhere', and Weeks et al. commented that the decline of freshwater quality over the previous quarter century had been rapid and 'alarming'—not just in support for biodiversity but also in a high proportion of pastoral and urban sites carrying high pathogen levels and nutrient loads. Pollution has become high, reflecting intensification of agriculture and urbanisation. So-called 'diffuse pollution' from agricultural intensification is now regarded as one of the biggest threats to New Zealand's freshwater ecosystems.

Defining suitable 'targets' for conservation and promoting relevant management to achieve these are central concerns in freshwater conservation. The United Kingdom organisation Pond Conservation, developed since 1988 through concerns that ponds were substantially neglected in freshwater conservation considerations, has two main objectives: (1) to promote the conservation of ponds and other fresh waters by providing good technical information and advice; and (2) to implement that advice on the ground with practical projects based in good science (Biggs et al. 2005). Development of the organisation focused initially on surveys that progressively documented ponds, from their numbers, extents and physical/chemical features to characterising biotic assemblages, in order to build up an information data set relevant to assessments and practical management. The four main areas of activity over the first 15 years of Pond Conservation had the effects of (1) demonstrating the importance of ponds for biodiversity; (2) understanding the factors that influence conservation values of ponds; (3) developing methods to assess the status of ponds; and (4) providing the basis for a National Pond Monitoring Network.

A series of provisional categories for assessing the conservation values of macroinvertebrate assemblages of British ponds has wide potential values elsewhere. The broader 'Pond Manifesto' (EPCN 2008) emphasised the importance of ponds around the world, noting their values for cultural, educational, historical and ecological purposes, and their contributions to local and regional diversity—as well as needs for their protection and conservation. For Europe, and applicable widely elsewhere, the manifesto states as key messages: (1) collectively, ponds represent an exceptional freshwater resource, and (2) ponds are a varied habitat type occurring across all … landscapes. They are important assets for human welfare, agriculture and water management, as well as for biodiversity. Again collectively, ponds (defined in the manifesto by area of <10 ha) represent 30% of the global surface area of standing fresh waters and occur across all biogeographical regions. Both permanent and temporary ponds vary enormously in character, with a correspondingly high variety of threats. Recent attention to characterising and managing 'ponds as a whole' rather than 'single ponds' implies that ponds may be the richest harbours of macroinvertebrates of all freshwater categories (Fig. 11.1, for Britain, after EPCN 2008).

The principles underlying the major elements of a strategy for pond conservation in Europe (Table 11.3, EPCN 2008) transfer easily to Australia, and would be needed in future national policy. The logistic necessity of concentrating more on 'communities' than on individual species conservation can draw usefully on the five major tactics for conservation of aquatic insects listed by Foster (1991) as: (1) heightened awareness of the need to maintain complex communities by reference to high profile taxa; (2) heightened awareness of the values of wetland insects in identifying sites of conservation value that cannot be identified by reference to other groups; (3) a rationale for community analysis developed in terms understood and used by

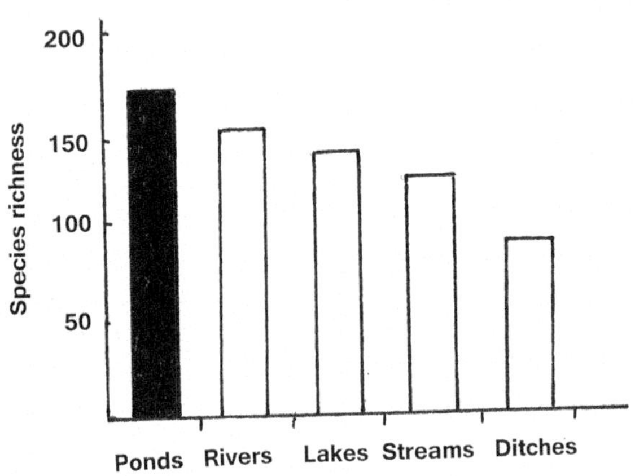


**Fig. 11.1** The regional diversity of macroinvertebrates in the River Cole catchment, southern England in ponds (black bars) and other waterbody categories (from EPCN 2008 with data from Williams et al. 2003, who cited the following figures for species richness in the running water systems noted: ponds 173, rivers 152, streams 124, ditches 90, overall total 249; ponds included 70%of total richness in the catchment)

**Table 11.3** Lessons from the European Pond Conservation Network: stated objectives and strategy (EPCN 2008)

| |
|---|
| Objectives of EPCN:<br>1. To exchange information on pond ecology and conservation between researchers, managers and practitioners<br>2. To promote understanding of pond ecology by encouraging the development and coordination of fundamental and applied research<br>3. To raise the profile of ponds and guide national and supranational policies for their protection<br>4. To promote effective practical pond conservation<br>5. To disseminate information on the importance, attractiveness and conservation of ponds to the people |
| Major objectives of strategy for conservation of ponds in Europe:<br>1. Policy and legislation. Work at the European level and with existing national networks and organisations to ensure that current and new policies, legislation and funding opportunities support the conservation of ponds<br>2. Research and monitoring. Identify and undertake applied and fundamental research work on pond biology and ecology needed to support best practice pond conservation and management, and pond monitoring<br>3. Communication and awareness raising. Develop and deliver a communication strategy to ensure that stakeholders, including policymakers, scientists, managers and practitioners and the general public are aware of the most up to date information about pond ecology, the importance of ponds, best practice techniques and conservation issues<br>4. Conservation of the pond resource. There is a critical need to work on the ground to halt the loss and degradation of the pond resource. Key sites need to be protected and new high quality ponds which will replace sites now destroyed or too damaged to restore need to be created. The role of ponds as 'stepping-stones' in the landscape also needs to be maintained |

conservation decision makers; (4) community studies generating plausible management suggestions; and (5) a 'key site' approach, as an inventory of wetland sites supporting internationally important species and communities for potential listing in the Ramsar Convention (p. 255).

The fates of many Australian aquatic insects devolve fully on the condition and accessibility of their habitats—a truism equally so for terrestrial taxa but perhaps far more difficult to address for water bodies faced with increasing human pressures superimposed on more natural fluctuations and that may be influenced by changes far from their apparent boundaries. Alterations far upstream from river reaches of interest, or beyond the immediate riparian zones of lakes or ponds can provide novel, sudden and drastic disturbances to waters that may already be subject to an array of more proximal and more obvious and managable influences on water flows, levels and quality. Problems of countering more distant impacts can often not be met by attention only to a focal site but demand concerted efforts over far larger areas. These may embrace aquatic and terrestrial environments, even across entire catchments of running waters and drainage basins of lakes or ponds. However intensive and detailed the management for a focal threatened insect species at a key site may be, success may not be assured without sufficient 'buffering' or attention to more widespread threats or changes.

Conservation of the naturally high functional connectivity of, especially, lotic waterbodies with wider biomes demands very careful and comprehensive planning.

Potentially, many more invertebrates may need to be considered in freshwater conservation programmes, reflecting the extensive areas and land uses that affect any given waterbody (Linke et al. 2011), and all of which may influence its condition and security. Thus, the 'drainage basin' is a functional regional unit in lotic ecology, and 'lake districts' have parallel roles in lentic ecology. The needs of assemblages of insects and others may validly be considered on this larger scale, rather than restricting attention to individual sites—even when those sites are the only ones known to be occupied by particular taxa, their management can only rarely be addressed fully in isolation (Heino 2009).

Nevertheless, many insect conservation projects necessarily devolve on individual sites or very limited regions, as the only place(s) where the species of interest is known to occur. 'Analogue sites' elsewhere may be sought, but many determined and well-targeted surveys have failed (1) to reveal the species elsewhere, even in apparently very suitable habitats and (2) not clarified reasons or the needed critical resources for occupancy or absence. In the 1980s–1990s, for example, many sites in central Victoria were explored unsuccessfully for the Ancient greenling damselfly (*Hemiphlebia mirabilis*, p. 219) following its rediscovery in the state. In many similar cases, initial fortuitous discovery of a notable species can stimulate or facilitate more detailed investigations. Site-orientated conservation is the most frequent approach in insect conservation, with priorities (when a choice between sites is available) determined on criteria such as residence of the focal species, wider richness, species endemism, and other parameters that contribute to some form of 'relative biodiversity value' or conservation importance and that may be overlain by urgency in the face of threats. However, even the resident or breeding status of a focal taxon may be difficult to confirm (Khelifa 2019), especially amongst taxa with a strong terrestrial phase in which they may move far from their natal aquatic habitats. Adults of some Odonata, for example, may be encountered in aquatic habitats where they do not reproduce—but those adults may also be far easier to detect than their larvae or exuviae. Likewise, biases in diversity assessments may arise when lotic and lentic taxa meet over part of their range, so that observed richness is 'artificially inflated' in some complex hydrological systems.

The wide principles of managing freshwater environments for conservation are only rarely devised specifically to conserve insects but, rather, to assure and restore the major ecological functions of those systems, their benefits to human use and values, and their aesthetic qualities. However, the reality of the diverse anthropogenic impacts on insects discussed in earlier chapters and the usually inadequate knowledge that can bear on their management implies the need for further focused management, especially to sustain the numerous ecologically specialised insect species that have demonstrablty or supposedly become threatened, or even lost. General management principles may need to be tailored for individual sites, species and contexts. The known array of threatened aquatic insects, although far from completely defined, is sufficient to indicate some approaches through which such refinements can be pursued. The breadth of principles was summarised for conserving habitats of Odonata by Corbet (1999) (Table 11.4), with the active steps predicated on the security (by deliberate protection or reservation) of the site(s) involved. Corbet grouped

**Table 11.4** Some considerations for management of freshwater ecosystems, based on the anthropogenic impacts that degrade habitats for dragonflies (Corbet 1999)

| Biotopes primarily affected | Impacts |
|---|---|
| Lotic and lentic | Lowering of water level by draining, extraction, or diversion |
| | Obliteration by earth moving, infilling, construction work, or removal of substrate |
| | Increase of sediment load by destruction of vegetation cover in catchment area |
| | Chemical or thermal contamination by runoff or discharge of agricultural, industrial or urban effluent |
| | Eutrophication, especially by seepage of manure or synthetic fertiliser applied to land |
| | Unrestricted human access to margins for recreation |
| | Progressive isolation of populations as biotopes are destroyed |
| Streams, rivers and canals | Destruction of heterogeneity of margins and rate of flow by canalization |
| | Destruction of vegetation and soil profiles by mechanical clearing |
| | Increase of predation on larvae, directly or indirectly, by stocking or propagation of fish, or by introduction of domestic ducks |
| Rivers | Erosion and physical disturbance of margins by inland water traffic |
| Streams and lakes | Acidification by afforestation with conifers |
| Lakes | Acidification by airborne industrial emissions |
| | Intermittent exposure of littoral zone during draw-down in lakes used for hydroelectric power |
| Bog pools | Obliteration by commercial peat extraction |
| Phytotelmata and ground litter | Obliteration by deforestation |

the principles into three categories, distinguishing the different emphases needed for lotic and lentic systems from more embracing general needs. Whilst Table 11.4 was derived especially for dragonflies, the measures listed have far wider relevance in aquatic entomology.

The recent establishment of the Alliance for Freshwater Life as a united call for protection of freshwater biodiversity and decreasing the global bias toward terrestrial conservation aims to provide a critical mass for representing that biodiversity and developing harmonious solutions to balance conservation and development needs (Darwall et al. 2018). Broad background to the issues involved across most major freshwater ecosystems (Cantonati et al. 2020) also emphasises that paucity of knowledge of invertebrates of some key geographical areas (notably within the Oriental, Neotropical and Afrotropical regions) hampers detection of some key 'hotspots' of endemism and richness that should receive high priority for protection. However, all major freshwater habitat categories are in some way threatened, with recognised or projected losses in their biodiversity. The 'Alliance' recognised five core areas for action, as (1) research, as the foundation of good conservation decisions and establishing the wider socio-economic context for sustainable use of water; (2) data and synthesis, rendering the available information accessible readily to conservation managers, researchers and other constituent interests, and building on existing schemes to facilitate wider access and use; (3) conservation and sustainable use of freshwater biodiversity, involving management and restoration to assure sustainability and to protect species and systems still in relatively good condition; (4) education and research, a core need at all levels to raise awareness problems and to garner support for action; and (5) policy, increasing awareness of the importance of freshwater species and ecosystems, and rendering policies truly beneficial.

## 11.2 Protected Areas

Freshwater protected areas are a vital component of many biodiversity conservation programmes, but much of the array of current reserves has had rather 'ad hoc' historical development, rather than arising from deliberate planning to achieve a representative suite of reserves that encompasses the full ecological variety present, and to ensure that all major ecosystem types have some formal protection for the future. An ideal for any such reserve system would be to assure such comprehensive protection and sympathetic management of the enveloping landscapes to protect them from further disturbance and major threats. However, as Kingsford and Nevil (2006) put it 'The history of freshwater protected areas in Australia is, in large part, a story of good intentions not carried through'. At that time, the Australian reserve system was not based on any systematic analysis of wetland types—despite some local exceptions (Fitzsimons and Robertson 2005). Measures to improve that representation and assure credibility for conservation recommendations were noted by Kingsford and Nevil, with those listed in Table 11.5 summarising the variety of interests and prospects from a workshop and additional support leading to more than 50 informed signatories to the document in which they are included.

Because most formal protected areas designated for biodiversity values have been based largely on charismatic vertebrates and vascular plants and predominantly emphasise terrestrial environments, their values fror aquatic biodiversity are often inadequately known or overlooked. Some focused investigations, such as in the Segura River Basin of Spain (Zamora-Marin et al. 2016), have revealed protected areas to have substantial gaps in freshwater faunal coverage. In that region, water beetles have been surveyed well and are recognised as valid surrogates for wider biodiversity (Abellan et al. 2005), with the incidence of 211 species assessed for 256 localities—namely protected areas (46 localities), protected areas and sites within the 'Natura 2000' network (118), and non-protected areas (138) with some of the last highly degraded. Beetle richness, in the above sequence of site categories, was 163, 201, and 171 species, with only 10 species not found in any of the protected areas.

**Table 11.5** Key avenues for action for progress toward an adequate representation of Australian freshwater ecosystems in protected areas, recognising the urgent need to establish comprehensive and representative freshwater protected areas and an effective management strategy that includes environmental issues (adapted extract from Kingsford and Nevil (2006), with permission from CSIRO Publishing)

1. Major rivers where ecosystems remain essentially intact should be protected
2. It is recommended that all Australian jurisdictions accelerate the development of freshwater protected areas
3. Ecosystem inventories also need accelerated development, both to underpin protected area identification and selection and to support 'sympathetic' management of biodiversity values within regional resource planning frameworks
4. Control of cumulative effects, particularly within catchment-scale management, needs much greater attention
5. Rehabilitation of significant aquatic sites should remain a priority
6. Stakeholders with common interests need to start building consensus and raising awareness, with their adequate consultation in selecting areas and building community support for management

Analyses presented by Zamora-Marin et al. implied that protected areas performed no better than non-protected areas or simply randomly selected areas for harbouring beetle richness. However, for the first two categories above, some tendency for higher richness was suggested, with three factors identified as deserving further investigation. Zamora-Marin et al. suggested that (1) because aquatic biodiversity may not have been considered in designating the protected areas, their aquatic communities may not show any especially rich faunas; (2) non-protected areas with low levels of human disturbance may support high richness; and (3) the Natura 2000 network included habitats that represent different rare or threatened European taxa rather than covering the most diverse or richest sites. However, almost all water beetles were present in at least one protected area, and none of the 10 other species is endemic or regarded as threatened. The importance of the Natura 2000 network in augmenting more formal protected areas includes both direct complementarity and likely roles as transition zones or buffers between more protected and more disturbed landscapes.

The priorities and problems of designating freshwater protected areas as a cornerstone of conservation endeavour, need and foundation for management have been discussed extensively, with realisation that any such measures beyond wholly local exercises (such as 'traditional' village ponds) are likely to pose economic and social controversy across sectoral interests with largely different priorities. Much of the practical difficulty is founded simply in the increased demands for water that already threaten many freshwater bodies across the world. Almost all of Earth's major rivers have in some way been overused, dammed, diverted or otherwise regulated, with degradation of their former variety and that of adjacent margins and landscapes. Anachronisms arise—many rivers, for example, form the boundaries of terrestrial protected areas, and are thus essentially protected only along one side whilst the other bank may be undergoing massive change. A widespread consideration and problem is simply the continuing conflicting demands and conflict between conservation needs and human use and dependence on aquatic resources. Many high priority water bodies are influenced strongly from far beyond their putative reserve boundaries and by disturbances from land uses over large areas. Fitzsimons and Robertson (2005) noted that many remaining wetlands in western Victoria are on private land and may not be available for reservation, even if tenure could be changed to increase long-term protection. Nevertheless, designation of a 'CAR' (Comprehensive, Adequate, Representative) protected area system for freshwaters is an urgent and continuing need. A strategic process to facilitate this has wide relevance (Fig. 11.2).

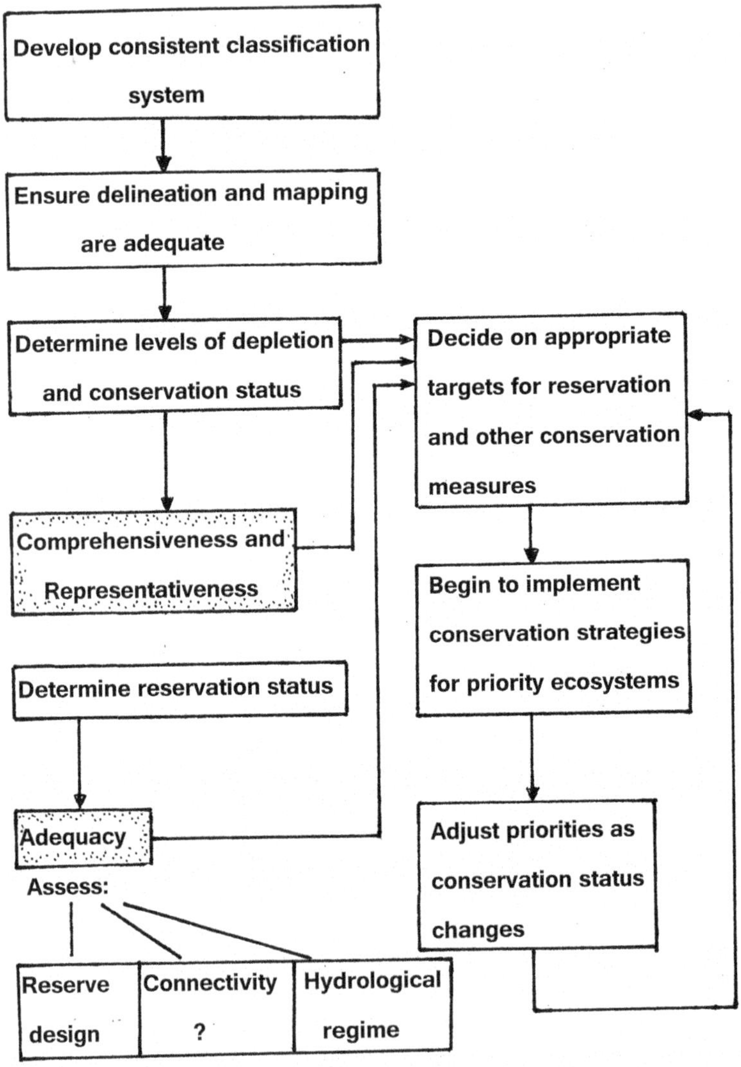


**Fig. 11.2** A strategic process to develop a comprehensive, adequate and representative system of freshwater reserves (Fitzsimons and Robertson 2005)

Ideally, protected area systems should comprise a 'set of conservation areas that best represents the native species, communities, and ecosystems of an ecoregion and the underlying processes that sustain them', as discussed by Higgins et al. (2005). Many, however, are designated or proposed for less-embracing purposes, such as protection of local populations or species for which security of a critical habitat is a predominant need for conservation management. Higgins et al. proposed a nested four-level hierarchical system for freshwater protected areas. These levels were (1) aquatic zoogeographic unit; (2) ecological drainage units in each of these; (3) aquatic ecological systems in each of these; and (4) macrohabitats in each, so ranging from overall planning units through finer scales to the more localised realms of ecological systems and habitats. The principle is founded largely on the concept of 'coarse filter conservation targets' as representative ecological units that will support many common species and communities and their environments, and so is more embracing than the more usual species-focused concentration on small individual sites. It also allows for detection of species that are not captured at this level and that may indeed need a fine-level approach, should that be possible.

In comparison with terrestrial and marine ecosystems, freshwater protected areas have received rather less informed attention, not least because transfer of the ideals and ideas derived from terrestrial areas to freshwater systems is often not easy (Abell et al. 2007). Outside of the Ramsar network (p. 255), Saunders et al. (2002) noted that few 'protected areas' had been created especially for freshwater systems, and most such habitats are included only fortuitously within terrestrial reserves in which many relevant concerns may not be addressed. Abell et al. noted that all freshwater bodies receive impacts from their surrounding environments, rather than just those generated internally, and a far wider catchment approach to protection is needed rather than simply focusing on the waterbody alone. Only then can the full variety of actual and potential threats be defined or predicted adequately for management to proceed. That recognition has led to realisation that the full protection of a given freshwater system necessitates considering both (1) its entire upstream catchment and (2) the downstream habitats of any key focal species. Activities such as building roads in reserved parks, for example, may have only trivial effects on the local terrestrial environment, but far more severe impacts by causing sedimentation and pollution of nearby aquatic ecosystems. Key habitats in existing protected areas may need some form of enhanced protection within these. For water beetles in Spain, Ribera (2000) recommended extension of protected areas in national parks to also include mid-elevation streams, in order to counter changes over whole catchments. Likewise, aquifers in Australia that support refugia for invertebrates may need a heightened level of protection (Davis et al. 2013). In some cases, refugia (p. 279) are the only hope for poorly dispersing species to persist as climate changes.

However, some protected areas have been established to protect freshwater species and habitats, and sometimes to promote tourism to scenic sites. When focal taxa for such reserves are designated, most are vertebrates. Only rarely are inventories of aquatic insects sufficient to assess their representation in protected areas, but many such insects may not occur there. In north-western Argentina, Nieto et al. (2017) noted that 'only an insignificant percentage' of the distribution of the region's freshwater invertebrates was in protected areas and pointed out that those protected areas had largely been raised for plants and vertebrates and that (as in many other places) neither invertebrates nor aquatic ecosystems were considered constructively. Well-planned protected area systems must work toward this more comprehensive consideration (Nieto et al. 2017).

Saunders et al. (2002) presented three 'strategies' for designation and conservation of priority freshwater protected areas, whilst also recognising that the probability of attaining any one of these was low, but also that the principles can be applied realistically in site-specific assessments, together with the interlaced threats the strategies address. Ideally, freshwater protected areas should (1) be in intact catchments, (2) have natural hydrological regimes, and (3) contain no non-native species. Saunders et al. (2002) noted also that in some contexts it may be possible to concentrate conservation effort on adjacent terrestrial areas and headwaters, whose condition can markedly influence adjacent or downstream waters. For headwaters, for example, unaltered upstream disturbance avoids impacts on downstream habitats. Indeed, the twin principles of protecting headwaters and land close to waterbodies may be combined with other management practices to provide some protection when the above three ideals cannot be met. However, and as Saunders et al. warned, 'the simple creation of freshwater protected areas … does not guarantee the long-term survival of these vital ecosystems'.

Using the protected area network of Michigan, United States, as a study case, Herbert et al. (2010) found this inadequate to safeguard important freshwater conservation features, not least because those areas had mostly been acquired or designated for other priorities. They were considered to provide only 'a weak safety net' for aquatic features. An allied theme is the vulnerability of small or isolated wetlands within larger reserves, and that may be deemed expendable. Based on surveys of amphibians of such small wetlands in South Carolina, Semlitsch and Bodie (1998) demonstrated that (1) small wetlands (as small as 0.2 ha) are valuable in maintaining biodiversity, and (2) their losses will also directly reduce connectivity between populations. In short, small isolated wetlands are not expendable.

It is impossible to protect all freshwater areas that harbour, in many instances very tenuously, threatened species, so that some rational form of setting priorities amongst the numerous deserving candidates is necessary. One approach, developed from a workshop that reviewed the then current protection schemes for freshwater, marine and terrestrial ecosystems in order to provide a tool for honing conservation efforts, was described by Darwall and Vie (2005). The approach drew on a seven-stage process for setting site priorities (Table 11.6), with assessments of species' conservation status and distribution, and a full range of habitats represented. Amongst the aquatic insects only Odonata are sufficiently well documented for species identification and conservation status assessments to be included, although availability of other data may enable more local inclusion of additional taxa. Thus, for Step 5 of Table 11.6, an 'important site' is identified if it satisfies one or more of four criteria, each with conditions (Table 11.7). Identifying priority sites for possible protection is a vital first step—but in many cases in practice will also mark the end of a process that for various reasons does not proceed to effective conservation.

**Table 11.6** The seven steps proposed for setting priorities among sites in selecting them for conservation of freshwater biodiversity (Darwall and Vie 2005)

1. Define the geographical boundaries within which to identify important Sites
2. Define the wider ecological context of the designated assessment area
3. Identify and map the distribution of inland water habitat types
4. Catalogue the distribution and conservation status of priority aquatic taxa
5. Apply species-based selection criteria (Table 11.7) to identify sites
6. Ensure full representation of inland water habitats among those sites selected
7. Ensure inclusion of keystone species

**Table 11.7** The species-based selection criteria proposed for identifying sites as 'important' for freshwater biodiversity conservation (Darwall and Vie 2005). A site is identified as important if it satisfies one or more of the four following criteria

| |
|---|
| *Criterion 1.* A site is known or thought to hold a significant number of one or more globally threatened species or other species of conservation concern, where (1) globally threatened status is determined according to the IUCN Red List categories and criteria; (2) the term 'significant numbers' is taxon-specific; (3) assessment may be based on infra-specific taxa; and (4) 'other species of conservation concern' allows for inclusions of taxa of regional or national concern even if they are common elsewhere |
| *Criterion 2.* A site is known or thought to hold non-trivial numbers of one or more species (or infraspecific taxa as appropriate) of restricted range, where (1) 'non-trivial numbers' is interpreted to exclude vagrant individuals, and the threshold number is taxon-specific; (2) guidelines for defining a 'restricted range' will be taxon-specific |
| *Criterion 3.* A site is known or thought to hold a significant component of the group of species that are confined to an appropriate biogeographical unit or units, where (1) non-native species are excluded from analysis; (2) the biogeographical unit is taxon specific—freshwater ecoregions may be appropriate units for many taxa; and (3) species-rich areas are identified, not as areas rich in the total number of species, but as areas rich in species restricted to that biome |
| *Criterion 4a.* A site is known or thought to be critical for any life history stage of a species—such as refugia, feeding areas, migratory sites, etc. |
| *Criterion 4b.* A site is known or thought to hold more than a threshold number of individuals of a congregatory species, where threshold number is taxon-specific (in practice largely used for birds) |

Expansion of freshwater protected areas in much of Australia is restricted by economic and social factors, even when major gaps in ecosystem coverage are evident. Sympathetic management of rivers outside protected areas is thus a central need. Measures to approach this include (1) provision of environmental flows; (2) reduced sedimentation; (3) improved water quality; (4) protection and restoration of refugia for aquatic species; (5) establishment of corridors to promote connectivity; and (6) translocation of sensitive species if natural range changes such as needs for southward movement in response to climate are not possible (Turak et al. 2011).

The Convention on Wetlands of International Importance (known widely as the 'Ramsar Convention') was designed to halt the worldwide loss of wetlands and to conserve and manage those that remain. At the time it was launched (2 February 1971), countries were encouraged to nominate sites that contained representative, rare or unique wetlands or that are important to conserve biodiversity toward a 'list' of wetlands of truly international significance. Australia was among the first group of Contributing Partners, with many differing wetlands—ephemeral or permanent, and freshwater to marine, nominated.

Management to maintain the 'ecological character' that describes each wetland is needed under the Convention. The strategic approach in Australia necessarily involves the Commonwealth government cooperating with State and Territory governments and agencies in establishing this baseline, against which future change can be monitored. In 2005, the definition of wise use of wetlands was modified to 'the maintenance of the ecological character, achieved through the implementation of ecosystem approaches within the context of sustainable development', again ensuring that the activities that affect wetlands will not lead to losses of biodiversity or other ecological, cultural or social values. Detailed Australian policy for this (Ramsar Convention 2008) flows from the Commonwealth Environment Protection and Biodiversity Conservation Act 1999.

The criteria for identifying Australian wetlands of international importance are wide-ranging, and many incorporate aspects of invertebrate biodiversity conservation, even though not stated specifically, and in contrast to more explicit focus on birds and fish (Table 11.8). The generic threats to wetlands noted in Table 11.9 were compiled from consideration of the threats identified for each of Australia's then 64 Ramsar sites and the management plans available for these. No aquatic insect species are noted in the taxon list accompanying the reference to 'invasive and other problematic species and taxa', with the only insects mentioned being the Yellow crazy ant (*Anoplolepis gracilipes*) and bees (unspecified taxa).

**Table 11.8** The criteria used to determine 'nationally important wetlands' as candidates for listing under the Ramsar convention (Ramsar Convention 2008)

A wetland may be considered nationally important if it meets at least one of the following criteria:

1. It is a good example of a wetland type occurring within a biogeographic region in Australia
2. It is a wetland which plays an important ecological or hydrological role in the natural functioning of a major wetland system/complex
3. It is a wetland which is important as the habitat for animal taxa at a vulnerable stage in their life cycles, or provides a refuge when adverse conditions such as drought prevail
4. The wetland supports 1% or more of the national populations of any native plant or animal taxa
5. The wetland supports native plant or animal taxa which are considered endangered or vulnerable at the national level
6. The wetland is of outstanding historical or cultural significance

**Table 11.9** The categories of threats to Australia's Ramsar sites, as listed by theme and major contributing activities following an extensive review (from Ramsar Convention 2008)

| Threat theme | Threat activity |
|---|---|
| Agriculture and aquaculture | Agriculture and aquaculture |
| | Annual and perennial nontimber crops |
| | Livestock farming and grazing |
| | Marine and freshwater aquaculture |
| | Wood and pulp plantations |
| Biological resource use | Fishing and harvesting aquatic resources |
| | Habitat shifting and alteration |
| | Storms and flooding |
| | Temperature extremes |
| Ecosystem/community stresses | Ecosystem degradation |
| Energy production and mining | Energy production and mining |
| | Mining and quarrying |
| Human intrusions and disturbance | Human intrusions and disturbance |
| | Recreational activities |
| | War, civil unrest and military exercises |
| Invasive and other problematic species and taxa | Invasive non-native/alien species |
| | Problematic native species |
| | Invasive and other problematic species and taxa |
| Natural system modifications | Dams and water management/use |
| | Fire and fire suppression |
| | Other ecosystem modifications |
| Pollution | Agricultural effluents |
| | Airborne industrial/commercial pollutants |
| | Excess energy |
| | Garbage and solid waste |
| | Household sewage and urban waste water |
| | Industrial and military effluents |
| | Pollution |
| Residential and commercial development | Commercial and industrial areas |
| | Housing and urban areas |
| | Residential and commercial development |
| | Tourism and recreation areas |
| Species stresses | Indirect species effects |
| | Species mortality |
| Transportation and service corridors | Shipping lanes |
| | Roads and railroads |
| Uncategorised | Uncategorised |

## 11.3 Management and Restoration

In marked contrast to many terrestrial insects of equivalent conservation concern in Australia, species-focused conservation plans for aquatic insects are few, and generally couched in rather general terms. Needs for conservation may be strongly evident from losses, declines, habitat alterations and defined impending changes, and more-or-less reliable designations of species as 'endangered' or other recognised category of threat flow from those perceptions. Translating those needs into detailed plans for action and conservation management, however, has lagged considerably. With very few exceptions, recommendations in any subsequent conservation plans, action statements and similar outcomes obliged under listing legislations are necessarily couched in rather general terms. Most of those documents emphasise needs for further surveys and documentation of the species' biology, whilst protecting known habitats and preventing further disturbances to all sites, including historical sites, from where the species has at some time been reported. Similar lack of detail applies to most other aquatic invertebrates, whilst the more detailed proforma produced for the Giant freshwater crayfish (*Astacopsis gouldi*, p. 167) in Tasmania is exceptional in the greater detail provided for this relatively well-known species. That plan includes series of costed recommendations and activities under each of five 'strategies' (Table 11.10), with the realisation that many will be difficult to achieve within any reasonable time, especially before a scheduled 5-year review of that recovery plan.

For the Hawkesbury-Nepean River system near Sydney, Chessman and Williams (1999) recommended a study covering the two major approaches evident more generally in insect conservation—individual attention to key threatened or 'listed' species and a more holistic strategy to conserve the wider communities and habitats and to manage threatening processes. Networks of 'subcatchment reserves' could be given high priority, these based on areas that retain much natural vegetation and have undergone little disturbance. However, because of varying tenure/ownership of lands, much of the area in private hands, opportunities to achieve such outcomes comprehensively are very limited.

The wider management of freshwater ecosystems for conservation necessitates consideration of a wide range of values. The list generated by Likens et al. (2009)

**Table 11.10** The five strategies identified to achieve the objectives of the recovery plan proposed for the Giant freshwater crayfish, *Astacopsis gouldi* (Australian Government 2017)

1. Conduct population trend monitoring and research to assess the current status of the species and evaluate the effectiveness of recovery actions
2. Mitigate key threats impacting upon Giant freshwater crayfish habitat
3. Increase the reservation status and improve the quality of identified key locations for the Giant freshwater crayfish
4. Reduce the impacts of illegal fishing on the Giant freshwater crayfish
5. Engage with the general public, local government and non-government organisations in developing and delivering conservation measures

(Table 11.11) provided a perspective of the principles for incorporation into management, and emphasised the needs for integration across numerous themes, in calling for a national programme of aquatic ecosystem science and management 'embedded in a common vision for the future of Australian landscapes and rivers and for their services to society'.

The formidable complexities of conserving and managing the Murray-Darling Basin ecosystems, as those given the greatest attention and need, have not eased since the critique provided by Pittock and Finlayson (2011) in which they portrayed a soberingly bleak future as conditions for biodiversity continue to decline with climate changes and increasing human demands for water. At that time, 13 river valleys were rated as in very poor condition, seven were poor, two moderate and only one good in combined assessments based on hydrology, fish and macroinvertebrates. Large areas of floodplain forests were changing to more terrestrial conditions, and urgent restoration measures to restore river flow and longitudinal connectivity, accompanied by increased protection and re-establishment of riparian vegetation, were among the priorities suggested.

Urban streams are an important conservation focus, with an overarching need for conceptual change in attitude so that management is developed to treat streams as 'natural assets' and 'ecosystems', rather than simply as 'drains'. Five main groups of restoration goals for these streams were listed by Walsh et al. (2005) as (1) riparian vegetation; (2) instream habitat enhancement; (3) end of pipe stormwater treatment; (4) elimination of allied stressors (such as sewer overflows and pollution sources); and (5) dispersed stormwater treatment (reducing catchment runoff to near the pre-urban levels). These measures can be used together in various combinations. Extent of remediation needed is often severe and, again following Walsh et al., integration of ecological, social, behavioural and economic research is necessary in conserving and restoring urban streams. Preservation of riparian vegetation, and its active restoration where needed, are major goals of stream restoration. However, the benefits

**Table 11.11** A statement of values of Australia's inland aquatic ecosystems, formulated by a group of experts in Bungendore, New South Wales, in 2008 (Likens et al. 2009)

1. Freshwater ecosystems are a foundation for our social, cultural, spiritual and economic wellbeing
2. Aquatic ecosystems have intrinsic values which go beyond economic value
3. The economy is part of the ecosystem and the health of the two are co-dependent
4. Climate change, population shifts and associated resource use have increased the need for urgent action
5. Globally, aquatic ecosystems are under serious threat and urgent action is needed to protect them
6. Many of Australia's aquatic ecosystems are badly degraded; however, they are worth restoring
7. Pathways for change occur through scientific engagement with communities and governments
8. The loss of species can reduce critical ecological functions. This is highly undesirable
9. There is a major need to ensure intergenerational equity
10. There is an urgent need to ensure that advice based on good aquatic-ecosystem science is available to inform land and water management in Australia

may depend on the wider landscape context and the levels of upstream disturbance. Naiman and Decamps (1997), supported by Roy et al. (2006), suggested that riparian forests might be able to maintain high quality aquatic habitats only if upstream effects do not have predominating severe impacts on the biota present.

Less than two decades ago, Paul and Meyer (2001) urged greater attention to the roles of urban streams and their restoration and noted that 'little attention is given to restoration of a native stream biota or the ecological services streams provide'. These are linked and necessitate expanding the field of urban stream restoration beyond the conventional roles of engineers and physical scientists in meeting the challenges of harmonising physical and chemical processes with biological themes. Approaches to river restoration in Australia (Cottingham et al. 2004) are very varied, but many exercises are difficult to evaluate retrospectively, because of different techniques, scales, duration, available resources and objectives. Collectively, they have demonstrated the complexity of holistic rehabilitation exercises that depend on the effectiveness of the approaches and methods used. Measures used commonly (Table 11.12) can each contribute to achieving ecosystem outcomes, but large-scale factors (such as prevailing climate or catchment conditions) may restrict, reverse or obviate the gains from smaller-scale restoration attempts. Not least, adequate monitoring and longer-term evaluation is often neglected—even in cases involving targeted attempts to protect or rehabilitate endangered species.

Much restoration of freshwater ecosystems is undertaken with a stated aim of promoting increased heterogeneity in providing, as examples, greater and more varied food resources, more physical refuges, greater surface area, more connectivity and more regular flow patterns, and so on—that fulfill the presumption that this increased variety fosters greater biodiversity. Physical changes to habitat structure are both easier to undertake and more convincing as 'restoration' than more diffuse measures that directly address ecological functions. However, as reviewed by Palmer et al. (2010), relatively few such exercises showed increased similarity of treatment sites to reference sites in their biodiversity complements. Across the studies evaluated, involving 78 independent river/stream restoration exercises in which invertebrate richness was addressed, only a third showed any positive relationship between biodiversity and in-stream heterogeneity. Most of those exercises involved measures leading to increasing substrate complexity (such as by adding

**Table 11.12** Some representative rehabilitation measures and objectives used for rivers in eastern Australia (from Cottingham et al. 2004)

| |
|---|
| Meeting the environmental water requirements of rivers |
| Reintroducing flow-related habitat for fish and invertebrates |
| Reintroducing wood (e.g. resnagging) and other physical habitat features |
| Rehabilitating riparian vegetation |
| Introducing riffle habitat to streams in urban areas |
| Managing sediment delivery and dynamics, both in-stream and in the riparian or catchment zone |
| Restocking species and tracking recovery |

boulders or wood), creating more riffle-pool complexes, regrading banks or altering the flow channel and were undertaken in response to agriculture, urbanisation or channelisation. For greater and more beneficial biodiversity outcomes, Palmer et al. recommended that greater attention should be directed to the factors that affect the stream and include invertebrates in remedying the problems that are most likely to limit restoration success. Whole watershed or catchment perspectives may indeed help to improve restoration outcomes. Restoration efforts need a clearly defined target 'goal' or condition, where possible compared with parallel reference or 'control' sites (Geist and Hawkins 2016). In many cases, restoration to any putative original state is impossible, with partial restoration toward that condition often termed 'remediation' or 'rehabilitation'—essentially terms that are often interchanged with 'restoration' as processes that may lead to novel habitats without strict natural analogues. Active restoration is more practicable in freshwater than in marine systems (where fostering natural recovery may be the better approach: Geist and Hawkins 2016), in part reflecting that many freshwater species have far more restricted distributions and can become of conservation concern as drivers for wider efforts. However, it is still complicated immensely by imprecise (or no) knowledge of the biology of most species present, and that even closely related insects and others can have very different needs. Predicting outcomes of restoration attempts, even those embedded in well-focused species conservation programmes, is usually highly uncertain and—as in many other aspects of conservation—publication of 'negative outcomes', however unsatisfactory, can provide valuable insights for future work (Geist 2015). The variety of factors leading to degradation, and their multiple interactions, can render definition of the precise steps needed for restoration of any freshwater habitat very tentative, and exacerbated further by the different time scales over which disturbances operate. Geist (2015) proposed seven steps that could help to improve the success of freshwater conservation (Table 11.13), collectively indicating need for sound management to be designed, and some of the problems that can hamper that progress.

One problem that arises in attempting to compare 'reference sites' to 'restoration sites' in aquatic ecosystems to establish a standard for emulation is simply that the physical scale of restoration may differ greatly from that of the disturbances in wider riverine landscapes. As Becker and Robson (2009) noted, the response by any single site to restoration management is unlikely to produce an invertebrate assemblage that resembles one at any former less-disturbed site. 'Reference sites' are also likely to each represent some form of disturbance history and individual location, so that each restoration site is within a mosaic of varied catchment-scale disturbances.

In some cases, active restoration may prove undesirable. Post-mining subsidence pools in the Czech Republic supported more dragonflies if they were allowed to undergo natural succession than if they were restored or reclaimed, with those imposed changes affecting bank structure and water levels (Harabis and Dolny 2015). Large disturbances led to habitat degradation and extinctions of dragonflies in the pools. Lack of resemblance to natural disturbance regimes occurred through stabiising the prevailing mining-influenced habitat conditions and leading to poor colonisation by sensitive Odonata species. Many of the restoration steps in this example led to steady decline of the Dragonfly Biotic Index (p. 183), and did not restore the natural

**Table 11.13** Seven steps toward improving freshwater conservation, suggested and discussed by Geist (2015)

| |
|---|
| 1. Decisions on conservation objectives |
| Define targets and priorities in terms of feasibility and time scales, with 'subobjectives' that can be used for subsequent evaluation. Need perspective that extends beyond simple population considerations |
| 2. Determination of the status quo |
| Based on above targets, evaluate present state of ecosystems, communities, populations (etc.), together with physical and chemical properties of habitats. Use standardised monitoring to evaluate pre- and post-restoration stages, with inclusion of reference sites |
| 3. Identification of bottlenecks and problems |
| Identify, from the systematic monitoring data, the key bottlenecks and problems related to conservation targets |
| 4. Setting priorities and making decisions on conservation action |
| After '3', need balanced decisions on suitable conservation actions, ideally with inputs from stakeholders/sponsors. Need approaches that have clear focus, ecologically suitable options and realistic timelines. Include socioeconomic themes of how community support will be forthcoming |
| 5. Conservation action |
| May benefit from transdisciplinary approach and effective communication. Perhaps trial measures on limited areas and monitor to determine desired effects and undesired side-effects, and expand if successful |
| 6. Evaluation and adaptive management |
| Systematic and scientific assessments of conservation actions, and use of that information in adaptive management, often minimised in favour of more subjective judgements, or not undertaken. Strong need for evidence-based conservation actions based on objective criteria |
| 7. Exchange and publication of results |
| Essential to publish both positive and negative outcomes |
| In many cases, expected success does not occur, and awareness of these can help avoidance of repeated 'mistakes' and modify approaches to similar projects in the future |

succession patterns needed by pioneer species. Species of high conservation value were not supported in the absence of natural levels of heterogeneity and disturbance. However, finding that such 'secondary habitats' as the products of industrial activity can substitute, at least in part, for lost natural wetlands and foster dragonfly assemblages through the heterogeneity they do provide has considerable potential for design of related future projects (Dolny and Harabis 2012). Different successional stages in ponds can influence dragonfly richness, so that increasing successional variety over the shores of a pond or across different ponds in the same area may be a useful management strategy (Steytler and Samways 1995). The principle is widespread. For constructed ponds (p. 270) along an urban land-use gradient in England, macroinvertebrate assemblages reflected differences in macrophyte cover, nutrient status, riparian shading, and the form of the pond edge, in addition to the surrounding landscape and proximity of other wetland habitats (Thornhill et al. 2017). Local physical factors were thereby highly influential, and four priority themes were suggested to improve conservation values of such urban ponds: (1) improve conditions for macrophyte growth, such as by reducing shading and by increasing heterogeneity of pond margins; (2) reduce nutrient inputs, such as by reducing drainage and increased pre-entry treatments of urban water; (3) assess the options within 100 m to allow naturalised vegetation such as scrub to develop; and (4) plan on a landscape scale to produce resilient networks of ponds.

Restoration can be focused to the habitat needs of a particular species or higher taxon, rather than seeking more general remediation. Thus, a translocation or re-introduction of significant species can succeed only if that species' requirements are fulfilled at the release site, and prior restoration management may be needed to assure this. Re-introduction of the local endemic rare damselfly *Ischnura gemina* (Coenagrionidae) into an urban park in San Francisco, an area in which most of its natural habitat had succumbed to urbanisation, is one such example (Hannon and Hafernik 2007). The release site was the only local breeding area for the species before its earlier local extinction. Restoration involved clearing the aquatic vegetation that had grown in and over the water channel, to leave open sun-exposed sites within the context of a larger habitat restoration project that involved removal of riparian trees and shrubs from a large seep previously occupied by *Ishnura*, and construction of three nearby pond habitats, each of at least 5 × 2 m. Introduction was by release of mating pairs, as one of the first attempts to introduce Odonata to a new site by the adult stage alone, and was initially successful. However, despite short–term success, re-establishment eventually failed—probably due to failure to suppress rapid growth of aquatic vegetation that filled the channel. Because urbanisation had removed the natural processes that maintained open water habitat, continuing or periodic management was needed to remove excess sediment and vegetation and thus to mimic past regimes. Habitat specialist Odonata tend to be amongst the most vulnerable representatives in more general declines of the order in such changed environments (Ball-Damerow et al. 2014), with damselflies often more sensitive than the larger true dragonflies. More broadly, as discussed elsewhere (p. 176) the population dynamics of Odonata can be influenced strongly by management of the nearby terrestrial areas (Harabis 2017), and those species with a long pre-reproductive period may depend on the terrestrial environment for up to several months. However, the aquatic breeding habitats must remain the primary focus for conservation, with provision for continuing management.

A long-term perspective is needed for many aspects of freshwater conservation, and one major drawback in planning is simply that long-term surveys of aquatic insects have historically been rather rare. As Jackson and Fureder (2006) commented 'long-term studies have led to new discoveries and unexpected insights for freshwater macroinvertebrates and ecosystems that could not have been evident from more temporally limited studies'. For the future, those insights depend on a combination of short-term and long-term surveys. The latter will almost inevitably be few, restricted by lack of funds, continuing agency commitment and political goodwill—as well as the limited career spans and longevity of scientists who drive such exercises. The longest studies noted by Jackson and Fureder included up to 30 years of data. However impressive, Jackson and Fureder also noted that this is still much less than the time needed to detect impacts of climate change or major land-use modifications, or of highly irregular and unpredictable disturbances such as '50-year' or '100-year' droughts or floods.

**Table 11.14** Five principles for effective research and management of lotic fishes proposed by Fausch et al. (2002)

| |
|---|
| 1. Choose appropriate scales—maintain flexibility to think and work at multiple scales |
| 2. Processes will interact across scales—embrace the ecological complexity of lotic systems rather than try to force simplicity upon them |
| 3. Unique features can have overriding effects—discrete habitat features in space or rare events in time can have important effects on fishes in lotic systems |
| 4. Unintended consequences of habitat degradation will occur in all directions, including upstream—anthropogenic disturbances may have strong effects at long distances from their source |
| 5. Match observations and predictions to the scales at which managers effect change |

As well as need for long-term information, Fausch et al. (2002) emphasised the need for a 'continuous' view of rivers and streams for their conservation to be pursued effectively—far different from the most common approach and restriction of observations only on small stretches of these long linear habitats, but with each such reach influenced by changes elsewhere along that length. Their essay focused on fish, but the five principles for effective research and management they proposed for lotic fishes emphasise parallel needs for many aquatic invertebrates (Table 11.14). These principles led to four areas in which large-scale 'riverscape' approaches to management are needed: (1) managing habitat for endangered or other threatened species; (2) countering invasions by alien species; (3) managing ecosystems to sustain fish populations for sport and commercial needs; and (4) addressing climate change. The third of these acknowledges that the scale at which natural habitat is created and used by fish may be far greater than that addressed by conventional management and can be extrapolated easily to invertebrate needs. In practice, conservation of stream fish diversity will require much innovative endeavour to 'address questions and design management strategies at scales that match the life histories of the species involved' (Fausch et al. 2002). So, also, for insects for which the effective environment may span both aquatic and terrestrial ecosystems, each with particular features and integration across catchments and their landscapes. The approach of treating streams as 'networks' that support dynamic metacommunities may help to convey this wider need, as discussed for New Zealand (Campbell and McIntosh 2017).

## 11.4 Fish Conservation

The conservation programmes for individual threatened fish species can easily overlook the needs of equally deserving invertebrates found in the same areas. In some cases those invertebrates have not even been detected, and may become threatened unintentionally and unknowingly. In others, they may simply be disregarded, or presumed to benefit from measures such as site protection and restoration. The latter may commonly be the case but is not invariably so for ecologically specialised taxa. As noted earlier, conservation of aquatic vertebrates generally has a far higher 'profile' than any equivalent intent for insects. In practice, this gives native fish (and others, such as amphibians and waterbirds) an important umbrella role for aquatic insect conservation, although management measures may not always be compatible, and the principle of more holistic conservation of inland water biota has yet to be developed fully in Australia. Conservation of individual wetlands for selected frog species, for example, very rarely acknowledges co-occurring aquatic invertebrates.

A North American aquatic bug, the Ash Meadows naucorid (*Ambrysus amargosus*) exemplifies some of the dangers that can arise. This was the first, and for long, only, aquatic insect listed under the United States Endangered Species Act, and is known only from the thermal Point of Rocks Springs, Nevada, where specialists had found it at intervals (Polhemus 1993). It was declared as 'threatened', and the population then 'protected' within a created Ash Meadows Natural Wildlife Refuge. However, by 1992 Polhemus found that the spring outflows that supported the species during his previous visit (in 1984, when the area had already been changed considerably from its earlier hospitable conditions) had been modified considerably. The later changes occurred during another conservation campaign, the construction of a refuge for the endangered Devil's Hole Pupfish (*Cyprinodon diabolis*)—in particular by placement of large rocks to secure a plastic water pipeline for the fish, but which essentially caused the loss of naucorids from one of their remaining habitat patches. The structural change was associated with spread of alien vegetation and increased trampling by wild horses seeking water. Polhemus used this dramatic example to demonstrate the problems of even signaled aquatic insects in conjunction with greater concerns for threatened vertebrates that gain higher public and political profiles.

Parallel examples are not uncommon. In general, profiles of even the most threatened aquatic insect species are very secondary to those of any co-occurring vertebrates—considered more 'worthy' but whose conservation may not provide any sound 'umbrella protection' for invertebrates in the same water bodies. More broadly, the principle of invertebrate conservation 'capitalising' on the greater conservation interest in endangered aquatic vertebrates, so that insects 'piggy-back' on the umbrella vertebrate interests, is an important avenue for development. Wetlands visited by large numbers of migratory waterbirds, including both common and rare taxa are a key focus in some bird conservation programmes and such sites can have high priority for conservation. Recommended in other contexts—for example, for mayflies in Montana, United States (Stagliano 2016), where some species share critical habitats (substrates of large sand or gravel) with the Pallid sturgeon fish (*Scaphirhynchus albus*) and Piping plover (*Charadrius melodius*), both targeted for active conservation, these measures may benefit species that have declined considerably from their former state as optimal habitat has been lost or degraded. Also in Montana, conservation efforts for the Idaho giant salamander (*Dicamptogon aterrimus*) may help to safeguard small steep forested streams that are critical for some invertebrates. Those streams and wetlands without salmonid fish appeared to be associated with the highest levels of mayfly rarity, in part paralleling the accepted values of stream ecosystems in Australia that lack introduced trout (p. 114).

'Over the last 25 years, the regular stocking of native fish into impoundments and rivers has established large, popular recreational fisheries and contributed significantly to conservation of these species' (Rowland and Tully 2004). Three species, all endemic to the Murray-Darling river system, have been especially important, and two (Murray cod, *Maccullochella peelii peelii*, and Silver perch, *Bidyanus bidyanus*) have been listed for conservation concern. The third species (Golden perch, *Macquaria ambigua*) had also declined. All three species had suffered local extirpations, so that the large-scale hatchery rearing programmes indeed have an important conservation role. However, the fish hatchery industry is a major contributor to supply for recreational angling, with management of genetic stocks, disease avoidance and suppression, and production of 'fit' stock all components of reducing/obviating poor hatchery practices whilst (under continually refined 'quality assurance' conditions) continuing to produce millions of fish annually for release. Several dragonfly species can become abundant in aquaculture ponds, where they can also be major predators of fish fry. Ingram et al. (1997) cited *Hemianax papuensis*, *Hemicordulia tau* and *Orthetrum caledonicum* as the most commonly encountered species in the south-east. *H. tau* has been reported to reach larval densities exceeding $2.5/m^2$, inducing attempts to supress their numbers. One tactic noted by Ingram et al. was to prevent oviposition by removing marginal vegetation used as egg deposition sites. It is also recognised that some larger Hemiptera prey on small fish, but their numbers only rarely reach levels of economic concern; water scorpions (Nepidae), for example, are often found in aquaculture ponds. Hemiptera, together with aquatic beetles and others, may also compete with small fish for prey.

Fish stocking, to enrich lakes and rivers for recreational angling, involves both native species (such as the above species in the south-east, and barramundi in northern Australia) and alien species (such as trout), and can also involve translocations of species of conservation interest. Stocking of farm dams (p. 270) is also common. Two major contexts are (1) introductions to streams that previously lacked species desired by anglers, and (2) supplementation of existing fish stocks to boost catch rates. In Queensland, native fish species are the primary focus for translocations, many of which are controlled and documented programmes, but others are more casual and undocumented (Burrows 2004). In some areas, re-stocking has been the basis of increased tourism industry and a continuing boon to local economy through angling as primarily 'a major recreational, social and economic activity'. Organised state-supported hatcheries and stocking groups are widespread—the latter occur, for example, in most catchments in Queensland's Wet Tropics Region (Burrows 2004). The industry continues to expand, with little historical consideration for wider impacts on local aquatic biodiversity. Nevertheless, concerns have indeed been voiced for some time.

A resolution from the Australian Society for Fish Biology in 1977, repeated in 1984, and circulated to all relevant agencies across Australia, was largely overlooked and unheeded. That resolution urged restriction or prevention of movements of native freshwater fish into drainage systems in which they were at that time not known to occur naturally.

A major concern in the Queensland Wet Tropics is the vulnerability of many of the insects of upland streams, many of them above waterfalls that have been effective barriers to natural fish movements upstream and allowed the upland waters to remain uninvaded. Those previously fish-depauperate streams may be especially vulnerable to fish translocations. Burrows (2004) commented that 'Many invertebrates in upland Wet Tropics streams are not adapted to coexist with translocated fishes to which they are now subjected'. Although in some places with less evidence, this comment is likely to apply widely in Australia. And, also as in the Wet Tropics, more detailed investigations of ecological impacts of translocated native fish, as well as of alien introductions, are needed. Those studies could usefully include more popular taxa that may also suffer increased vulnerability. One recommendation by Burrows, indeed, was to undertake such evaluations also for frogs and other fish.

## 11.5 Education and Involvement

Education programmes and publicity for promoting insect conservation have very varied targets—from young people to politicians, managers, and potentially interested and supportive people in many walks of life. This breadth necessitates a correspondingly varied range of activities, limited only by imagination and the resources available, but unified by the need to disseminate accurate information, clear rationale for conservation need and action, and to demonstrate how people can help in this endeavour. Perhaps only for dragonflies has the potential for any aquatic insect group been approached, with local cultural and proprietal interests for this global flagship order giving them a uniquely prominent role as ambassadors for wider aquatic insect conservation needs—not least as foci for an increasing array of 'dragonfly reserves' (p. 250). Some years ago, the attendees at a major International Symposium on Conservation of Dragonflies in Japan issued the 'Kushiro Appeal' (Corbet et al. 1995) as a statement of belief in, and primary needs for, conservation of Odonata and their habitats. That statement (Table 11.15) remains entirely relevant, and includes clear endorsement of the value of wide education.

As in so many aspects of conservation, agency and other 'official' expertise and logistic capability is commonly far short of what is needed for effective assessments and management of freshwater environments, and wider avenues of recruiting interest—such as through volunteer groups and 'citizen scientists'—can make substantial contributions at all stages. Effective communication and education can engender both awareness of needs and the urgency for action, and practical help. Ponds and streams are inherently attractive to many young people, and 'school trips' incorporating these into science curricula, and basic introductions to their biota by local enthusiasts and managers can do much to foster enduring interests. 'Awareness trails' and good signage that includes informative comment and illustration of notable insects and their roles and needs can contribute to countering the 'extinction of experience' of natural history discussed by Cheesman and Key (2007) as largely indoor activities continue to displace such interests among young people

**Table 11.15** The text of the Kushiro Appeal on conservation of Odonata (Corbet et al. 1995)

1. Because humans depend for, practical as well as aesthetic reasons, on diversity of habitats and their occupants, humans living today have an obligation to conserve such diversity for future generations
2. Destruction of dragonfly habitats is increasing, almost entirely due to human impact, powered by growth of human numbers and economic activity; where habitat restoration is possible, it should reflect natural dynamics and preferably keep intact the integrity of the natural systems
3. In order to conserve dragonfly habitats, the following actions should be accorded high priority:
   3.1. Conservation of aquatic and terrestrial habitats, and their surroundings, including preservation as parks and reserves
   3.2. Expansion of ecological research, including research on the impact of recreation on dragonfly habitats, as a basis from improved management of areas being conserved
   3.3. Ecological management of natural and restored habitats; and
   3.4. Monitoring the successes, or failures, of conservation programmes and, where necessary, modifying such programmes
4. The future success of dragonfly conservation depends on education, especially of nonspecialists and children
5. Destruction of dragonfly habitats is a global problem; so wherever possible conservation strategies should be coordinated and reinforced both nationally and internationally

and, in many instances, also displace any nascent natural curiosity over the 'real world'. One of the four issues identified by Morse (2009) as 'especially deserving of immediate and long-term attention' is a trend that 'young people tend to know less about species' and be less ready to pursue careers intent on discovering them and augmenting knowledge about them and their ecological needs. This is, most simply, a component of reduced education about natural history, especially field-based study, in favour of more 'technological' topics and indoor computer-based learning. Morse believed, in a view held much more widely, that a significant change in direction to emphasise biodiversity education and discovery (whilst this is still possible in the face of 'the unprecedented rate at which [aquatic] species are being lost') is truly urgent. That need does not diminish the relevance of, for example, modern molecular and statistical approaches to species diagnosis, evolutionary divergence, and determining extinction rates—but conservation founded on sound biological knowledge and observation of the species and environment of concern remains a paramount need and focus.

Recreational angling remains one of the few socially acceptable activities benefiting from sound knowledge of freshwater habitats. Fly-fishing, in particular, has a strong foundation in (and legacy from) knowledge of the life cycles and seasonality of freshwater insects, and conveys strong need for natural conditions to persist.

Nevertheless, community volunteers (often referred to as 'citizen scientists') are a key resource potentially contributing considerably to basic documentation of aquatic insects through participation in surveys and monitoring activities. Such involvements are not new: Firehock and West (1995) noted that the involvement of citizens in river protection can be traced back over several centuries to the English position of hiring a 'Riverkeeper' charged with safeguarding interests of anglers and protecting the river's welfare. In North America, the Izaak Walton League was formed in 1922 with a parallel purpose, and efforts to survey and counter water pollution from 1926 and activities proliferating as a nation-wide volunteer network was grown. Their first exercises in volunteer monitoring involved EPT as key indicator orders—if all three were present a stream was rated 'excellent'; if only Ephemeroptera and Trichoptera occurred, it was 'good'; and if only Trichoptera were found, it was adjudged 'fair', with all sampling based on a prescribed standard technique.

Benthic insects have figured highly in volunteer assessments, together with adult Odonata. Whilst many dragonflies and damselflies can be identified by non-specialist volunteers to species level from relatively conspicuous features and aided by well-illustrated field guides, most EPT larvae are diagnosed to only order and, sometimes, family level. Particular characteristic taxa can sometimes be recognised consistently. Thus, volunteers in North Carolina used (1) philopotamid caddisflies, with larvae having bright orange heads; (2) any caddisfly larvae with cases made of stones; and (3) large perlid stonefly larvae (Penrose and Call 1995), all regarded as good indicators of water quality as being intolerant of pollution.

Volunteer programs can foster local interests and can themselves be encouraged by activities such as formation of 'Friends' groups for local waterbodies or focal species support. A very wide range of contributions is possible, and volunteer agendas can complement and extend activities initiated by local authority. It follows that any guiding 'committee of management' or parallel overseeing body should include representatives of any volunteer interest group (whose expertise as 'naturalists' may by far exeed that available within officialdom) and be seen as a community activity in which all interested stakeholder groups are involved. Firehock and West (1995) listed several features that all such projects should include, as (1) managers and those who will use the data should be involved from the early stages of project design, with the initial needs also including a process for recruiting and training volunteers and reporting information; (2) volunteers should have clear goals, and understand clearly why and how they are monitoring or otherwise contributing to clearly established goals and practical outcomes; (3) every partner sector should have clearly defined roles, and all groups should work in partnership rather than 'go it alone'; and (4) projects should be self-sustaining rather than fully dependent on government or agency funding, and include a mechanism for review and possible re-direction as needed in the future. Many volunteer projects are driven by the enthusiasms of single individuals and can be easily abandoned if that person can no longer participate. Emphasis on fostering public awareness can sometimes counter this and, as Penrose and Call (1995) commented, volunteer monitoring programs 'enhance public stewardship of our natural resources' in leading to cooperations between volunteers and government agencies.

Expressions of appreciation and acknowledgement of all such participation should also be a routine and sincere activity, without which—as, regrettably, often the case—willingness to contribute further can be dissipated easily. Volunteer interest and participation in practical conservation exercises can never be 'taken for granted'.

Many investigations, whether more formal or embracing wider community inputs, on aquatic insects or other freshwater ecology projects may give rather inconclusive results, or do not satisfy the predictions that led to the study. In many instances those outcomes are never communicated or publicised, and Kefford et al. (2003, 2010) urged the wisdom and values of publishing 'negative' results. In their example, they noted the paucity of information on effects of suspended sediments on lowland stream invertebrates and that the possible general lack of effects (1) may be genuine or (2) flows from lack of communication of other studies. As they noted, in a perspective of far wider value in insect conservation studies 'finding no effect of something is generally perceived as less exciting' and may never be written up or communicated further. Those outcomes, however, may constructively inform policy or management.

## 11.6 Artificial Water Bodies

Many human activities have led to creation of new water bodies—from the large water storage dams formed by impoundments, to the variety of smaller wetlands, ornamental ponds and lakes created as urban wetlands, and in parks and home gardens. Collectively, the latter, in particular, may create networks of novel water bodies across landscapes, with the outcomes that present-day distributions of many mobile insects and others reflect both (1) loss from natural habitats as those diminish, and (2) colonisation and expansion into anthropogenic water environments (Bilton et al. 2001). In some areas, such as the Cape Floristic Region of South Africa, natural ponds are scarce but numerous artificial ponds have been constructed for farm irrigation (Simaika et al. 2016). They are now an integral feature of the local landscape, and incidences of rare local water beetles and bugs imply their value for conservation. Sampling of Odonata over three years in natural and artificial ponds showed six of the 28 species found only at the latter. Constructs such as farm ponds and dams, aquaculture ponds, water-filled quarries or gravel pits, storm drains and many others augment the availability of aquatic habitats and can also promote connectivity for dispersing insects. Farm dams in the Drakensberg Range of Kwa-Zulu Natal, South Africa were considered to be 'nature reserves' for dragonflies (Samways 1989).

Farm dams merit special consideration in Australia as a significant landscape feature over much of the continent. Australia has more than half a million farm dams, categorised as small (<5 ha) anthropogenic or natural waterbodies and used for the two main purposes of irrigation and watering stock (Brainwood and Burgin 2009). A variety of subsidiary purposes, such as fish farming and agricultural waste water storage, occurs. However, changes in water policy, affecting the proportions of surface water that can be harvested by farm dams, have led to decommissioning of many dams, together with many in the expanding urban and periurban areas being filled in to enable development. As Brainwood and Burgin commented, the biodiversity values of farm dams have received little direct attention. Their surveys in two New South Wales sites compared the macroinvertebrates of dams and stream sites in the same areas, using artificial substrate samplers deployed for six-week periods in spring and summer. The dams supported a variety of taxa. However, those dams had existed for several decades, with opportunity to accumulate species (not named in Brainwood and Burgin's account)—but more species occurred in the stream sites than in all three dams together. Assemblages in each dam across seasons were more similar within the dam than across different dams, with heterogeneity of any dam, reflecting vegetation and general structure affecting the species richness.

The invertebrate communities of new ponds confirm their importance as complementary habitats of considerable value in maintaining regional diversity (Coccia et al. 2016). In Spain, 6–7 years from restoration led to such ponds supporting communities which differed from those of reference ponds, adding further to their conservation importance.

From another point-of-view, artificial water bodies in Australia might even be a major threat to native biodiversity, beyond aquatic biota. Landsberg et al. (1997) noted that few areas of pastoral rangeland in the arid and semi-arid regions are more than about 10 km from an artificial watering point constructed for use by stock. One consequence is that, as well as stock, those water sources are exploited by feral animals (p. 118) and kangaroos, leading to additional changes in grazing intensity on the rangelands and possible adverse impacts on terrestrial insects from overgrazing.

Farm dams in South Africa have been credited with 'considerable inadvertent dragonfly conservation' in providing additional habitats for some very localised species (Samways et al. 1996), and as stepping stones for mobile taxa (Simaika et al. 2016). Those dams vary greatly in the assemblages they can support but, as lessons for future constructions of dams or ponds, it was believed essential to provide a wide range of biotopes and microhabitats to foster rare specialist species and also sustain the wider array of more generalised eurytopic species. The most important components of habitat heterogeneity for Odonata are the amount and composition of aquatic and semi-aquatic vegetation, shoreline features, hydrological factors, and sun exposure (Schindler et al. 2003). Restoration can thereby involve relatively simple measures once such basic parameters are clarified. Thus, Janssen et al. (2018) proposed that affecting the light regime (such as by removing shading terrestrial vegetation) can promote aquatic and riparian vegetation that is important for different dragonfly life stages, and so increase their diversity. In Germany, the key drivers for dragonfly diversity were cover of tall sedges (affected indirectly by available sunlight) and submerged macrophytes (affected directly through restoration). Those measures also increased habitat suitability for several species that would not otherwise occur. Counts of exuviae confirmed a strong correlation between tall riparian vegetation and dragonfly larval density. These broad restoration measures—with details reflecting characteristics of individual ponds—focused on removal of woody vegetation, together with dead wood and sludge, from the aquatic environment.

Many landowners do not specifically acknowledge the values of their farm dams as habitats for native biota until this is pointed out to them. In southern Queensland, such dams are an integral feature of the landscape as reliable water sources for cattle (the primary need influencing dam management), with impacts of cattle including erosion by trampling, nutrient enrichment by dunging, and removal of palatable riparian plants. The benthic fauna of a series of those ponds comprised 14 orders and 45 families of invertebrates (Markwell and Fellows 2008). Most families were of Hemiptera, Coleoptera, Odonata and Diptera, commonly the most diverse orders in pond ecosystems. Diversity varied across the dams, with the number of families correlated significantly with percentage of macrophyte cover. Despite the priority use and need for their dams, some of the landowners interviewed by Markwell and Fellows were keenly aware of dam biodiversity values and management to conserve these, by measures such as tree planting and fencing parts of the dam perimeter to restrict and localise cattle access.

Most such dam or pond constructs are not predicated on insect conservation considerations or needs, but a major exception is for Odonata, for which 'dragonfly ponds' are a significant and strongly-promoted measure for conservation. Initiated most strongly in Britain, where long-term surveys (such as that by Moore 1991) demonstrated their conservation values and capability to attract and support diverse assemblages of species, they are now familiar in many parts of the world. Many Odonata thrive in such artificial environments. These may thereby be less vulnerable to loss in connectivity or changed hydrology of natural waters (Simaika and Samways 2015). However, stream specialists are more vulnerable and may become more so as those habitats warm, especially in conjunction with losses of riparian vegetation. Simaika and Samways exemplified two nationally endangered South African taxa, *Chlorolestes fasciatus* (at higher elevations) and *C. tesselatus* (at lower elevations) as both vulnerable to such changes, but in largely different environments. Artificial ponds may not always compensate for natural declines of such specialist taxa, but many Odonata can indeed benefit from their availability. One fundamental principle is that the construction of new ponds or other waterbodies does not justify the continuing destruction of those already present—new constructs are ideally augmentations to natural habitats rather than replacements for them. However, with increasing interests in 'offset habitats' to compensate for development activities, considerable effort may be needed to achieve this.

More generally, wetlands constructed in agricultural landscapes to filter agricultural runoff have considerable potential as aquatic invertebrate habitats by augmenting the remaining natural habitats in the area. To be efficient filterers and maximise nutrient retention, a combination of a deeper area for sedimentation and binding of phosphorus and a shallower vegetated area that allows denitrification and nitrogen uptake by plants has been advocated, and a combination of good water quality and vegetational variety can provide good habitats for Odonata (Huikkonen et al. 2019). Many of the species found were widespread generalists which are nevertheless important in sustaining ecosystem functions, and a key need in conservation.

With very large numbers of constructed ponds already in Britain, a recent scheme (initiated by Pond Conservation from 2010) aims to create a million ponds in the British countryside within 50 years, providing unprecedented opportunity for conservation of aquatic biota. Primack et al. (2001) demonstrated the role of pond construction and restoration in promoting further the entrenched cultural values of dragonflies in Japan. They noted that artificial dragonfly ponds had become 'a significant addition to Japanese landscapes', with siting them near schools an important component of encouraging interest in natural history and conservation amongst young people—a lesson reflected also elsewhere. They are also an important counter to the large-scale losses of wetland habitats due to urbanisation.

Constructed ponds are novel features and may not fully replace natural ponds, but marginal vegetation characteristics can enhance that resemblance considerably. Artificial reservoirs in the Greater Cape Floristic Region of South Africa, constructed primarily for water storage, can closely resemble natural ponds once their marginal vegetation has developed. They shared about 40% of the 45 sampled lentic insect species with natural ponds and supported some rare endemic species over periods of drought (Deacon et al. 2019), when the more natural habitats became unavailable. Conservation values may be increased further by managing reservoir shorelines to increase their resemblance to more natural water bodies. Deleterious effects can also occur. Reservoirs on Table Mountain constructed to supply water for Cape Town had dramatic negative impacts on the highly endemic local water beetle fauna, a high proportion of which were lost, together with many of those of high conservation interest. These included species of Hydraenidae, Dryopidae and Elmidae, and Turner (2007) regarded the impact of reservoirs as 'devastating'. Losses largely reflected conversion from riverine to lacustrine habitat, to which many running-water specialists are susceptible. Fuller evaluation of water beetle assemblages was needed before any future parallel modifications are made.

Much thought and practical trialing has gone into optimising the desirable features for artificial dragonfly ponds, with the primary purpose of attracting a variety of species that can establish there. Parameters such as water depths, substrate textures, extent and heterogeneity of vegetation (both riparian and aquatic macrophytes), and slope of edges (from vertical to shallow gradients) are all manipulable, within the basic premise that heterogeneity may favour species richness of the colonising Odonata whilst particular focal species may require more specific habitat structures. Similar parameters apply in restoration efforts for both lotic and lentic waterbodies and, indeed, resource heterogeneity is a critical consideration in any parallel context in which insect richness and variety is a primary conservation focus, together with eliminating or minimising harmful disturbances. Continuing management may be needed to maintain the needed heterogeneity of habitat structure. However, it is clear that ponds for dragonflies or for wider biodiversity can be tailored to minimise unfavourable features and, in many cases, to ensure that particular features needed for key focal species are provided. The key requirements listed by Hassall (2014) (Table 11.16) suggest many of these. As an important community adjunct, many

**Table 11.16** Some key requirements for consideration in constructing artificial ponds, such as for dragonflies (after Hassall 2014)

1. Avoid vertical walls that prevent amphibians leaving the water
2. Maintain submerged and emergent plant communities with light management
3. Situate to maximise connectivity with existing ponds
4. Be aware of human access that may influence species introductions
5. Use for functions other than biodiversity, such as stormwater management, may not reduce diversity but may influence the composition of communities that occur
6. Plan a variety of pond types, as different conditions support different communities in a wide range of taxa

ponds can also host amphibians, and this may in some cases help to foster interest in their worth.

Many constructed aquaculture ponds in southern Australia are relatively small and shallow. Ingram et al. (1997) noted a common pattern of them being 0.5–2.5 m deep, with banks having about a '1 in 3' slope. Placement commonly reflects local topography—so that if ponds are built into a depression or gully the excavated soil or clay can be used in wall/bank construction. Many are permanent but others (such as rearing ponds for fish fry) may be filled seasonally from a local water supply. Drainage can be facilitated by a gently sloping floor leading to a single 'deep drain point'. Both static ponds, most commonly with earthern floors, and 'raceway' ponds with a continuous water flow and used for rearing salmonids occur. All such constructions are likely to accumulate aquatic macroinvertebrates and may thereby (1) augment pond networks in the area and (2) fulfill dual roles of fish husbandry and wider insect conservation.

The spatial scales at which ponds are constructed (with their connectivity a prime conservation issue) in continuing agrienvironment schemes can strongly influence their worth in Odonata conservation (Raebel et al. 2012). Counts of exuviae were used to compare Odonata across 29 ponds in southern England, covering a range of different variable features—each of which can be assessed easily (and without needs for complex equipment) by managers (Table 11.17). Vegetation was categorised more extensively, as indicated, and indeed proved important. Ponds with floating and submerged vegetation yielded the largest numbers of exuviae, whilst ponds with limited or no vegetation yielded relatively few. Abundance also decreased with

**Table 11.17** Variables used to classify ponds in lowland England in assessing suitability for Odonata in relation to agrienvironment schemes (Raebel et al. 2012)

| Variable | Description |
|---|---|
| Pond age | Pond created 3 or less years ago or >3 years ago |
| Water area variation | Yearly variation in water surface area: 4 classes—0–10%, 10–20%, 20–50%, >50% |
| Buffer | Unmown, grassy margin 2 or more m wide; 3 classes: absent, partially surrounding, fully surrounding |
| Cattle | Cattle trampling intensity: 3 classes—zero, low, high |
| Distance | Distance to nearest viable pond: 5 classes—0–100 m, 100–200 m, 200–300 m, 300–400 m, >400 m |
| Dry | Pond did or did not dry out the summer before sampling |
| Fish | Fish present or absent |
| Transparency | Water transparency: 4 classes—transparent >0.5 m from bank, transparent on shallow bank only, low turbidity, high turbidity |
| Trees | Percentage of deciduous trees within an area of 5 m around the pond: 7 classes—0–5, 5–10, 10–30, 30–40, 40–50, 50–70, >70 |
| Vegetation type | Pond vegetation type: 7 classes, based on species composition and structure |

increased water turbidity (with high transparency most evident in well-vegetated ponds, and turbidity possibly linked functionally with reduced photosynthetic activity and limiting the ability of visually-hunting larvae to forage: Corbet, 1999). Recently-created ponds characteristically produced more exuviae than older ponds, but also had lower species richness. Presence or absence of fish apparently had little effect in this study.

Raebel et al. (2012) suggested that key factors to consider when planning ponds for dragonflies in agricultural landscapes included (1) having low agrichemical inputs as buffers and maximising pond catchment area covered by those buffers; (2) encouraging native submerged and floating vegetation; (3) maintaining open pond surface areas; and (4) creating varied ponds within close proximity. The importance of wider 'pondscapes' in which artificial and natural ponds are considered together led Hill et al. (2018) to propose five policy-based thematic recommendations for incorporating them into wider conservation considerations, as (1) environmental context: with logistic/cost advantages in adopting groups of ponds rather the individual ponds as management units; (2) urban planning: with mitigation for pond losses from urbanisation based on landscape scale rather than individual habitat creation; (3) flood management: integrating ponds into water storage strategies; (4) agriculture: recognising the significance of many farmland ponds for conservation; and (5) education: emphasising greater practical attention to ponds and their plight to both urban and rural communities.

Related modifications of water bodies may or may not incorporate ponds. Measures to promote awareness of dragonflies in the wider community vary considerably, but one pioneering example is a 'dragonfly awareness trail' in Pietermaritzburg, South Africa (Suh and Samways 2001), in which water body modifications, to foster high diversity of Odonata, and informative signage allow visitors to appreciate the insects in a series of environments. Such exercises can build on numerous attempts to create dragonfly reserves, including artificial water bodies, in Europe, Japan, and North America. Well-designed trails can encourage many people to see dragonflies, and learn about them from personal experiences, whilst at the same time not damaging the local environment by straying from paths into sensitive areas, or their numbers becoming excessive or unregulated.

More broadly, most ponds in urban areas are 'artificial' as constructed or modified by people, almost all are also managed primarily for purposes other than biodiversity conservation and isolated from more natural waterbodies. Numbers continue to increase—thus, the number of urban ponds in the Melbourne area has increased five-fold in only about 20 years (Hale et al. 2015), providing abundant opportunity to contribute to biodiversity conservation. Prospects for doing so (Oertli and Parris 2019) must harmonise conservation with the designated primary roles of those ponds. Even when, as is commonly the case, urban ponds support fewer species than rural ponds, they are still valuable habitats for many invertebrates—including those of conservation concern. A management framework toward management of urban ponds to optimise biodiversity conservation could include four major key points: Oertli and Parris proposed these as (1) the pondscape with a diverse and environmentally varied suite of characteristics, such as pond age, size, depth, shade, primary productivity and management practices; (2) pond quality high at the local scale, to provide a range of habitats and avoid creation of ecological traps (p. 138), such as by having good water quality, gently sloping banks, a range of aquatic vegetation, and lacking invasive non-native taxa; (3) high habitat quality in the area of several hundred metres surrounding the ponds, such as by large green spaces and low levels of impervious surfaces; and (4) high density of high quality ponds within the urban matrix, with high connectivity to facilitate movements of species between them. Importantly, Oertli and Parris (2019) stressed that urban ponds need to be actively protected and their worth promoted as active components of biodiversity conservation strategies: those strategies may need to be designed especially for urban environments or modified from those optimal for more natural pondscapes.

## 11.7 Temporary Ponds

Temporary ponds have received far less conservation attention than permanent ponds, but surveys in Britain have confirmed that these common formations can provide valuable habitats for considerable arrays of macroinvertebrates (Nicolet et al. 2004). Almost three-quarters of the 71 temporary ponds surveyed contained at least one 'nationally scarce' macroinvertebrate. Assemblages were dominated by Coleoptera, comprising on average more than half (56%) of the species in ponds, and recorded from every pond examined, followed by Hemiptera (12% of species, from 92% of ponds) and Trichoptera (11% of species). Invertebrate richness was correlated strongly with plant richness, and that study also confirmed that many of the invertebrate species occurred in both temporary and permanent ponds. The mobility of adult Coleoptera and Hemiptera allows them to leave temporary pools as these dry out, whilst many of the Trichoptera, and other orders, were species that emerge before a seasonal dry period or (more rarely) have larvae that can survive in substrate refuges. Management should retain decaying vegetation and leaf litter on the substrate during dry periods, and Nicolet et al. (2004) also recommended constructions of temporary ponds as worthwhile augmentation to more permanent pond mosaics. Similarly, seasonal pasture wetlands can contribute to biodiversity in agricultural landscapes which are often rather depauperate (Robson and Clay 2005), and also merit management to assure their persistence and protect them from drainage. Many temporary ponds support relatively few macroinvertebrates when compared with more permanent ponds, with some suggestion that richness of the former increases as the period over which water is present increases (Collinson et al. 1995).

Several categories of 'pond permanence' were recognised in comparative surveys of macroinvertebrates (mostly assessed at the family level) in Canada (Gleason and Rooney 2018), as (1) ephemeral (with water for only 1–2 weeks in spring), (2) temporary (water over the first four weeks of spring), (3) seasonal (water for about eight weeks, extending into summer), (4) semi-permanent (water drawn down only in drought years), and (5) permanent (water throughout the year). Each of the 62 macroinvertebrate taxa was allocated to a 'desiccation strategy', as tolerators, needing wet layers, needing only dry layers, and dispersers. The permanence categories showed a continuous gradient in community composition, with the largest differences being between temporary and permanent waterbodies, and influenced also by the open-water area, which reflected both hydrological and habitat zonation. Of the insects assessed, only Culicidae were more abundant in the temporary ponds.

In an earlier British example, water beetle larvae and adults in 77 ponds in northern England were related to the duration of water presence (Eyre et al. 1992), with occurrences examined separately to reveal different assemblages in ponds in relation to the length of their seasonal dry period. Some species could not tolerate any dry period, others preferred temporary water, and yet others were distributed in various ways between these regimes.

In south western Victoria, samples of invertebrates were taken from 19 wetlands over two weeks in spring—and with a perennial wetland included in each of the four 'clusters' sampled. Non-insect groups predominated in samples, but chironomids were also common, and Coleoptera (53 species) the richest taxon encountered. Each of the perennial wetland sites contained species not found in adjacent seasonal wetlands, and also more species. As in England and Wales (Nicolet et al. 2004, above), it is likely that diverse beetle assemblages may be common in seasonally flooded wetlands in Australia. However, in contrast, Robson and Clay (2005) found some other taxa (such as *Anisops* [Hemiptera] and three species of *Atalophlebia* [Ephemeroptera]) in perennial wetlands but not in seasonal waterbodies. Individual seasonal wetland faunas differed more than individual perennial wetland faunas, with some of the former suffering disturbance from cattle access. Again, management was recommended to protect seasonal pasture wetlands from threats such as drainage.

## 11.8 Intermittent Streams

The implications of unnaturally changed flow regime and periods of water absence are universal among concerns for stream invertebrates and, whilst many species are long-adapted to withstand some such variations, examples discussed earlier confirm that imposed changes to water permanence can have profound impacts. Considerations of effects of water drawdown or installation/construction of impoundments should include assessment of their effects on macroinvertebrates of the lotic systems affected.

## 11.9 Stormwater Retention Ponds

Constructed stormwater retention ponds are a means of protecting wider water resources from urban pollution and of moderating excessive water flows from storms. Those along motorways in France, where their construction has been mandatory since 1992, have three major objectives (Scher and Thiery 2005)—(1) to control water flow during storms; (2) to prevent waterbodies receiving chronic road pollution such as hydrocarbons and trace metals; and (3) to protect water from exceptional contamination, such as oil spills from traffic accidents.

Highway runoff storage ponds, as a category of stormwater ponds, have the additional feature that they often receive high concentrations of pollutants in road runoff, with one of their functions being to retain those pollutants and prevent their movements into the wider landscape. They can thus differ from 'natural ponds' in the same region, but comparisons of Coleoptera, Heteroptera and Odonata of the two pond categories in France (Le Viol et al. 2009), although interpreted only to family level (a total of 34 families represented), suggested that the communities of the two pond types did not differ greatly—95% of highway pond families occurred in natural ponds, and 99% of natural pond families were found in highway ponds. However, the abundance of 11 of the 28 families found in both categories differed considerably, seven being more abundant in highway ponds, and four in natural ponds. Despite the abiotic differences, the highway ponds had clear biodiversity values. The resemblances might in part reflect that the stormwater ponds studied were 'old' (34 years) and had natural bottom rather than the synthetic polyethylene membrane used in many more recently constructed ponds. In contexts such as this, highway ponds can augment the biodiversity values of road verges, with both serving as refuges for invertebrates now largely eliminated from the wider local landscapes.

Stormwater ponds can be colonised by animals, but their suitability for invertebrates has received little attention. Surveys of six retention ponds along motorways in south–eastern France over 1993–1996 included assessment of adult Odonata (Scher and Thiery 2005). Within a total of 29 species found (60% of those recorded from lentic waters in the region), 11 were also identified as larvae, and numbers at each pool were substantial (n = <u>18</u>, 14, <u>21</u>, <u>19</u>, 10, 13 species), and highest in those ponds with a more natural substrate (underlined in the above list) rather than those lined with a synthetic polyethylene membrane. Wider comparisons with the regional fauna showed that the retention ponds lacked two regionally typical taxa (*Orthetrum coerulescens*, *Coenagrion mercuriale*), both of which need high densities of plants for suitable breeding habitat. However, four other species were more common in the ponds than elsewhere and occurred there together with six of the ten more common regional species. The retention ponds were clearly used as breeding sites for many taxa, but some Odonata were recorded simply as 'visiting adults'.

Attitudinal change, to transform common perception of urban water as a nuisance to one of it being an asset for its environmental values, includes development of modified Sustainable Drainage Systems (SUDS), which harmonise more with conservation of both water quality and stream biota. Design of SUDS with such positive attributes remains 'a very challenging task in reality' (Zhou 2014), especially in incorporating considerations of future climate changes and increasing urbanisation with which those systems must contend.

Stormwater impacts are by no means restricted to urban areas. Severe runoff from crop fields into agricultural drainage ditches was considered the most severe threat to the ecological values of those ditches (Herzon and Helenius 2008).

## 11.10 Refuges

A refuge for aquatic insects must, by definition, be a physical entity secure from key disturbances and that can function as a source of colonists for wider habitats in the landscape after disturbances there have passed. They include a variety of scales, from individual habitat features to entire waterbodies or their remnants. Most such bodies are constructed for other purposes, such as water storage or drainage systems, or are simply remnants of abandonment from agriculture or industry. They can be thought of as part of a matrix of habitats within the landscape, but the more familiar 'natural refuges' are now augmented by numerous anthropogenic equivalents within disturbed landscapes, and whose importance is increasing. A variety of refuges against drought, for example, are potentially available to stream insects—damp sediment is used for aestivation by some taxa, whilst remnant perennial pools are also important. Fewer taxa have desiccation-resistant eggs and can persist in dry sediments, but within each higher taxon different species may exhibit restraint or preference for one of several refuge strategies available.

The twin terms of 'refuges' and 'refugia' are often not distinguished, and are used interchangeably, but Davis et al. (2013) argued that their separation is helpful to understanding. They considered 'refuges' as microhabitats that provide protection from contemporary spatial or temporal disturbances, while 'refugia' are more properly based on longer-term evolutionary time scales, of millennia. Semantic confusions can occur also in distinguishing 'macrorefugia' and 'microrefugia', with the former reflecting larger areas of favourable climate (for example), and the latter, small favourable areas within a largely unfavourable environment. Thus, meaning of such terms can differ in individual accounts, but conceptual differences between refuges and refugia in freshwater ecosystems are distinguished most commonly in relation to climate change. For aquatic habitats in Australia's arid biome, the variety was depicted by Davis et al. as in Fig. 11.3. Those habitat forms with the greatest separation of microclimate from regional climate are those most likely to act as evolutionary refugia, and some habitats function as both ecological refuges and evolutionary refugia depending on the dispersal traits of the taxa they harbour, their hydrological connectivity, and overall proximity.

The functional variety of natural refuges is illustrated well by considering those used by or available to insects that must be able to counter impacts of intermittent water losses in streams and are noted on p. 231. Their survival on floodplains, many of which are inundated at least once a year, for periods ranging from a few days (typified by some higher elevation sites) up to several months (more characteristic of lowland plains), necessitate a range of strategies—including refuges—for their characteristic invertebrates to persist. The partial analogy with plants, of 'invertebrate seedbanks' remaining in the substrate over dry periods, was explored by Tronstad et al. (2005). They collected soil divots (10 cm deep, surface area 60 × 30 cm) from three floodplain regions with different inundation periods (inundated for 285, 127 and 7 days over the previous season, reflecting an elevational gradient as above) along the Sipsey River, Alabama (United States), and rehydrated them individually with filtered river water.

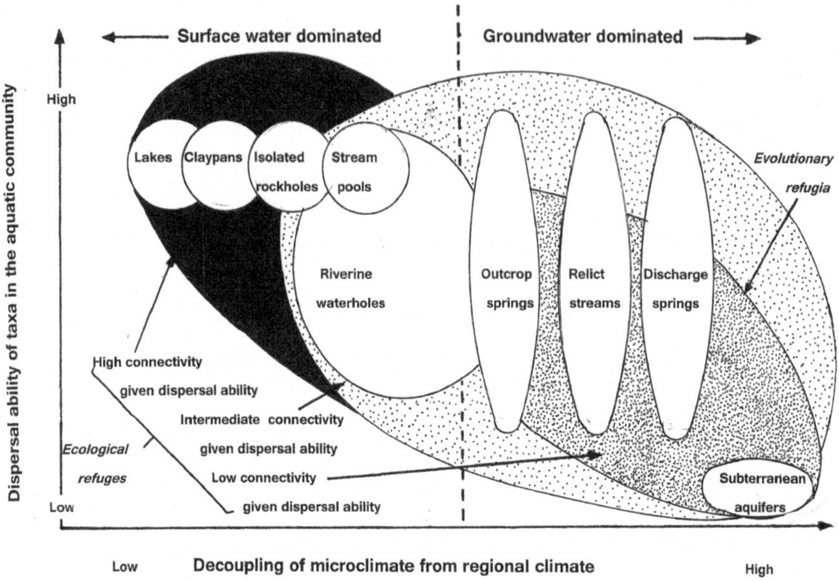


**Fig. 11.3** Conceptual differences between evolutionary refugia and ecological refuges based on levels of decoupling of microclimates from regional climate, with the greatest decoupling most likely to act as evolutionary refugia. Habitats with the least decoupling function as ecological refuges for only the most mobile aquatic taxa, whilst some habitats potentially act in both roles, depending on dispersal tendency of the taxa and their geographical proximity and hydrological conditions (Davis et al. 2013)

The insects emerging over the next ten weeks represented aquatic, semi-aquatic and terrestrial taxa. Nine orders of insects, as well as crustaceans and molluscs, emerged. About 89% of the >45,000 individuals were semi-aquatic and only about 6% aquatic. The latter were more abundant at low than at high or intermediate elevation sites, and Chironomidae (89%) were by far the most abundant adult insects emerging at low elevation sites, with lower numbers at the others. Many emerged from the soil within 10 days of wetting and were probably active, rather than dormant—in contrast, dormant eggs hatched only after about 10 weeks at the higher elevations. It seemed that regular flooding maintained the diverse invertebrate seedbank in the soil, not least because some chironomids and others did not become dormant in short non-inundated periods.

The protection and rehabilitation of refuges was listed as a key future direction for Australian river restoration (Cottingham et al. 2004), in emphasising long-term measures such as incorporating resilience to major disturbances such as floods, droughts and fires. Many refuges have been viewed traditionally as facilitating spread of alien species, rather than of primary conservation value, but even small constructs such as farm dams or stormwater drains, as above, may prove significant. The variety, reviewed by Chester and Robson (2013) confirms that very little is known in detail of the conservation and refuge values of many water bodies, and that more detailed surveys of all categories are needed: for many, any 'biodiversity value' is simply undocumented. Chester and Robson also distinguished between 'primary' and 'secondary' refuge habitats, with the former those in optimal primary habitat and the latter essentially suboptimal as being in other areas where primary habitat is no longer available or used only seasonally as conditions permit. When populations are small, secondary refuges may indeed be needed for species recovery to occur. Should drought occur at times when early instars of key invertebrates are present, these may benefit especially from structural refuges such as the moist interstices in the substrate—but those may not be available in waters where sand or sediment has filled those natural spaces (Boulton 2003). Local attributes associated with high refuge values were presence of macrophytes, absence of fish, natural substrate bed materials, and hydroperiod. At a landscape level, connectivity through proximity to other water bodies, and natural vegetation were important. Even relatively depauperate ponds may act as 'stepping stones' in enhancing connectivity. Several key management needs were identified through Chester and Robson's review: each raises a number of research needs that can enhance understanding and biodiversity vales. The listing in Table 11.18 includes much of value for aquatic insect conservation.

Localised deep pools are important hydrological refuges during drought. Chessman (2009) noted that, in parallel with their better-documented values for fish, they can also be significant for macroinvertebrates in arid or semi-arid areas. The variety of opportunities for refuge provided by the hyporheic zone demonstrates the needs and values of management to retain those roles, in addition to more usual stream rehabilitation approaches that tend to address the surface stream only. Increasing structural complexity of the substrate, such as by introducing wood into the streams, is a component of wider schemes that focus on heterogeneity, including local 'refugial hotspots' (Stubbington 2012). Part of the heterogeneity needed for refugia is to sustain the tolerance limits of the invertebrates that habitually use them (Storey and Quinn 2013), with implication from their study that sustaining the microclimates resulting from presence of riparian forest in New Zealand is an important part of protecting the aquatic community in intermittent streams. The wide use of

**Table 11.18** The key management needs for conservation management of potential anthropogenic freshwater refuges (Chester and Robson 2013)

1. Many artificial waterbodies may retain particular species because they do not have a natural water regime or vary in some other way from natural conditions. Therefore, the use of traditional restoration actions should be carefully considered by managers before implementation
2. Consideration should be given to enhancing connectivity between anthropogenic refuges and surrounding waterbodies and terrestrial landscapes
3. Artificial waterbodies should be managed together with surrounding natural waterbodies as a mosaic of habitats for freshwater species. Attention should be paid to beta-diversity patterns across waterbodies
4. Hydroperiod is easier to control in many artificial waterbodies than in natural ones. Hydroperiod may therefore be managed to benefit particular species or groups of species
5. Attention should be paid to potential threats to anthropogenic refuges such as dredging, piping, draining, and pesticide use

dry sediments as refugia was evident when 37 of the 53 invertebrate taxa found in the sediments of perennial streams became active when dry sediment samples were re-wetted. By comparison, local taxa were found in remnant pools likely to be the other main dry season refuge in that system (Storey and Quinn 2013). This example supported the protection or restoration of riparian vegetation in order to moderate summer streambed temperatures and ensure the refuge potential implied for shallow stream bed sediments.

Substrate characteristics can affect their values as refuges, by influencing the susceptibility of prey invertebrates to predation, through changing levels of prey detection and vulnerability. Thus, in laboratory studies based on New York stream insects, Fuller and Rand (1990) assessed the susceptibility of single and mixed prey taxa (comprising larvae of Ephemeroptera [*Baetis tricaudatus*, *Ephemerella subvaria*], Diptera [*Simulium vittatum*] and Trichoptera [several species of Hydropsychidae]) exposed to two predator taxa (larvae of *Agnetina capitata* [Plecoptera] and *Nigronia serricornis* (Megaloptera]), in each of a range of substrates—very coarse sand, gravel-pebble, artificial turf, or gravel with additional sand. Different substrates differentially affected the two predators, reflecting biology of the taxon combinations tested, in particular their foraging and prey detection behaviours. Mobility of the prey, together with substrate form, affects the refuge potential of a site for any particular predator–prey association (p. 87).

Remnant pools are also important in areas of seasonally intermittent river flow (p. 26). Many insects and others have adapted over long periods to relatively predictable seasonally intermittent flow as 'evolutionary cues' that foster diversity of their life style traits. This may not be so for recent and sudden anthropogenic transformations from perennial to intermittent flow. Datry et al. (2014) warned that some irreversible modifications to community structure and function may occur, as trends that could include loss of species with low tolerance to desiccation, and habitat fragmentation limiting dispersal and chances of re-colonisation. Community structure and resilience may be influenced also by changed interactions, such as by losses of predatory fish. In emphasising the needs for a far wider ecological perspective of impacts of anthropogenic river drying, Datry et al. also noted the confounding impacts of simultaneous pollution, overfishing and wider habitat degradation, so that distinguishing impacts of each of a complex of 'threats' or 'stressors' may be very imprecise and largely site-specific. The wider consequences of such changes in relation to refuge capability extend to themes such as eutrophication, lowered oxgen content and high levels of chemical contamination, collectively providing environmental milieux far different from the perennial rivers previously present.

In the Cape Floristic Region of South Africa, many aquatic insects are well-adapted to withstanding droughts and in part achieve this through living in low quality habitats such as remnant pools that contain sufficient resources for them to survive over a dry period (Deacon et al. 2019). Artificial reservoirs in the region harboured many species found normally in natural ponds, and also yielded some rare endemic species not previously recorded in the area. Characteristics of the marginal vegetation of reservoirs were influential and approaches to render this similar to that of natural ponds may be a valuable management strategy.

**Table 11.19** Five primary functions of urban ponds, and their key characteristics: a typology used by Hassall (2014)

| Pond type | Characteristics |
|---|---|
| Garden pond | Small size; set within an impervious matrix; often stocked with fish; very rarely dry out; maintained to prevent succession |
| Industrial ponds | Medium size; urban or periurban, often away from residential areas; sometimes contaminated; constructed to hold water for use, or left after mineral extraction; rarely in use for original purpose |
| Ornamental lakes and ponds | Medium-large size; heavily managed for aesthetics; hypertrophic; fish and ducks encouraged and fed; access to public encouraged and uncontrolled; often with vertical sides that may influence animal movements |
| Drainage systems | Highly variable in size; primary function is hydrological management; diverse designs; wide variation in 'naturalness'; temporary (detention basins) or permanent (retention ponds) |
| Nature reserves | Medium-large size; managed primarily for biodiversity (often birds); either co-opted natural ponds or created to appear natural; access to public encouraged but controlled |

The variety of 'urban ponds', used as a single category by Chester and Robson, was explored further, when Hassall (2014) listed five categories with different primary functions (Table 11.19), concentrating on the most frequent categories—but noting also unusual systems such as bomb craters, swimming pools and monumental fountains, each of which has also received some attention.

The presence of some form of 'refuge' or shelter can strongly influence the diversity and survival of many macroinvertebrates in a variety of waters. In subterranean cave streams in Brazil, effectiveness of refuges related to the structure of the stream channel (Pellegrini et al. 2018)—only in some areas were conditions suitable for materials such as pieces of wood to accumulate. However, comparison of the insects in several cave streams implied that different factors may influence the assemblages: water flow, temperature and general availability of organic materials were all significant. As in New Zealand cave streams (McNie and Death 2017), river silting from terrestrial activities such as deforestation or mining can lead to both increased needs for refuges for invertebrates and difficulties in maintaining their roles.

## 11.11 Woody Debris

Submerged woody debris was for long a focus for removal from streams and rivers, as 'rubbish', but progressive recognition of its ecological values in generating structural complexity and providing shelter, substrates and food for animals—from invertebrates to fish—has led to reconsideration and, in many cases, deliberate introductions or regulatory controls of removal. The importance of larger branches ('snags') can be considerable. More broadly, allochthonous riparian forest debris is a major source of energy for both benthic invertebrates and fish (Reid et al. 2008).

However, the form and placement of wood placed deliberately for conservation may markedly influence the projected benefits, as shown by comparisons of invertebrates associated with dead wood in various sites in the Hunter River, New South Wales (Sceally et al. 2007). Assemblages were compared on 'simple' (single trunk logs of eucalypt wood) and 'complex' structures (groups of sticks and branches of eucalypts) to give two levels of structural complexity in pools and riffles at three sites along the river. Inhabitants were compared with the 'natural assemblages' sampled by sweep-netting (pools) or kick-sampling (riffles) and samples from 'natural' dead wood at each site. Forty-five taxa were collected and, at all three sites, mean richness was lowest on simple pool substrates and highest on the complex substrates in both pools and riffles. Abundances were increased in substrates that trapped organic matter, with far more such material in riffles than in pools, and complex substrates always contained more organic material. Site-specific features also occurred. More generally, the substrate(s) available for colonisation by benthic invertebrates markedly affect the communities that can develop. Comparison of various substrates in Alabama, United States, revealed considerable differences across the total 96 taxa obtained (Rinella and Feminella 2005), with particular taxa associated variously with sandy or woody substrates.

In Victoria, 'Removal of woody debris from Victoria's rivers and streams' is listed formally as a Potentially Threatening Process, for which an Action Statement (Heron and Doeg 2003a, b) has been prepared under the Flora and Fauna Guarantee Act 1988 (FFG). Previously, 'desnagging' was considered an essential component of river management, but the accepted formal grounds for listing this as a threat under FFG were wide-ranging, as (1) posing a significant threat to a range of flora and fauna; (2) posing a significant threat to the survival of two or more taxa; and (3) posing a significant threat to the survival of a community. The major expressed concern was the decline of many native fish species. Many live in lowland streams, where desnagging has been most intensive in the past—but also where woody debris may provide the most (or, even, only) stable environments. Restoration of native riparian vegetation is linked intricately with future self-sustaining supply of 'natural' dead wood.

## 11.12 Riparian Zones

Vegetation succession in riparian zones is often amenable to management, whether to modify structure or change species composition such as by removal of alien trees or shrubs (p. 100). Odonata attending a constructed reservoir in Pietermaritzberg, South Africa, were monitored to examine changes over 13 years after the founding impoundment in 1988 (Suh and Samways 2005). Assemblage changes were evident soon after impoundment, with only 12 species beforehand and 26 species (of a total of 30 species over the full survey period) in 1993. Much of this increase was attributed to vegetation changes fostering a wide range of adult dragonfly behaviour, and coexistence of species with differing habits and requirements. The reservoir was part of a wider dragonfly awareness trail (p. 275), but Suh and Samways cautioned that allowing the succession to proceed further would lead to overgrowth of the water margins and increased siltation. They recommended several practical management steps to prevent those effects and maintain the diversity of Odonata present, as (1) rotational clearing of marginal vegetation; (2) removal of alien plants; (3) encouraging native plants in the diversity present, whilst also maintaining good sunlight penetration; (4) maintaining a good level of open water; and (5) dredging the reservoir at the inlet from the stream. The principles in those recommendations have far wider relevance. More generally for newly constructed wetlands, measures to increase plant abundance and reduce water turbidity are likely to enhance invertebrate abundance and diversity, as well as ecological functions (Stewart and Downing 2008).

## 11.13 Perspective and Prospects

The themes discussed above are by no means a complete listing of relevant topics but illustrate the variety of considerations and (often complementary) approaches needed for effective conservation of and in inland water environments. For many water bodies, simply preventing further despoliation and change through effective protection and buffering will provide the greatest benefits for their inhabitants. For many others, active reduction of threats and proactive measures to counter further potential threats at an early stage of conservation planning are a priority, with assessment of environmental impacts of any planned disturbances (and including those not directly on the water body itself but likely to affect it from adjacent terrestrial influences) ideally including routine appraisal of impacts on aquatic insects. All kinds of waterbodies contribute to regional biodiversity, each harbouring taxa and associations not found in other categories (Williams et al. 2008). It is inevitable that human needs for water will continue to increase, and to compromise the wellbeing of freshwater environments and their inhabitants throughout the world, and that conservation of insects and their relatives will continue to garner only limited support and publicity despite their increasingly-acknowledged functional importance in managing the health of freshwater ecosystems (Cantonati et al. 2020). Need to conserve both human needs and freshwater biodiversity is acknowledged widely in Australia, and the foundation knowledge currently available on the insect fauna—clarifying its global significance as largely endemic, localised and often vulnerable—provides abundant opportunity for enhancing their conservation through modifying both policy and management of freshwater environments.

Whilst all native aquatic biota should in principle be sustained, most insects continue to be largely overlooked in detailed planning and are unlikely to achieve greater prominence in the near future in the face of conflicting interests and needs for water use. Needs for 'wise' and practical water management to support Australia's population, industry and wider environment have never been greater, and such practical targets are a major priority. Nevertheless, the examples and contexts discussed in this book (1) display the richness, variety and unique nature of vulnerable Australian insect taxa and (2) suggest how inland water management may increasingly sustain these and the intrinsically fragile environments on which they depend. Without that progress, many aquatic insects are likely to disappear, and Australia's biological heritage be diminished. The few 'flagship species' noted earlier are simply the most obvious and preliminary manifestations of far wider needs, but illustrate the variety of environments of concern and the complex biology and needs of the insects themselves. Harmonising needs of endemic biota, here largely the overlooked 'meek inheritors' that characterise those environments, with wider demands of humanity remains a formidable undertaking—but one without which the future of many inland water insects seems bleak. Those insects, in addition to being globally significant components of Australia's biota, are also integral to the continued health, functioning and sustainability of the wider inland water environments they occupy throughout the continent.

## References

Abell R, Allan JD, Lehner B (2007) Unlocking the potential of protected areas for freshwaters. Biol Conserv 134:48–63

Abellan P, Sanchez-Fernadez D, Velasco J, Millan A (2005) Assessing conservation priorities for insects: status of water beetles in southeast Spain. Biol Conserv 121:79–90

Arthington AH, Naiman RJ, McClain ME, Nilsson C (2010) Preserving the biodiversity and ecological services of rivers: new challenges and research opportunities. Freshw Biol 55:1–16

Arthington AH, Kennen JG, Stein ED, Webb JA (2018) Recent advances in environmental flows science and water management–innovation in the anthropocene. Freshw Biol 63:1022–1034

Ball-Damerow JE, M'Gonigle LK, Resh VH (2014) Local and regional factors influencing assemblages of dragonflies and damselflies (Odonata) in California and Nevada. J Insect Conserv 18:1027–1036

Barmuta LA, Linke S, Turak E (2011) Bridging the gap between 'planning' and 'doing' for biodiversity conservation in freshwaters. Freshw Biol 56:180–195

Becker A, Robson BJ (2009) Riverine macroinvertebrate assemblages up to 8 years after riparian restoration in a semi-rural catchment in Victoria, Australia. Mar Freshw Res 60:1309–1316

Biggs J, Williams P, Whitfield M, Nicolet P, Weatherby A (2005) 15 years of pond assessment in Britain: results and lessons learned from the work of Pond Conservation. Aquat Conserv Mar Freshw Ecosyst 15:693–714

Bilton DT, Freeland JR, Okamura B (2001) Dispersal in freshwater invertebrates. Ann Rev Ecol Syst 32:159–181

Boulton AJ (2003) Parallels and contrasts in the effects of drought on stream macroinvertebrate assemblages. Freshw Biol 48:1173–1185

Brainwood M, Burgin S (2009) Hotspots of biodiversity or homogeneous landscapes? Farm dams as biodiversity reserves in Australia. Biodiv Conserv 18:3043–3052

Burrows DW (2004) Translocated fishes in streams of the Wet Tropics Region, North Queensland: distribution and potential impact. Cooperative Research Centre for Tropical Rainforest Ecology and Management, Rainforest CRC, Cairns

Campbell RE, McIntosh AR (2017) Space-time interactions and invertebrate assemblage change in stream networks. Aust Ecol 42:227–237

Cantonati M, Poikanes S, Pringle CM, Stevens LE, Turak E (and 18 other authors) (2020) Characteristics, main impacts, and stewardship of natural and artificial freshwater environments: consequences for biodiversity conservation. Water 12(1):260. https://doi.org/10.3390/w12010260

Cheesman OD, Key RS (2007) The extinction of experience: a threat to insect conservation? In: Stewart AJA, New TR, Lewis OT (eds) Insect Conservation Biology. CAB International, Wallingford, pp 322–348

Chessman BC (2009) Climatic changes and 13-year trends in stream macroinvertebrate assemblages in New South Wales, Australia. Glob Change Biol 15:2791–2802

Chessman BC, Williams SA (1999) Biodiversity and conservation of river macroinvertebrates on an expanding urban fringe: western Sydney, New South Wales, Australia. Pacif Conserv Biol 5:36–55

Chester ET, Robson BJ (2013) Anthropogenic refuges for freshwater biodiversity: their ecological characteristics and management. Biol Conserv 166:64–75

Coccia G, Vanschoenwinkel B, Brendonck L, Botero L, Green AJ (2016) Newly created ponds complement natural waterbodies for restoration of macroinvertebrate assemblages. Freshw Biol 61:1640–1654

Collinson NH, Biggs J, Corfield A, Hodson MJ, Walker D, Whitfield M, Williams PJ (1995) Temporary and permanent ponds: an assessment of the effects of drying out on the conservation value of aquatic macroinvertebrate communities. Biol Conserv 74:125–133

Corbet PS (1999) Dragonflies. Behaviour and ecology of Odonata. Harley Books, Colchester

Corbet PS, Dunkle SW, Ubukata H (eds) (1995) In: Proceedings of the international symposium on conservation of dragonflies and their habitats, Kushiro, Japan. Japanese Society for Protection of Birds, Kushiro

Cottingham P, Walsh C, Rooney G, Fletcher T (eds) (2004) Urbanization impacts on stream ecology–from syndrome to cure? (Outcomes of workshops held at the symposium on Urbanization and Stream Ecology, Melbourne). CRC for Freshwater Ecology, Canberra

Darwall WRT, Vie J-C (2005) Identifying important sites for conservation of freshwater biodiversity: extending the species-based approach. Fish Manage Ecol 12:287–293

Darwall W, Bremerich V, De Wever A, Dell AI, Freyhof J (and 18 other authors) (2018) The Alliance for Freshwater Life: a global call to unite efforts for freshwater biodiversity science and conservation. Aquat Conserv 28:1015–1022

Datry T, Larned ST, Tockner K (2014) Intermittent rivers: a challenge for freshwater ecology. Bioscience 64:229–235

Davis J, Pavlova A, Thompson R, Sunnucks P (2013) Evolutionary refugia and ecological refuges: key concepts for conserving Australian arid zone freshwater biodiversity under climate change. Glob Change Biol 19:1970–1984

Deacon C, Samways MJ, Pryke JS (2019) Aquatic insects decline in abundance and occupy low-quality artificial habitats to survive hydrological droughts. Freshw Biol 64:1643–1654

Dolny A, Harabis F (2012) Underground mining can contribute to freshwater biodiversity conservation: allogenic succession forms suitable habitats for dragonflies. Biol Conserv 145:109–117

EPCN (European Pond Conservation Network) (2008) The pond manifesto. Geneva

Eyre MD, Carr R, McBlane RP, Foster GN (1992) The effects of varying site-water duration on the distribution of water beetle assemblages, adults and larvae (Coleoptera: Haliplidae, Dytiscidae, Hydrophilidae). Arch Hydrobiol 124:281–291

Fausch KD, Torgersen CE, Baxter CV, Li HW (2002) Landscapes to riverscapes: bridging the gap between research and conservation of stream fishes. Bioscience 52:483–498

Firehock K, West J (1995) A brief history of volunteer biological water monitoring using macroinvertebrates. J N Am Benthol Soc 14:197–202

Fitzsimons JA, Robertson HA (2005) Freshwater reserves in Australia: directions and challenges for the development of a comprehensive, adequate and representative system of protected areas. Hydrobiologia 552:87–97

Foster GN (1991) Conserving insects of aquatic and wetland habitats, with special reference to beetles. In: Collins NM, Thomas JA (eds) The conservation of insects and their habitats. Academic Press, London, pp 237–262

Fuller RL, Rand PS (1990) Influence of substrate type on vulnerability of prey to predacious aquatic insects. J N Am Benthol Soc 9:1–8

Geist J (2015) Seven steps towards improving freshwater conservation. Aquat Conserv Mar Freshw Ecosyst 25:447–453

Geist J, Hawkins SJ (2016) Habitat recovery and restoration in aquatic ecosystems: current progress and future challenges. Aquat Conserv Mar Freshw Ecosyst 26:942–962

Gleason JE, Rooney RC (2018) Pond permanence is a key determinant of aquatic macroinvertebrate community structure in wetlands. Freshw Biol 63:264–277

Hale R, Coleman R, Pettigrove V, Swearer SE (2015) Review: identifying, preventing and mitigating ecological traps to improve the management of urban aquatic ecosystems. J Appl Ecol 52:928–939

Hannon ER, Hafernik JE (2007) Reintroduction of the rare damselfly *Ischnura gemina* (Odonata: Coenagrionidae) into an urban California park. J Insect Conserv 11:141–149

Harabis F (2017) Does the management of surrounding terrestrial habitats increase the tendency of odonates to leave aquatic habitats? Biodivers Conserv 26:2155–2167

Harabis F, Dolny A (2015) Odonates need natural disturbances: how human-induced dynamics affect the diversity of dragonfly assemblages. Freshw Sci 34:1050–1057

Hassall C (2014) The ecology and biodiversity of urban ponds. WIREs Water 1:187–206

Heino J (2009) Biodiversity of aquatic insects: spatial gradients and environmental correlates of assemblage-level measures at large scales. Freshw Rev 2:1–29

Herbert ME, McIntyre PB, Doran PJ, Allan JD, Abell R (2010) Terrestrial reserve networks do not adequately represent aquatic ecosystems. Conserv Biol 24:1002–1011

Heron S, Doeg T (2003a) Degradation of native riparian vegetation along Victorian rivers and streams. Action Statement no 183. Department of Sustainability and Environment, Victoria, Melbourne

Heron S, Doeg T (2003b) Removal of woody debris from Victorian rivers and streams. Action Statement no 129, Flora and Fauna Guarantee Act 1988, Department of Sustainability and Environment, Victoria, Melbourne

Herzon I, Helenius J (2008) Agricultural drainage ditches, their biological importance and functioning. Biol Conserv 141:1171–1183

Higgins JV, Bryer MT, Khoury ML, Fitzhugh TW (2005) A freshwater classification approach for biodiversity conservation planning. Conserv Biol 19:432–445

Hill MJ, Hassall C, Oertli B, Fahrig L, Robson BJ (and six other authors) (2018) New policy directions for global pond conservation. Conserv Lett 11:e12447. https://doi.org/10.1111/conl.12447

Huikkonen I-M, Helle I, Elo M (2019) Heterogenic aquatic vegetation promotes abundance and species richness of Odonata (Insecta) in constructed agricultural wetlands. Insect Conserv Divers. https://doi.org/10.1111/icad.12396

Ingram BA, Hawking JH, Shiel RJ (1997) Aquatic life in freshwater ponds. A guide to the identification and ecology of life in aquaculture ponds and farm dams in South Eastern Australia. Identification guide no. 9. Cooperative Research Centre for Freshwater Ecology, Thurgoona

Jackson JK, Fureder L (2006) Long-term studies of freshwater macroinvertebrates: a review of the frequency, duration and ecological significance. Freshw Biol 51:591–603

Janssen A, Hunger H, Konold W, Pufal G, Staab M (2018) Simple pond restoration measures increase dragonfly (Insecta: Odonata) diversity. Biodiv Conserv 27:2311–3232

Kefford BJ, Papas PJ, Nugegoda D (2003) Relative salinity tolerance of macroinvertebrates from the Barwon River, Victoria, Australia. Mar Freshw Res 54:755–765

Kefford BJ, Zalizniak L, Dunlop JE, Nugegoda D, Choy SC (2010) How are macroinvertebrates of slow flowing lotic systems directly affected by suspended and deposited sediments? Environ Poll 158:543–550

Khelifa R (2019) Sensitivity of biodiversity indices to the life history stage, habitat type and landscape in Odonata community. Biol Conserv 237:63–69

Kingsford RT, Nevil J (2006) Urgent need for a systematic expansion of freshwater protected areas in Australia. Pac Conserv Biol 12:7–14

Knight AT, Cowling RM, Rouget M, Balmford A, Lombard AT, Campbell BM (2008) Knowing but not doing: selecting priority conservation areas and the research-implementation gap. Conserv Biol 22:610–617

Landsberg J, James CD, Morton SR, Hobbs TJ, Stol J, Drew A, Tongway H (1997) The effects of artificial sources of water on rangeland biodiversity. Biodiversity Technical Paper no. 3. Environment Australia, Canberra/CSIRO.

Le Viol I, Mocq J, Julliard R, Kerbiriou C (2009) The contribution of motorway storm retention ponds to the biodiversity of aquatic macroinvertebrates. Biol Conserv 142:3163–3171

Likens GE, Walker KF, Davies PE, Brookes J, Olley J (and eight other authors) (2009) Ecosystem science: toward a new paradigm for managing Australia's inland aquatic ecosystems. Mar Freshw Res 60:271–279

Linke S, Turak E, Nel J (2011) Freshwater conservation planning: the case for systematic approaches. Freshw Biol 56:6–20

Markwell KA, Fellows CS (2008) Habitat and biodiversity of on-farm water storages: a case study in southeast Queensland, Australia. Environ Manage 41:234–249

McNie PM, Death RG (2017) The effects of agriculture on cave-stream invertebrate communities. Mar Freshw Res 68:1999–2007

Moore NW (1991) The development of dragonfly communities and the consequences of territorial behaviour: a 27 year study on small ponds at Woodwalton Fen, Cambridgeshire, United Kingdom. Odonatologica 20:203–231

Morse JC (2009) Biodiversity of aquatic insects. In: Foottit RG, Adler PH (eds) Insect biodiversity. Science and Society (1st edn) Wiley-Blackwell, Oxford, pp 165–184

Naiman RJ, Decamps H (1997) The ecology of interfaces: riparian zones. Annu Rev Ecol Syst 28:621–658

Nicolet P, Biggs J, Fox G, Hodson MJ, Reynolds C, Whitfield M, Williams P (2004) The wetland plant and macroinvertebrate assemblages of temporary ponds in England and Wales. Biol Conserv 120:261–278

Nieto C, Ovando XMC, Loyola R, Izquierdo A, Romero F, (and six other authors), (2017) The role of macroinvertebrates for conservation of freshwater systems. Ecol Evol 2017:1–12

Oertli B, Parris KM (2019) Review: toward management of urban ponds for freshwater biodiversity. Ecosphere 10(7):https://doi.org/10.1002/ecs2.2810

Palmer MA, Menninger HL, Bernhardt E (2010) River restoration, habitat heterogeneity and biodiversity: a failure of theory or practice? Freshw Biol 55:205–222

Paul MJ, Meyer JL (2001) Streams in the urban landscape. Annu Rev Ecol Syst 32:333–365

Pellegrini TG, Pompeu PS, Ferreira RL (2018) Cave benthic invertebrates in south-eastern Brazil: are there 'key' factors structuring such communities? Mar Freshw Res 69:1762–1770

Penrose D, Call SM (1995) Volunteer monitoring of benthic macroinvertebrates; regulatory biologists' perspectives. J N Am Benthol Soc 14:203–209

Pittock J, Finlayson C (2011) Australia's Murray-Darling Basin: freshwater ecosystem conservation options in an era of climate change. Mar Freshw Res 62:232–243

Poff NL, Matthews JH (2013) Environmental flows in the Anthropocene: past progress and future prospects. Curr Opin Environ Sust 5:667–675

Polhemus DA (1993) Conservation of aquatic insects: worldwide crisis or localized threats? Amer Zool 33:588–598

Primack R, Kobori H, Mori S (2001) Dragonfly pond restoration promotes conservation awareness in Japan. Conserv Biol 14:1553–1554

Raebel EM, Merckx T, Feber RE, Riordan P, Macdonald DW, Thompson DJ (2012) Identifying high-quality pond habitats for Odonata in lowland England: implications for agri-environment schemes. Insect Conserv Divers 5:422–432

Ramsar Convention (2008) National framework and guidance for describing the ecological character of Australian Ramsar wetlands. Module 2 of the National guidelines for Ramsar wetlands–implementing the Ramsar Convention in Australia. Department of the Environment, Water, Heritage and the Arts, Canberra

Reid DJ, Quinn GP, Lake PS, Reich P (2008) Terrestrial detritus supports the food webs in lowland intermittent streams of south-eastern Australia: a stable isotope study. Freshw Biol 543:2036–2050

Ribera I (2000) Biogeography and conservation of Iberian water beetles. Biol Conserv 92:131–150

Rinella DJ, Feminella JW (2005) Comparison of benthjc macroinvertebrates colonising sand, wood, and artificial substrates in a low-gradient stream. J Freshw Ecol 20:209–220

Robson BJ, Clay CJ (2005) Local and regional macroinvertebrate diversity in the wetlands of a cleared agricultural landscape in south-western Victoria, Australia. Aquat Conserv Mar Freshw Ecosyst 15:403–414

Rowland SJ, Tully P (2004) Hatchery quality assurance program for Murray cod, Golden perch and Silver perch. New South Wales Department of Primary Industries, Grafton

Roy AH, Freeman MC, Freeman BJ, Wenger SJ, Meyer JL, Ensign WE (2006) Importance of riparian forests in urban catchments contingent on sediment and hydrologic regimes. Environ Manage 37:523–539

Samways MJ (1989) Farm dams as nature reserves for dragonflies (Odonata) at various altitudes in the Natal Drakensberg mountains, South Africa. Biol Conserv 48:181–187

Samways MJ, Caldwell PM, Osborn R (1996) Spatial patterns of dragonflies (Odonata) as indicators for design of a conservation pond. Odonatologica 25:157–166

Saunders DL, Meeuwig JJ, Vincent ACJ (2002) Freshwater protected areas: strategies for conservation. Conserv Biol 16:30–41

Sceally JA, Mika SJ, Boulton AJ (2007) Aquatic macroinvertebrate communities on wood in an Australian lowland river: experimental assessment of the interactions of habitat, substrate complexity and retained organic matter. Mar Freshw Res 58:153–165

Scher O, Thiery A (2005) Odonata, Amphibia and environmental characteristics in motorway stormwater retention ponds (southern France). Hydrobiologia 551:237–251

Schindler M, Fesl C, Chovanec A (2003) Dragonfly associations (Insecta: Odonata) in relation to habitat variables: a multivariate approach. Hydrobiologia 497:169–180

Semlitsch RD, Bodie JR (1998) Are small, isolated wetlands expendable? Conserv Biol 12:1129–1133

Simaika JP, Samways MJ (2015) Predicted range shifts of dragonflies over a wide elevation gradient in the southern hemisphere. Freshw Sci 34:1133–1143

Simaika JP, Samways MJ, Frenzel PP (2016) Artificial ponds increase local dragonfly diversity in a global biodiversity hotspot. Biodiv Conserv 25:1921–1935

Stagliano DM (2016) Mayflies (Insecta: Ephemeroptera) of conservation concern in Montana: status updates and management needs. Western N Am Nat 76:441–451

Stansberry DH, Stein CB (1971) Stream channelization and the preservation of biological variability: why naiads (pearly freshwater mussels) should be preserved. Stream channelization, Part 4. Committee on Government Operations, United States House of Representatives 1971, pp 2177–2179

Stewart TW, Downing JA (2008) Macroinvertebrate communities and environmental conditions in recently constructed wetlands. Wetlands 28:141–150

Steytler NS, Samways MJ (1995) Biotope selection by adult male dragonflies (Odonata) at an artificial lake created for insect conservation in South Africa. Biol Conserv 72:381–386

Storey RG, Quinn JM (2013) Survival of aquatic invertebrates in dry bed sediments of intermittent streams: temperature tolerances and implications for riparian management. Freshw Sci 32:250–266

Stubbington R (2012) The hyporheic zone as an invertebrate refuge: a review of variability in space, time, taxa, and behaviour. Mar Freshw Res 63:293–311

Suh AN, Samways MJ (2001) Development of a dragonfly awareness trail in an African botanical garden. Biol Conserv 100:345–353

Suh AN, Samways MJ (2005) Significance of temporal changes when designing a reservoir for conservation of dragonfly diversity. Biodiv Conserv 14:165–178

Thornhill I, Batty L, Death RG, Friberg NR, Ledger ME (2017) Local and landscape scale determinants of macroinvertebrate assemblages and their conservation value in ponds across an urban land-use gradient. Biodiv Conserv 26:1065–1086

Tronstad LM, Tronstad BP, Benke AC (2005) Invertebrate seedbanks: rehydration of soil from an unregulated river floodplain in the south-eastern U.S. Freshw Biol 50:646–655

Turak E, Marchant R, Barmuta LA, Davis J, Choy S, Metzeling L (2011) River conservation in a changing world: invertebrate diversity and spatial prioritisation in south-eastern coastal Australia. Mar Freshw Res 62:300–311

Turner CR (2007) Water beetles associated with reservoirs on Table Mountain, Cape Town: implications for conservation. J Insect Conserv 11:75–83

Walsh CJ, Roy AH, Feminella JW, Cottingham PD, Groffman PM, Morgan RP (2005) The urban stream syndrome: current knowledge and the search for a cure. J N Am Benthol Soc 24:706–723

Weeks ES, Death RG, Foote K, Anderson-Lederer R, Joy MK, Boyce P (2016) Conservation science statement. The demise of New Zealand's freshwater flora and fauna: a forgotten treasure. Pac Conserv Biol 22:110–115

Wells SM, Pyle RM, Collins NM (comp) (1983) The IUCN Invertebrate Red Data Book. IUCN, Gland and Cambridge

Whitten T, Holmes D, MacKinnon K (2001) Conservation biology: a displacement behaviour for academia? Conserv Biol 15:1–3

Williams P, Whitfield M, Biggs J (2008) How can we make new ponds biodiverse? A case study monitored over 7 years. Hydrobiologia 597:137–148

Zamora-Marin JM, Guttierez-Canovas C, Abellen P, Millan D (2016) The role of protected areas in in representing aquatic biodiversity: a test using alpha, beta and gamma diversity of water beetles from the Segura River basin (se Spain). Limnetica 35:179–192

Zhou Q (2014) A review of sustainable urban drainage systems considering the climate change and urbanization impacts. Water 6:976–992

# Appendix A

List of Australian aquatic insects that have been listed formally under Commonwealth or State/Territory legislation as of conservation concern, as at late 2018 (Taylor et al. 2018) (species listed alphabetically under each order).

**Ephemeroptera**

*Tasmanophlebia lacuscoerulei*, Large Blue Lake mayfly.

**Odonata**

*Acanthaeschna victoria*, Thylacine darter.

*Agriocnemis dobsoni*, Tropical wisp.

*Agriocnemis kunjina*, Pilbara wisp.

*Antipodophlebia asthenes*, Terrestrial evening darner.

*Archaeophya adamsi*, Adams emerald dragonfly.

*Archaeosynthemis spiniger*, Spiny tigertail.

*Archiargiolestes parvulus*, Midget flatwing.

*Archiargiolestes pusillissimus*, Tiny flatwing.

*Armagomphus armiger*, Armourtail.

*Austroaeschna ingrid*, Grampians darner.

*Austroaeschna muelleri*, Carnarvon darner.

*Austroaeschna speciosa*, Tropical unicorn darner.

*Austroagrion pindrina*, Pilbara billabongfly.

*Austroargiolestes alpinus*, New England flatwing.

*Austroargiolestes elke*, Azure flatwing.


T. R. New, *Insect Conservation and Australia's Inland Waters*,
https://doi.org/10.1007/978-3-030-57008-8

*Austrocordulia leonardi*, Sydney hawk.

*Austrocordulia territoria*, Top End hawk.

*Austrogomphus divaricatus*, Fork hunter.

*Austropetalia tonyana*, Alpine redspot.

*Austropetalia victoria*, Alpine redspot.

*Austrophlebia subcostalis*, Northern giant darner.

*Austrophya mystica*, Rainforest mystic.

*Calliagrion billinghursti*, Large riverdamsel.

*Cordulephya bidens*, Tropical shutwing.

*Cordulephya divergens*, Clubbed shutwing.

*Cordulephya montana*, Mountain shutwing.

*Diphlebia hybridoides*, Giant rockmaster.

*Eurysticta coolawanyah*, Pilbara pin.

*Eusynthemis deniseae*, Carnarvon tigertail.

*Griseargiolestes bucki*, Turquoise flatwing.

*Griseargiolestes metallicus*, Metallic flatwing.

*Hemicordulia koomina*, Pilbara emerald.

*Hemigomphus cooloola*, Wallum vicetail.

*Hemigomphus magela*, Kakadu vicetail.

*Hemiphlebia mirabilis*, Ancient greenling damselfly.

*Huonia melvillensis*, Forestwatcher.

*Ictinogomphus dobsoni*, Pilbara tiger.

*Indolestes obiri*, Cave reedling.

*Lestoidea barbarae*, Large bluestreak.

*Lestoidea brevicauda*, Short-tipped bluestreak.

*Lestoidea lewisiana*, Mount Lewis bluestreak.

*Lithosticta macra*, Rock narrow-wing.

*Micromidia convergens*, Early mosquitohawk.

*Nososticta koongarra*, Citrine threadtail.

*Nososticta pilbara*, Pilbara threadtail.

*Nososticta taracumbi*, Melville Island threadtail.

*Orthetrum balteatum*, Speckled skimmer.

*Orthetrum boumiera*, Brownwater skimmer.

*Petalura gigantea*, Giant dragonfly.

*Petalura litorea*, Coastal petaltail.

*Petalura pulcherrima*, Beautiful petaltail.

*Pseudocordulia circularis*, Circle-tipped mistfly.

*Pseudocordulia elliptica*, Ellipse-tipped mistfly.

*Telephlebia tillyardi*, Tropical evening darner.

*Telephlebia tryoni*, Coastal evening darner.

*Tonyosynthemis claviculata*, Clavicle tigertail.

*Tonyosynthemis ofarrelli*, Slender tigertail.

*Zephyrogomphus longipositor*, Rainforest hunter.

**Plecoptera**

*Eusthenia nothofagi*, Otway stonefly.

*Leptoperla cacuminis*, Mount Kosciuszko wingless stonefly.

*Leptoperla kallistae*, Kallista flightless stonefly.

*Riekoperla darlingtoni*, Mount Donna Buang wingless stonefly.

*Riekoperla intermedia*, stonefly.

*Riekoperla isosceles*, stonefly.

*Thaumatoperla alpina*, Alpine stonefly.

*Thaumatoperla flaveola*, Mount Stirling stonefly.

**Diptera**

*Arachnocampa* sp., Mount Buffalo glowworm.

*Edwardsina gigantea*, Giant torrent midge.

*Edwardsina tasmaniensis*, Tasmanian torrent midge.

**Trichoptera**

*Archaeophylax canarus*, caddisfly.

*Costora iena*, Great Lakes caddisfly.

*Ecnomina vega*, Macquarie River caddisfly.

*Hydrobiosella sagitta*, St Columba Falls caddisfly.

*Hydroptila scamandra*, Upper Scamander River caddisfly.

*Oecetis gilva*, South Esk River caddisfly.

*Orphninotrichia maculata*, Wedge River caddisfly.

*Orthotrichia adornata*, Derwent River caddisfly.

*Oxyethira mienica*, Ouse River caddisfly.

*Ramiheithrus kocinus*, Corinna caddisfly.

*Stenopsychodes lineata*, Bluff Hill Creek caddisfly.

*Tasimia drepana*, Picton Rivers caddisfly.

*Taskiria mccubbini*, McCubbin's Lake Pedder caddisfly.

*Taskiria otwayensis*, caddisfly.

*Taskiropsyche lacustris*, Lake Pedder caddisfly.

## Reference

Taylor GS, Braby MF, Moir ML, Harvey MS, Sands DPA (and10 other authors) (2018) Strategic national approach for improving the conservation management of insects and allied invertebrates in Australia. Austral Entomol 57:124–149

# Index




T. R. New, *Insect Conservation and Australia's Inland Waters*,
https://doi.org/10.1007/978-3-030-57008-8

## W



## Y



## Z

The manufacturer's authorised representative in the EU is Springer Nature Customer Service Centre GmbH, Europaplatz 3, 69115 Heidelberg, Germany. If you have any concerns regarding our products, please contact ProductSafety@springernature.com

Printed and bound by CPI Group (UK) Ltd, Croydon, CR0 4YY
23/09/2026
02221547-0001